ROUTLEDGE LIBRARY EDITIONS:
GEOLOGY

Volume 25

PERIGLACIAL GEOMORPHOLOGY

PERIGLACIAL GEOMORPHOLOGY

Binghamton Geomorphology Symposium 22

Edited by
JOHN C. DIXON AND ATHOL D. ABRAHAMS

LONDON AND NEW YORK

First published in 1992 by John Wiley & Sons Ltd

This edition first published in 2020
by Routledge
2 Park Square, Milton Park, Abingdon, Oxon OX14 4RN

and by Routledge
52 Vanderbilt Avenue, New York, NY 10017

Routledge is an imprint of the Taylor & Francis Group, an informa business

British Library Cataloguing in Publication Data
A catalogue record for this book is available from the British Library

ISBN: 978-0-367-18559-6 (Set)
ISBN: 978-0-429-19681-2 (Set) (ebk)
ISBN: 978-0-367-46450-9 (Volume 25) (hbk)
ISBN: 978-0-367-46465-3 (Volume 25) (pbk)
ISBN: 978-1-00-302890-1 (Volume 25) (ebk)

Publisher's Note
The publisher has gone to great lengths to ensure the quality of this reprint but points out that some imperfections in the original copies may be apparent.

Disclaimer
The publisher has made every effort to trace copyright holders and would welcome correspondence from those they have been unable to trace.

PERIGLACIAL GEOMORPHOLOGY

Proceedings of the 22nd Annual Binghamton Symposium in Geomorphology

Edited by

JOHN C. DIXON
Department of Geography
University of Arkansas, USA

and

ATHOL D. ABRAHAMS
Department of Geography
State University of New York at Buffalo, USA

JOHN WILEY & SONS
Chichester · New York · Brisbane · Toronto · Singapore

Other Wiley Editorial Offices

John Wiley & Sons, Inc., 605 Third Avenue,
New York, NY 10158-0012, USA

Jacaranda Wiley Ltd, G.P.O. Box 859, Brisbane,
Queensland 4001, Australia

John Wiley & Sons (Canada) Ltd, 22 Worcester Road,
Rexdale, Ontario M9W 1L1, Canada

John Wiley & Sons (SEA) Pte Ltd, 37 Jalan Pemimpin #05-04,
Block B, Union Industrial Building, Singapore 2057

Library of Congress Cataloguing-in-Publication Data:

Binghamton Symposium in Geomorphology (22nd : 1991)
Periglacial geomorphology : proceedings of the 22nd Annual Binghamton Symposium in Geomorphology / edited by John C. Dixon and Athol D. Abrahams.
p. cm.
Includes bibliographical references and index.
ISBN 0-471-93342-2 (ppc)
1. Frozen ground—Congresses. 2. Geomorphology—Cold regions–Congresses. I. Dixon, John Charles. II. Abrahams, A. D. (Athol D.), *1946–* III. Title.
GB642.B56 1991 91-39799
551.3′8—dc20 CIP

British Library Cataloguing in Publication Data:

A catalogue record for this book is available
from the British Library.

ISBN 0-471-93342-2

Typeset in 10/12 Times by Acorn Bookwork, Salisbury, Wiltshire.
Printed and bound in Great Britain by Biddles Ltd, Guildford and King's Lynn.

Contents

Contributors

Chris Burn, Department of Geography, University of British Columbia, Vancouver, British Columbia V6T 1W5, CANADA.
David R. Butler, Department of Geography, University of Georgia, Athens, GA 30602, U.S.A.
T. Nelson Caine, Department of Geography, University of Colorado, Boulder, CO 80309, U.S.A.
G. Michael Clark, Department of Geological Sciences, University of Tennessee, Knoxville, TN 37996, U.S.A.
Joseph L. DeMorett, Ocean Drilling Program, Texas A&M University, College Station, TX 77843, U.S.A.
James S. Gardner, President's Office, University of Manitoba, Winnipeg, Manitoba R3T 2N2, CANADA.
John R. Giardino, Department of Geography and Geology, Texas A&M University, College Station, TX 77843, U.S.A.
Kevin J. Hall, Department of Geography, University of Natal, P.O. Box 375, Pietermaritzburg 3200, SOUTH AFRICA.
Bernard Hallet, Quaternary Research Center, University of Washington, Seattle, WA 98195, U.S.A.
James Hedges, Big Cove Tannery, PA 17212, U.S.A.
Kenneth M. Hinkel, Department of Geography, University of Cincinnati, Cincinnati, OH 45221, U.S.A.
Antoni G. Lewkowicz, Department of Geography, Erindale College, University of Toronto, Mississauga, Ontario L5L 1C6, CANADA.
George P. Malanson, Department of Geography, University of Iowa, Iowa City, IA 52242, U.S.A.
Frederick E. Nelson, Department of Geography, Rutgers University, New Brunswick, NJ 08903, U.S.A.
Samuel I. Outcalt, Department of Geological Sciences, University of Michigan, Ann Arbor, MI 48109, U.S.A.
Francisco L. Pérez, Department of Geography, University of Texas, Austin, TX 78712, U.S.A.

Wayne H. Pollard, Department of Geography, McGill University, Montreal, Quebec H3A 2K6, CANADA.
Dan J. Smith, Department of Geography, University of Saskatchewan, Saskatoon, Saskatchewan 57N 0W0, CANADA.
Colin E. Thorn, Department of Geography, University of Illinois, Urbana, IL 61801, U.S.A.
John D. Vitek, Graduate College and School of Geology, Oklahoma State University, Stillwater, OK 74078, U.S.A.
Edwin D. Waddington, Geophysics Program, University of Washington, Seattle, WA 98185, U.S.A.
Steven J. Walsh, Department of Geography, University of North Carolina, Chapel Hill, NC 27599, U.S.A.
Robert O. van Everdingen, The Arctic Institute of North America, University of Calgary, Calgary, Alberta T2N 1N4, CANADA.

Preface

This book contains the proceedings of the 22nd Annual Geomorphology Symposium. It is appropriate that the theme of this year's symposium should be periglacial geomorphology, as recently there has been a resurgence of interest in cold climate research. In the past few years a number of conferences have been organized by the IGU Periglacial Commission, and several sessions on periglacial geomorphology were held at the 12th INQUA meeting. The resulting publications (Clark, 1988; French and Koster, 1988; Koster and French, 1988), however, have emphasized periglacial research in Europe and have contained relatively few contributions from North America. Accordingly, one of the aims of the 22nd Annual Geomorphology Symposium was to highlight the considerable quantity and diversity of periglacial geomorphic research being undertaken in Arctic and alpine environments by Canadian and American geomorphologists.

Global climatic and accompanying environmental change represents a contemporary focus of research in the natural sciences. The periglacial environment is one of the regions of the world where such changes will undoubtedly have their greatest impact. Any assessment of the effects of global change clearly requires a thorough understanding of contemporary geomorphic processes operating in the periglacial realm. The papers contained in this volume explore a variety of contemporary geomorphic processes and begin to examine some of the potential impacts of global change on the nature and extent of permafrost and seasonal ice phenomena.

An understanding of periglacial geomorphic processes in both Arctic and alpine environments is essential to assessing the impact of human development and recreation. In both Canada and the United States pressure for natural resource development and extraction has, and will continue to have, significant impact in high latitude and high altitude environments. Alpine environments in particular are under increasing pressure from recreational use as adjacent urban populations grow. This increasing recreational demand results in growing concerns over enhanced soil erosion, decreased hillslope stability, and potential adverse effects on recreational facilities.

In the past half-decade or so, Washburn (1985) and French (1987) have attempted to identify research problems in periglacial geomorphology and to highlight research themes and trends in the field. These authors have recognized some two dozen topics that fall within the active research arena of periglacial geomorphology, including frost-related processes, pingos, palsas, mass wasting processes, nivation, ground ice, and paleoenvironmental studies. This year's symposium and the resulting proceedings examine some of these areas of research.

The papers in this volume are largely divided into two sections. The first half of the volume, chapters 1 through 9, focuses primarily on periglacial processes in alpine environments, while the remaining chapters deal with Arctic periglacial processes. The volume begins with a theoretical paper by Thorn in which he defines periglacial geomorphology in terms of a single unifying concept, that of the geomorphic role of ground ice. In the process of justifying his definition, he provides some contemporary insights into many of the periglacial research problems outlined by Washburn (1985) and French (1987).

Operationalization of Thorn's theoretical approach to periglacial geomorphology in the paleo record is tested in the Appalachian Highlands south of the glacial limit in chapter 2 by Clark and Hedges. These workers investigate the evolution of broad accordant summits and their associated landforms. They adduce a range of conspicuous periglacial forms, such as summit tors, upland flats, risers, and terraces, to support their hypothesis that the summits are relict cryoplanation terraces.

Contemporary alpine periglacial processes are the subject of the next two chapters. Caine analyzes spatial patterns of geochemical denudation in the Colorado Rocky Mountains. He shows that although solute yields from a small alpine catchment are low, in favorable locations they exceed rates of suspended sediment transport by an order of magnitude. Geochemical processes are particularly important in maintaining small-scale landforms, such as nivation hollows. Gardner, working in the Canadian Rocky Mountains, examines the concept of enhanced freeze-thaw weathering in the vicinity of the glacier-cirque headwall margin. He reports on freeze-thaw temperature regimes in a restricted zone at the glacier-headwall contact and argues in favor of a zone of enhanced mechanical weathering in the vicinity of the randkluft lip.

A very different perspective on mechanical weathering processes is provided from the maritime Antarctic environment by Hall. Hall isolates a diversity of components intimately associated with freeze-thaw weathering and distinguishes these mechanisms from those associated with salt crystallization and insolation weathering.

Rock glaciers have been of recurring interest to periglacial geomorphologists since the 1950s. A recent comprehensive review of these landforms is provided by Giardino, Shroder, and Vitek (1987). Despite a large volume of research on rock glaciers, much remains to be learned about their origin, mobility, and role in overall hillslope evolution in alpine environments.

Giardino, Vitek, and DeMorett examine the pattern of water movement and its associated geochemical characteristics in three rock glaciers in the Sangre de Cristo Mountain Range, southwestern Colorado. From their observations and measurements they formulate a model of throughflow water movement. This paper represents an important contribution to the understanding of geochemical processes operating on alpine hillslopes.

Butler, Malanson, and Walsh examine the morphology of snow-avalanche paths in the Rocky Mountains of northwestern Montana and southwestern Alberta. Their paper highlights the significance of these mountain landforms as conduits for the transfer of sediment from alpine to nonalpine environments. The significance of this sediment transfer is examined in terms of its impact on human activities at lower elevations.

The theme of mass wasting is continued in the chapter by Smith, who presents an extended record of rates of solifluction in the Rocky Mountains of Canada. He finds that movement is the result of the combined effect of creep and shearing. Results of the study indicate that the work performed by these processes is of comparable magnitude to that of rockfall and snow avalanches.

The nature and origin of patterned ground remains a major theme in periglacial geomorphic research. Pérez investigates the origin of miniature stone stripes on hillslopes in the Venezuelan Andes and attributes the formation of these small-scale features mainly to needle ice development.

The mechanisms of patterned ground formation are explored from a broader and more theoretical perspective by Hallet and Waddington. They examine the role of buoyancy forces generated in association with freeze-thaw processes, and they explore the implications of these forces for diapiric and soil circulation processes. They argue that although buoyancy processes may not specifically be responsible for the generation of new patterned ground, they may be significant in maintaining features such as sorted circles.

Mass wasting processes in Arctic environments are analyzed in the chapter by Lewkowicz. He examines the distribution and causes of active-layer detachment slides on Ellesmere Island. Recent slides are related to the susceptibility of soils to frost action and rates of advance of the thaw plane at critical depths within the active layer.

The action of ice has long been a central theme in periglacial geomorphology and is examined in the final three chapters of this volume. Pollard and van Everdingen discuss the formation of seasonal ice bodies in the Yukon Territory, Mackenzie Valley, and Axel Heiberg. This chapter focuses on the development of icings and seasonal frost mounds and their subtypes. The theme of ground ice is further explored by Nelson, Hinkel, and Outcalt, who address the question of palsa-scale frost mound development. Their paper examines both aggradational and degradational processes of formation.

Nelson, Hinkel, and Outcalt point out the potential of palsa-scale frost mounds to provide information on the paleoecology of the areas in which they occur, as well as the potential impact of global climate change on these landform features. The theme of environmental change on frost-related features is addressed in the final paper by Burn. Results of geothermal modeling near Mayo, Yukon Territory, show that over the past 20 years there has been progressive warming of the permafrost. This trend is attributed to increases in both snowfall and winter air temperatures.

The papers presented at the 22nd Annual Geomorphology Symposium, and subsequently assembled in this volume, clearly indicate the healthy and active state of periglacial geomorphic research in North America and demonstrate continued progress in the understanding of periglacial processes in both Arctic and alpine environments. We hope that this volume will serve as a stimulus for further research in cold climate geomorphology in North America, and that the papers will promote new perspectives on, and new techniques within, the discipline. Finally, we trust that the papers in the volume demonstrate that progress has been made in addressing some of the research problems outlined by Washburn (1985) and French (1987).

References

Clark, M. J. (ed.), *Advances in Periglacial Geomorphology*, 481pp., John Wiley, Chichester, England, 1988.

French, H. M., Periglacial geomorphology in North America: current research and future trends, *Ecological Bulletins*, **38**, 5–16, 1987.

French, H. M., and E. A. Koster (eds), Periglacial processes and landforms, *Zeitschrift fur Geomorphologie Supplement Band*, **71**, 1988.

Giardino, J. R., J. F. Shroder, Jr., and J. D. Vitek (eds), *Rock Glaciers*, 355pp., Allen and Unwin, Boston, Mass., 1987.

Koster, E. A., and H. M. French (eds), Periglacial phenomena: Ancient and Modern, *Journal of Quaternary Science*, **3**, 110pp, 1988.

Washburn, A. L., Periglacial problems, in *Field and Theory: Lectures in Geocryology*, edited by M. Church and O. Slaymaker, pp. 166–202, University of British Columbia Press, Vancouver, B.C., 1985.

Acknowledgments

The organization of the 22nd Annual Geomorphology Symposium and the subsequent preparation of this book would not have been possible without the assistance of numerous institutions and colleagues. We wish to thank the Departments of Geography at both the University of Arkansas, Fayetteville, and the State University of New York at Buffalo for their financial support. A grant to the symposium was provided by the Conferences in the Disciplines program at the latter institution.

We wish to thank the participants and authors of the papers in this volume. Without their contributions there would be no symposium and no book. The high quality of papers is also due in part to the conscientious efforts of the manuscript reviewers, namely Jim Benedict, Duane Braun, David Butler, Nel Caine, Jim Gardner, Rick Giardino, Ken Hinkel, Toni Lewkowicz, Brian Luckman, Ben Marsh, Scott Morris, Fritz Nelson, Greg Olyphant, Sam Outcalt, Francisco Pérez, Wayne Pollard, Larry Price, Olav Slaymaker, Lincoln Washburn, and Stephen Watts. Their assistance is greatly appreciated.

1 Periglacial Geomorphology: What, Where, When?

C. E. Thorn
Department of Geography, University of Illinois

Abstract

Periglacial geomorphology embraces concepts ranging from the fuzzy to those that are rigorously grounded in theory and have been empirically and/or field corroborated. As a consequence of this admixture, the discipline has an unsatisfactory blend of core concepts. Many of the older concepts stem from climatic geomorphology and are not amenable to operational definition, while many of those rigorously developed by process geomorphologists are restricted to specific micro-environments, and it is difficult to assess their general applicability. This paper is an attempt to define periglacial geomorphology from a narrow perspective in which all of the constituent processes and forms are theoretical and/or empirically founded. The objective of this exercise is not to authorize some concepts while proscribing others, but to emphasize where the theoretical ground appears firm and where attention needs to be directed. The conclusion reached is that periglacial geomorphology is most appropriately defined as follows:

> Periglacial geomorphology is that part of geomorphology which has as its primary object physically based explanations of the past, present, and future impacts of diurnal, seasonal, and perennial ground ice on landform and landscape initiation and development. Additional components of the subdiscipline include similar investigations of the geomorphic roles of snowpacks (but not glaciers) and fluvial, lacustrine, and marine ice.

Such a definition contains a unifying concept, the geomorphic role of ground ice, and consequently offers a number of advantages over previous defini-

Periglacial Geomorphology. Edited by J. C. Dixon and A. D. Abrahams

tions. These include: the possibility of sound operational definitions; an appropriate focus on ground temperature/ground moisture interaction; better opportunities to evaluate the periglacial (i.e., ground ice) versus nonperiglacial contributions to landform/landscape development.

Introduction

The purpose of this paper is to construct the skeleton of a periglacial geomorphology based strictly on theoretically and/or empirically derived, field-verified processes. Such an endeavor will produce an admittedly truncated version of what is commonly considered to constitute the discipline. However, the result should highlight the rigorously scientific domain of what is arguably the primary approach to periglacial geomorphology in the English-speaking world.

I define geomorphology as the scientific explanation of natural landforms or assemblages of landforms (landscapes hereafter). In addition, I assume that the term periglacial[1] is used to identify a definable subset of processes, landforms, or landscapes within geomorphology which exhibit distinctive and substantive attributes. By suggesting that periglacial geomorphology should follow scientific standards, I am setting its practioners a difficult task because the size and complexity of the literature on scientific methodology clearly indicates that a definitive statement on the scientific method is impossible (Thorn, 1988a). However, fundamental scientific prerequisites mandate that the researcher investigate a problem emanating logically from an existing body of theory and one designed to extend, rectify, and/or replace that theory. The problem should be formulated as a hypothesis; this implies not only that the issue be firmly imbedded in theory but that it also be a substantive and quantifiable one. The hypothesis must then be tested for the purpose of falsification or perhaps corroboration, but it cannot be verified (Haines-Young and Petch, 1986).

The creation, testing, and modification of a hypothesis invariably requires the use of at least one "operational definition" and identification of "necessary and sufficient conditions." In a science such as geomorphology, where theory and fieldwork should be symbiotic, operational definitions and necessary and sufficient conditions should be expressed, if possible, in field terms. Consequently, it is upon the field expression of necessary and sufficient conditions that I will try to build my case of the what, where, and when of periglacial geomorphology.

Objectives

Technically, objectives lie outside the domain of science, as they are not subject to logical objection. Any objective(s) which a group of individuals holds in common is potentially a field of scientific endeavor provided that it is investigated in accordance with accepted scientific methodologies. There appear to be five primary objectives within the commonly accepted domain of contemporary periglacial geomorphology. These are:

(1) to identify the chemistry, physics and/or mechanics of periglacial processes;
(2) to identify periglacial landforms wherever they occur;
(3) to identify the presence of permafrost wherever it occurs;
(4) to investigate the properties and behavior of permafrost and active layer, and identify the associated processes;
(5) to reconstruct paleo-environments (an activity which may focus on either the past presence of periglacial regimes or, more narrowly, the past presence of permafrost).

It is immediately apparent that the above list contains no item which overlaps completely with Łoziński's (1909) original definition of "periglacial facies," and his subsequent (Łoziński, 1912) restatement of the concept. Jahn (1954, p. 118) cites two key sentences from Łoziński (1909):

> The peculiarity of periglacial facie consists in the mechanical splitting of the rocks in situ; and freezing is the most important factor of periglacial climate. Under periglacial climate, the action of freezing must have attained the intensity it shows at present in the neighborhood of sub-polar ice sheets.

According to Jahn's account, the two necessary and sufficient attributes of Łoziński's (1912) concept of things periglacial are adjacency to a Pleistocene ice sheet (simultaneously a temporal and spatial requirement) and the presence of rubble or rubble mantles (intrinsically assumed to be of freeze-thaw origin). Łoziński himself was aware of many of the difficulties created by such an operational definition: for example, the existence of presently active alpine rubble sheets; the correlation of such rubble sheets with resistant rocks; the seeming periglacial nature of loess; and the introduction by Andersson (1906) of the concept of solifluction.

While recognizing limitations to his periglacial concept, Łoziński still anchored it firmly on his initial criteria. Placed under careful scrutiny it is clear that it is not possible to fabricate a fully operational definition of Łoziński's concept. This is because his stated criteria fail to permit identification of all frost-dominated environments, nor do they permit a satisfactory measure for distinguishing his essentially "paleo" concept from areas exhibiting similar active processes and forms.

There is no intrinsic reason to abandon a term whose current usage differs from its original meaning; both everyday language and technical terminology are replete with examples of evolving terms. However, it is essential to find for the term periglacial a meaningful content, which is common to all the objectives the term is used to describe. The only logical options are (1) to restrict the term to a subset of objectives with common content and add other terms to cover the remainder of the objectives, or (2) to abandon the term periglacial altogether and establish an entirely new set of terms such as those rooted in geocryology. The choice between these two approaches is one which is frequently faced in science; neither approach is mandated nor is either proscribed.

The second salient point to emerge from the list of contemporary objectives in periglacial geomorphology is the interaction or overlap between things periglacial and those directly associated with, and restricted to, permafrost. A number of geomorphologists (e.g., Péwé, 1969; Harris, 1988) have defined periglacial geomorphology as synonymous with the occurrence of permafrost. I would reject this view. Not only is such a definition overly restrictive, it is illogical to define geomorphology by an exclusively thermal criterion.

If Łoziński's definition of periglacial is regarded as unamenable to operational definition, and correlation of the term periglacial with permafrost alone is deemed too restrictive and illogical, then the objective of this paper is given a context. This objective is a scientific explanation of that portion of the world's landscapes that can be meaningfully identified as the product of cold, but not glacial, processes and meaningfully identified by a unifying operational definition.

Theoretical frameworks

The definition, retention, or rejection of any term in geomorphology should each be predicated upon the clarity garnered. Terminological issues are an important facet of the inherent conflict in science between the need for meaningful generalization and the need for precision. I believe this issue is colored as much by emotions as by intellect and is reflected in the common tendency to be either a lumper or a splitter. However, whether a lumper or a splitter, it behoves every scientist to be clear-minded. Clarity of thought is dependent upon the small artifacts of the process (i.e., words) as well as the grandiose (i.e., overarching theory). I have expressed my thoughts on this particular matter elsewhere (Thorn, 1988a), but I would emphasize once again the problems stemming from retention of poorly defined morphogenetic terms. I would cite Washburn's (1956) seminal paper on patterned ground as a classic appreciation of this issue.

Washburn (1980 p. 157) restated his 1956 perspective as follows:

> An adequate explanation of patterned ground is a multivariate problem (Caine, 1972, p. 56). The following discussion . . . is based on . . . the following premises: (1) Patterned ground is of polygenetic origin; (2) similar forms of patterned ground can be due to different genetic processes; (3) some genetic processes may produce dissimilar forms; (4) there are more genetic processes

> than there are presently recognized terms for associated forms, and (5) it is desirable to maintain a simple and self-evident terminology.

The thrust of Washburn's case is to disaggregate poorly understood processes and forms, while establishing a comprehensible technical vocabulary. However, as Washburn stated himself, his classification scheme is a research, or investigative vehicle; it does not represent a definitive, physically based explanation of all patterned ground forms. Moreover, it is quite conceivable that ultimately features with similar morphology will be viewed as genetically unrelated. In separating process and form[2] it is essential to retain a clear and firm grasp on the fundamental objectives in geomorphology. Geomorphologists have little direct interest in form alone; recognition of forms is merely one convenient tactic in our strategy to explain landscapes. Ultimately, forms are of limited interest because we want to understand the origin of things, and there appears to be no one-to-one correspondence between process and form (i.e., between cause and effect). Haines-Young and Petch (1983, p. 464) discussed these issues in their usual lucid fashion, concluding that:

> the origin of landforms . . . cannot be defined *except* from evidence of how they have functioned or how they are functioning.

The emptiness of a preoccupation with form alone is well illustrated by Evans and Cox's (1974) report on cirque morphology and amplified by Sauchyn and Gardner's (1983) discussion of open rock basins.

Several conclusions may be drawn from analysis of the papers by Washburn (1980) and Haines-Young and Petch (1983). Most importantly, form and process cannot be fully disaggregated in geomorphology because form alone has no explanatory power. However, although form alone has no explanatory power, explained form provides us with our only opportunity to link the past, present, and future. Consequently, we must treat process and form symbiotically, while recognizing that process is pervasive. However, the form we identify and treat discretely is only a portion of a much larger related continuum.

Landform "definitions" are normally a scientific shorthand; more formally they represent "nominalist definitions" (Haines-Young and Petch, 1983, p. 463). They represent convenience more than anything else. If form represents the access point to explanation by providing the evidence for process, it follows that contemporary processes that produce ephemeral or no form are always going to tantalize geomorphologists, as there is no basis for evaluating their past or future significance.

It is now apparent that a focus upon form and a focus upon process do not really represent discrete approaches. The supposed differences actually represent preferences in starting points from which to approach similar or identical goals. Such choices stem from different theoretical perspectives and must form the crux of the next step.

CLIMATIC GEOMORPHOLOGY

In seeking to explain landforms, the two obvious starting points are either to begin by examining geomorphic processes or to begin by identifying the landforms themselves. The most enduring attempt to approach geomorphology using landforms as the starting point was undoubtedly that of William Morris Davis (1899). Unhindered by modern demands for quantitative rigor, Davis was able to offer explanations of landscapes at a regional scale. However, Davis quickly recognized flaws in his general model and as early as 1905 published his first paper (Davis, 1905) identifying a "climatic accident." Recognition of discrete geographic (geomorphic) cycles based on climatic zones represented a step back from over-generalization but a continuation of an approach in which form change over time had primacy. The formal and comprehensive restatement of this secondary Davisian position constitutes the starting point of what is commonly called climatic geomorphology. The French and German schools offer the most highly developed statements of climatic geomorphology (e.g., Tricart and Cailleux, 1972; Büdel, 1982), with Peltier (1950) and Tricart (1970) providing detailed treatments of periglacial geomorphology.

The conceptual linchpin of climatic geomorphology is the sequence: climatic regime controls geomorphic process(es) and process(es) controls landform(s). It is an approach which is also founded on the concept of the development of characteristic forms (Brunsden and Thornes, 1979). Priesnitz (1988, p. 64) expressed this succinctly as "constant climates cause characteristic forms." Under ideal circumstances such a view requires one-to-one correspondence between climate and geomorphic process as well as between process and form. A less satisfactory alternative is to claim that uniqueness occurs either in the intensity of a single process or in the relative intensities within a complex of processes.

The shortcomings of the climatic geomorphology approach from the perspective of process geomorphologists have been identified on many occasions, most notably by Stoddart (1969) and Derbyshire (1976). Among the more obvious flaws cited are: the disparate and inappropriate criteria used to establish climatic zones; failure to establish climate-process links; inadequate corroboration of process-form links; and uncertainty of temporal relationships between meteorological and/or climatic inputs and geomorphic responses. Furthermore, climatic geomorphologists have generally failed to establish their case holistically, for as Stoddart (1969, p. 174) noted:

> the recognition of distinct landform assemblages has depended less on total landscape morphometry than on the occurrence of less frequent but more spectacular type-landforms, such as inselbergs or pediments.

Indeed, geomorphologists who have a preference for studying process are likely to agree with Cooke and Warren's (1973, pp. 6–7) assessment of climatic geomorphology in a totally different realm:

> In our view, the formulation of simple and comprehensive generalizations about the nature of desert geomorphology, especially if they are based on the recognition of relations between desert landforms and desert climates, will be difficult, and in the present state of knowledge, is impossible.

Superficially, all periglacial geomorphologists would appear to be protagonists of climatic geomorphology. However, the combination of a preoccupation with process in periglacial geomorphology in the English-speaking world and the failure of climatic geomorphologists to pursue process research vigorously means that, in reality, there are at least two distinct groups of periglacial geomorphologists. This alleged dichotomy ignores, for example, some very real differences within the community of climatic geomorphologists (Stoddart, 1969).

In declining to pursue periglacial geomorphology from the theoretical framework of climatic geomorphology, I do not wish to imply that it is a worthless methodology. Climatic geomorphologists sacrifice precision for generalization. This is not an error providing it is done consciously. They also pursue their geomorphic investigations from a top-down perspective (literally and figuratively), as they obviously operate primarily at a regional level and encounter difficulties produced by mixed signals when moving down to smaller spatial scales. If one scientific approach (in this case climatic geomorphology) produces tenuous results, it is likely that another methodology will develop. In explaining landforms and landscapes, the alternative starting point to form is process, to which I now turn.

PROCESS GEOMORPHOLOGY

If climatic geomorphology may be crudely characterized as an extension of Davis's methodology, process geomorphology may be equally crudely characterized as an outgrowth of one attempt to correct Davis's methodology. Once isotope dating began to reveal flaws in several Davisian regional histories, geomorphologists sought firmer ground upon which to base their concepts of landform and landscape development. The relative lack of attention to process in the Davisian scheme made it an obvious candidate for investigation. No one person can truly be labeled as creator of process geomorphology, but certainly Horton (1945) and Strahler (1952) deserve identification as leading figures.

There is little doubt that process geomorphologists initially believed that they were going to provide the missing link in the Davisian scheme—that is, a process component in the regional development of landscapes. Preoccupied with the need for ever-increasing refinement, process geomorphologists have pursued (or perhaps retreated down) a reductionist path. If climatic geomorphology is bedeviled by fuzzy precepts and an inability to identify a clear climatic signal at small scales, process geomorphology is constrained by an inability to expand its claimed surety beyond the small scale and to aggregate meaningfully discrete results. It is exploration of these limits to periglacial process geomorphology, while maintaining strict operational definitions, that forms the focus of this paper.

Periglacial processes

ASSUMPTIONS AND RULES

In order to produce a tightly drawn and defensible product, it is necessary to specify some assumptions and rules. The first assumption is that geomorphology, and specifically process geomorphology, is a continuum, but one in which there is utility in subdivision. The second assumption is that air climates are insufficiently directly related to ground climates (surface and subsurface) to specify the latter from the former in a satisfactory manner.

The first rule to be specified is that the criterion or criteria used to define the concept "periglacial" must be consistent and demarcate something substantive. The second rule is that only processes which have a theoretical explanation (or, failing this, an empirical foundation), for which an operational definition can be established, and which have been corroborated in the field may be invoked.

CRITERIA FOR A PERIGLACIAL ENVIRONMENT

Łoziński's identification of freeze-thaw weathering as the fundamental periglacial process was refined by Washburn (1980, p. 6):

> By far the most widespread and important periglacial process is frost action. Actually, frost action is a "catch-all" term for a complex of processes involving freezing and thawing including, especially, frost cracking, frost wedging, frost heaving, and frost sorting.

Washburn's statement represents the beginning of identification of a sufficient process to define the term periglacial. However, frost action is not a necessary process by Washburn's (1980, p. 6) own admission:

> In addition to frost action, certain aspects of mass wasting, nivation, fluvial action, lacustrine action, marine action, and wind may produce periglacial features.

Using the frost action[3] criterion, a region should be defined as periglacial by virtue of frost processes being unique, or restricted, to it. However, the prerequisite of an operational definition expressed in geomorphic terms precludes this option. This is because frost processes must occur at an intensity which is sufficient to generate a detectable result (form). There are many regions where frost action is simply overwhelmed by other processes and consequently remains, in operational terms, undetectable. An operational definition that cannot be falsified or corroborated in the field is self-evidently worthless, for as Medawar (1967) characterized it, science is "The Art of the Soluble."

In choosing to spotlight the freeze-thaw cycle, a purely thermal event, periglacial geomorphologists have inevitably left ground moisture, rock lithology, soil texture, and regolith particle-size distribution in the shadows. It is critical not to overlook the fact that both air and ground freeze-thaw cycles are actually surrogate measures, and to appreciate that the production, presence, and melting of ground ice are the real focal issues. Consequently, definitive studies (as opposed to much needed supporting studies) must involve direct observation of ground ice and associated geomorphic responses. Long-term monitoring projects of ground ice behavior are likely to incorporate ground temperatures as a convenient surrogate measure. In these instances it is important to appreciate that air temperatures are an inadequate measure of ground conditions, and also that ground temperatures may be a questionable indicator of ground moisture state in the temperature range immediately above and below zero degrees Celsius.

Requirements for definitive field studies must reflect contemporary technological capabilities. If ground temperature data are generated it should be by thermistors-thermocouples connected to data loggers producing either continuous or high frequency sampled records. The nature of host material, be it rock lithology or regolith/soil fabric or texture, must be described comprehensively using appropriate standardized measures. With respect to ground moisture, continuous or high frequency monitoring is not yet feasible for many researchers (but see Outcalt and Hinkel, 1989, for the use of electric potential in monitoring moisture). Ignorance of dynamic soil moisture behavior is a significant problem because it is a reasonable simplification to assume that the interaction between ground temperature, moisture content, and the mechanical properties of the host material will provide basic answers to most questions. Nevertheless, it appears that definitive studies will presently have to continue to embrace much simpler ground moisture measures than temperature measures.

An area of frequent concern is the distinction between landform initiation and landform growth (or maintenance), both of which are encompassed in the normal use of the word origin. Washburn (1980) identified this issue in his discussion of patterned ground, and there are certainly other examples, such as rock glaciers (Johnson, 1987), where landform initiation may be quite distinct from development. This issue is too complex to resolve en passant, but it is worth remembering that any landform labeled periglacial may be periglacial only in origin, growth, or maintenance, or may be periglacial throughout its development.

Prior to attempting extension of the periglacial concept beyond an operational definition embracing only frost processes, it is useful to examine this restrictive version. This will be done by reviewing the four frost mechanisms identified by Washburn (1980, p. 6).

Frost wedging (weathering)

From its inception, periglacial geomorphology has been dominated by the concept of frost wedging, a synonym for weathering by freezing and thawing. The story is one of casual empiricism gathering respectability by repetition until it attained the stature of an article of faith. Despite lacking detailed theory, the concept gained some scientific credence in the 1950s and subsequently by extensive laboratory testing of small rock samples under conditions that may or may not mimic natural ones.

There were early warnings by Grawe (1936) of the limitations and requirements which might constrain frost weathering of bedrock, but Walder and Hallet's (1985) paper provided the first convincing theoretical treatment. The tenor of this paper is extremely encouraging because it developed a model analogous to that used to depict soil heaving (Taber, 1930; Gilpin, 1980). Furthermore, the model presented requires a remarkably modest set of prerequisites. Rapid crack growth can be generated in the temperature range −4 to −15°C and is favored by low cooling rates in an open system lacking high pore-water pressures as well as by prolonged freezing. Apart from its theoretical rigor, the primary appeal of Walder and Hallet's paper is the unexceptional conditions invoked. However, as they themselves point out, other frost-related mechanisms are possible, Michaud, Dionne, and Dyke's (1989) explanation of frost bursting being an apparent example.

Laboratory research into frost weathering has been quite extensive; however, it has lacked a number of important attributes. Foremost among these problems has been uncertainty concerning the thermal and moisture regimes which actually prevail within natural bedrock and regolith fragments. Consequently, freeze-thaw cycles used in laboratory samples may or may not reflect natural temperature ranges (Thorn, 1979). A similar problem has overshadowed the moisture issue, and most laboratory experiments have embraced very crude approaches to moisture conditions and supply. This critical issue has only begun to receive the attention it merits (e.g., Fukuda, 1983; Hall, 1988). As noted above, laboratory studies have been dominated by investigation of very small rock fragments, whereas the field situation obviously includes much larger rock fragments and bedrock faces. The potential for more than one relevant mechanism in this context was signaled by Tricart's (1956) early recognition of microgelivation (frost weathering restricted to pore spaces) and macrogelivation (frost weathering in which fractures are exploited) and is presently sustained by Walder and Hallet's (1985) recognition of the distinction between single- and multisided freezing. Although laboratory experimentation represents by far the largest portion of the frost weathering data base (e.g., Lautridou, 1988), it is difficult to view this experimentation very positively. Given almost total ignorance of both field conditions and applicable theory, the experiments have not even precluded the possibility of other mechanisms (e.g., Hudec, 1973).

Field corroboration is something of a misplaced concept with respect to frost weathering. At present there is no adequate field criterion to establish that bedrock weathering or further comminution of rock fragments has been dominated by freeze-thaw weathering. Nevertheless, it is clear that the majority of periglacial researchers believe that freeze-thaw weathering of bedrock is an established fact, and that it is an acceptable premise upon which to base many secondary concepts (e.g., cryoplanation). In some specialized contexts there does appear to be adequate evidence for this assumption (e.g., Mackay and Slaymaker, 1989). However, the most common argument to substantiate the importance of freeze-thaw weathering is both circumstantial and circular. The concept that freeze-thaw weathering dominates cold regions gained respectability long before there was the ability to test it in the field. As angular rock fragments are common in cold environments they were assumed to be the product of the dominant process, namely freeze-thaw weathering. Today, it is common to assume that angular rock fragments are definitive evidence of frost weathering. Nevertheless, the angularity of comminuted bedrock must always be strongly influenced by lithology and is certainly likely to stem from processes other than freezing and thawing in many circumstances (e.g., McGreevy and Whalley, 1982; Whalley, 1984).

It is my belief that what periglacial geomorphologists need more than any other single item is a way to determine in the field whether or not bedrock fragments have been frost weathered. Development of a rigorous test will be a difficult task, particularly given the problems of conducting empirical research at scales larger than the laboratory experiment. At present it seems that concentration on small-scale attributes offers the most promising avenue of advancement. A possible approach would be to conduct scanning electron microscope studies (e.g., Dowdeswell, 1982; Dowdeswell, Osterman, and Andrews, 1985) with a view to establishing criteria which might be determined eventually at the hand lens scale. This issue is one whose time is long overdue, but it will require rigorous distinction of frost-induced indicators among the myriad of features produced by other weathering mechanisms as well as careful linkage between laboratory and field studies.

Frost heaving

Frost heaving of soil or regolith is produced by two primary mechanisms. First, there is an approximately 9% expansion of volume during the phase change from liquid to ice. Second, there is a much more complex, as well as a potentially much larger, effect produced by the migration of unfrozen soil moisture to the freezing front (Taber, 1930). The latter mechanism will produce segregation ice, a phenomenon that has been observed experimentally (e.g., Taber, 1930), modeled numerically (Outcalt, 1976), and observed in the field (e.g., Mackay, 1971, 1972). While observed directly over a long period, a theoretical explanation of the formative process(es) of segregation ice has remained difficult. Much of the research on the topic has been summarized by Washburn (1980, pp. 68–70). Although not widely cited in frost heave studies, Gilpin's (1980) paper would make an interesting starting point if it could be shown that a single theoretical explanation could embrace both a portion of freeze-thaw weathering (Walder and Hallet's 1985 paper is founded on Gilpin's 1980 paper) and frost heaving.

The critical assumption in Gilpin's (1980) model is the presence of an attractive force toward the solid at a solid-liquid interface. There is presently no definitive explanation of such a force, but evidence for it is widespread. Derivative from this assumption is the presence of a disjoining force that tends to separate the faces of a growing ice crystal from the abutting solids (soil particles in this case). Water is thus able to migrate to the liquid layer (ca. 10–20 μm thick) between an ice crystal and the abutting soil particles, enter this layer, freeze, and contribute to the growth of a segregated ice lens.

Frost heaving has been corroborated in the field on a number of occasions (e.g., Washburn, 1967; Mackay, 1981) and may not only be measured surficially by fairly simple instrumentation (e.g., Washburn, 1967; Fahey, 1973, 1974) but also at shallow depths (Mackay and Leslie, 1987). Conceptually, on an inclined surface frost heave occurs orthogonally to the slope, but during the thaw phase the ground subsides in the vertical plane due to gravity. This combination produces what Washburn (1967) labeled potential frost creep. In fact, Washburn identified a slight retrograde movement during collapse, such that potential frost creep was not fully realized. A third element in Washburn's (1967) field measurements was the presence of a horizontal movement during melting; it was this third component he labeled gelifluction. Washburn's meticulous fieldwork in Mesters Vig, northeast Greenland, was combined with a concerted attempt to clarify the terminology surrounding the general issue of the qualitative concept of "solifluction" (Andersson, 1906).

Mackay's (1981) equally impressive fieldwork on Garry Island in the Northwest Territories has led him to distinguish between creep processes in permafrost and nonpermafrost contexts. The essential difference between the two situations is created by the presence of freezing from below, as well as from above, in the active layer. Mackay made several points in his paper, but the most interesting appears to be the identification of summertime frost heave in permafrost situations. This phenomenon is caused by water in the unfrozen active layer migrating downward and refreezing in the frozen active layer and/or in the permafrost below. Mackay's findings, now abundantly verified (see Mackay, 1983), necessitate reexamination of Washburn's distinction between the gelifluction and frost creep components of mass wasting above permafrost (Lewkowicz, 1988).

The allusion to geomorphic processes associated with the thaw phase of frost heave merits amplification. Under the categorization defined by Washburn (1967), gelifluction is associated directly with the thaw phase of frost heave: it is a horizontal increment of movement beyond that produced by potential frost heave minus the retrograde component. More recent work (e.g., McRoberts and Morgernstern, 1974) explains many periglacial mass movements, including gelifluction, in terms of the thaw consolidation concept. The concept, which can actually be traced back to Taber (1943), invokes generation of excess pore water pressures when ground ice melts faster than drainage can remove the resultant meltwater.

A distinctive form of ice segregation is needle ice. Most of the critical components in needle ice formation appear to have been identified by Outcalt (1971, 1979a,b), and it has been effectively reproduced in the laboratory (Soons and Greenland, 1970). Although some interesting facets of needle ice (e.g., raking) still appear to be open to debate (Mackay and Mathews, 1974), its general effectiveness as a heaving mechanism and its common occurrence in the field (under admittedly constrained contexts) are widely recognized.

The sum of these research projects is illustration of the extreme complexity present in the field and our growing ability to recognize it. A substantive issue stemming from this sophistication is whether or not such complex process regimes leave clear imprints of their presence on associated forms. It is difficult to envisage that the discrete process components recognized by Mackay could ever be detected from a purely static contemporary study, let alone in an investigation of a paleo-form. Consequently, we probably need to recognize formal hierarchical process suites in which field processes are subdivided into primary components, which leave detectable evidence, and secondary mechanisms, which are identifiable when active but leave no lasting trace.

Frost sorting

Frost sorting is the differential movement of particles of varying size during frost heave and/or subsequent settling. Eakin (1916) distinguished between vertical movement (heaving) and horizontal movement (thrusting), although it is the ejection of coarse fragments that has received most attention from field researchers (e.g., Morawetz, 1932; Schunke, 1974). Such behavior, if established, necessitates identification and explanation of a general forcing mechanism. However, the existence of a variety of sorted patterned ground forms also introduces secondary questions focused upon their spacing, regularity, and shape.

Laboratory research on the topic began early, with Taber (1929) establishing the general principle that particle migration is at 90° to the freezing front. Corte (summarized 1966) demonstrated that particles of differing size may all respond to an advancing freezing front. Nevertheless, it is the movement of coarse fragments that has been the focus of most laboratory experiments. This line of research has endowed the discipline with the twin concepts of frost push and frost pull to explain the preferential upward movement of coarse debris. Frost pull invokes the notion that, once uplifted during frost heave, coarse particles fail to sink fully to their prefreezing position—a situation which Kaplar (1965) has verified experimentally. Frost push is dependent upon the preferential development of ice lens at the base of coarse particles due to their higher thermal conductivity or thermal diffusivity. Washburn (1980, p. 91) suggested that the frost-pull mechanism is probably more widespread.

Both frost pull and push were developed to explain upward movement in the presence of a downward moving freezing front. Mackay (1984) investigated frost sorting driven by a freezing front moving upward from the permafrost table. In addition to reviewing the existing concepts underpinning frost sorting, Mackay (1984) introduced the notion that stones, like soil, may be considered "frost susceptible." He considered the presence of a "network of fine pores sufficiently large to permit the through-flow of water but sufficiently small to retard the advance of pore ice into them" (Mackay, 1984, p. 441) to be the factor determining frost susceptibility.

At the present time theoretical understanding of frost sorting is far from complete, but laboratory experiments have corroborated the feasibility of the frost-derived sorting mechanism invoked by field researchers. Although the size, shape, and juxtaposition of some sorted patterned ground phenomena are fairly easy to appreciate (e.g., those exhibiting diapiric type behavior), others present serious intellectual challenges.

A potentially comprehensive explanation for the surficial geometry and width-to-depth ratio of sorted polygons and stripes has been proposed by Ray, Krantz, Caine, and Gunn (1983) and developed by Gleason, Krantz, Caine, George, and Gunn (1986). This model invokes the presence of convection cells driven by unstable density stratification of thaw water in the active layer. Such cells promote an uneven base characterized by peaks and troughs that predetermine preferential locations for sorting mechanisms. Statistical testing of theoretically determined spacing against field spacing produced remarkably good agreement. The appeal of this proposed explanation is reinforced by its applicability to sorted stripes as well as to polygons occurring beneath ponds in periglacial environments.

Recent work by Hallet and Prestrud (1986) and Washburn (1989) lends credence to an older notion of direct circulation of the soil itself (see Washburn, 1956, pp. 852–855, for a discussion). Hallet and Prestrud pointed out that Mortenson (1932) demonstrated that convective soil movement cannot be founded upon notions of thermal gradients within a soil. However, if soil bulk density variability derived from feasible differences in soil moisture content is considered, then convective soil motion becomes a viable mechanism. In their initial report from an ongoing study of sorted circles in western Spitsbergen, Hallet and Prestrud presented a number of pieces of evidence favoring soil convection as an explanation of the circles. Although their ideas are independent of the pore water convection suggested by Ray et al. (1983), they do not conflict with it. Neither do Hallet and Prestrud's concepts preclude roles for frost heaving and sorting. Recent results from movement studies within sorted circles at Resolute in the Canadian Arctic obtained by Washburn (1989) also may be interpreted as being derived from soil circulation, although he was conservative in this, and Washburn's experiments, like those of Hallet and Prestrud, cannot yet be considered definitive.

The diverse morphology and potentially complex origins of patterned ground are still adequately characterized by Washburn's (1956) seminal examination of the topic. Despite the undoubted complexities that remain to be unraveled, it is quite clear that there are now substantive advances being made to provide physically based explanations of the myriad of patterned ground forms which have long been reported in the field-based literature. It is worth emphasizing the roles that are being assigned to soil moisture differences in recent work, as this is a further example of the benefits derived from reducing the preoccupation with temperature alone.

Frost (thermal contraction) cracking

Frost cracking, a widespread phenomenon in regions underlain by permafrost, occurs when thermally derived stresses exceed the tensile strength of the soil (Mackay, 1986). Lachenbruch (1962) provided the most frequently cited theoretical explanation in English. The explanation offered by Lachenbruch was a comprehensive one embracing not only the cause of cracking but also its depth, pattern, and repetitive nature, as well as minor secondary features such as the shallow surficial troughs which frequently mark the locations of frost cracks occupied by ice wedges.

Given the acceptable nature of Lachenbruch's explanation of frost cracking, most of the ensuing fieldwork has focused upon the associated ice-wedges. Frost cracking provides a location for the concentration of surficial meltwaters during the melt season. Such water freezes when it seeps into cracks in the permafrost, and it is the repetitive cracking of the ice wedge and refreezing of the meltwater which makes the ice wedge grow. Much of the detailed fieldwork (e.g., Black, 1974, 1982) has concentrated upon calculation of ice-wedge growth rate and associated attempts at dating. More recently, Mackay's (1986) lake drainage experiment at Illisarvik has provided an opportunity to study frost-crack and ice-wedge initiation. Field data from Illisarvik has revealed very rapid crack growth initially (as much as an order of magnitude greater than previous estimates of long-term growth rates), but encroaching vegetation trapped sufficient snow to terminate the growth of some cracks. Furthermore, although some features of frost-crack theory were substantiated (e.g., upward growth of cracks from ice wedges), others were not (e.g., direction of ice crack propagation and the linear coefficient of ground expansion).

Although frost cracking is most commonly associated with permafrost, it is possible for seasonal frost cracks to occur. These features have not been as widely studied as cracks occurring within permafrost but represent a potentially difficult issue for researchers using cracks as a diagnostic characteristic of permafrost. In general, ice wedges are indicative of "wet" permafrost, in contrast to sand wedges (a type of soil wedge), which are indicative of dry permafrost; other types of soil wedge are not diagnostic of permafrost at all (Black, 1976).

At the present time our knowledge of the mechanics of ice-wedge initiation and development appears to meet all the criteria specified earlier in this paper. However, problems remain with respect to rates and patterns of development, particularly in the early states of growth.

Fluvial, lacustrine, and marine contexts

If frost processes are retained as the unifying theme in periglacial geomorphology, it is possible to develop operational definitions for fluvial, lacustrine, and marine contexts. The presence of seasonal ice cover on the waterbody becomes the necessary and sufficient condition to make the environment a periglacial one, and the impact of the ice on the abutting shoreline becomes the local geomorphic issue.

In fluvial situations the terrace-like riverbank features created by ice-drives (e.g., Smith, 1980) may be considered diagnostic of a periglacial regime. Nevertheless, there are many fluvial situations in cold climates which appear to lack diagnostic landforms. No aspect of valley form in and of itself appears to be sufficient to establish a periglacial environment. This statement is not intended to deny that some asymmetric valleys are created by differences in periglacial process regimes, but merely to emphasize that asymmetry in and of itself cannot be considered diagnostic.

Small-scale features associated with ice-shove (e.g., Dionne and Laverdière, 1972) and ice-rafting processes (e.g., Dionne, 1981) appear to be sufficient to establish both lacustrine (Dionne, 1979) and marine shorelines as periglacial. Indeed ice may play both erosional and depositional roles in some places such as a tidal flat (Dionne, 1988, 1989). However, inclusion of such features as extensive strandflats (e.g., Trenhaile, 1983; Dionne and Brodeur, 1988) quickly transports us from the realm of the defensible to that of the speculative.

The use of varves as a cold-climate indicator (implicitly embracing seasonal lake ice) is widespread. Nevertheless, it should be remembered that identification of varves is a complex issue. Rather than discuss it at length, I would merely note that only carefully documented identification should be considered a persuasive argument.

The presence of permafrost

Permafrost is widely defined as ground which has been less than 0°C continuously for two or more years (e.g., Washburn, 1980). This purely thermal definition is widely recognized as by-passing the issue of whether or not the ground moisture present is in the solid or liquid state. Although such an approach has obvious merit in defining precisely what constitutes permafrost, it offers much less utility from a geomorphic perspective and runs counter to the case being built herein.

The potential geomorphic significance of permafrost lies in its ability to generate unique processes and/or its ability to accelerate or intensify processes occurring in periglacial regions not underlain by permafrost. Permafrost does generate unique periglacial processes, but these are not ubiquitous. In contrast, some typical periglacial processes and derivative forms occur without regard to the presence or absence of permafrost. Consequently, it appears appropriate to regard permafrost as a sufficient cause to identify an area as periglacial, but inappropriate to regard it as a necessary attribute of periglacial regions (e.g., Péwé, 1969; Harris, 1988). These traits produce a very broad issue within the discipline and can only be treated selectively here.

UNIQUE PERMAFROST PROCESSES

Weathering

Accepting weathering to be the mechanical and/or chemical breakdown of sediment and rock at or near the surface, there do not appear to be any weathering processes which may be operationally defined, in accordance with the criteria specified earlier, as unique to permafrost. Nevertheless, there are a number of equivocal issues that emerge under a conventional definition of periglacial landscapes. One example is the role of permafrost in cryoplanation

(Reger and Péwé, 1976), which includes a weathering component because it is largely dependent upon the process of nivation. Reger and Péwé (1976) regarded cryoplanation terraces as permafrost indicators, and findings such as those of Mackay and Slaymaker (1989) certainly suggest that permafrost may locally concentrate bedrock weathering by frost shattering. However, other researchers (e.g., Demek, 1969) question the need for the presence of permafrost in the process of cryoplanation.

Both cryoplanation and nivation are examples of concepts that presently defy satisfactory operational definition because they embrace many component parts. It is the unquestioned acceptance of debatable concepts such as these which must be abandoned if the discipline is to progress.

Mass wasting

French (1976) identified ground ice slumping as one of the principal geomorphic processes in permafrost regions, while Mackay (1970) suggested that a careful distinction should be made between thermokarst subsidence and thermokarst erosion on the basis of the former being independent of flowing water while the latter requires it. The undoubted importance of thermokarst processes is obvious in the presence of degrading permafrost, but it may be extremely difficult to establish the past presence of permafrost once degradation is complete. Without careful measurement, other unique processes may be difficult to identify even when active, summer frost heave in sediments overlying permafrost (Mackay, 1981) being a case in point.

One important topic where the role of permafrost remains uncertain is that of rock glaciers. A rock glacier derived from a degenerate glacier (i.e., a glacier-debris-system rock glacier; Whalley, 1974) might arguably be considered glacial rather than periglacial. A rock glacier derived from talus (i.e., a talus-derived rock glacier, Johnson, 1975) might or might not contain ice and, consequently, might be unrelated to periglacial conditions as commonly defined or to any frost-related concept.

Recent theoretical modeling of transverse ridges on rock glaciers (Loewenherz, Lawrence, and Weaver, 1989) suggests that the significance of some surficial features with respect to what they indicate about flow behavior may need fundamental rethinking. In short, until the significance of the presence or absence of ice and/or the distribution of ice (i.e., ice core or ice cement) on flow characteristics can be determined, a meaningful assessment of what does or does not constitute a rock glacier is unlikely to be amenable to an operational definition.

Ice-cored features

Landforms unique to permafrost produced by ice cores are relatively common. Such phenomena span a wide range of sizes and include a number of forms: for example, ice wedges (Black, 1974; Mackay, 1986), palsas (Seppälä, 1982, 1988), and pingos (Mackay, 1978; Pissart, 1988). Some features are also ephemeral: for example, seasonal frost mounds (Pollard, 1988). In general terms this appears to be a sphere of research which is soundly based and, while careful distinctions are necessary (see earlier discussion of ice wedges), operational definitions meeting the specified criteria can be made.

Fluvial, lacustrine, and marine processes and environments

The presence of permafrost substantially modifies hydrological regimes (Woo, 1986) and is of undoubtable importance in fluvial, lacustrine, and marine contexts. In some instances ice patterns within the permafrost produce direct relations within thermokarst: for example, beaded drainage and baydjarakhs (Washburn, 1980, p. 270). However, it seems certain that the bulk of thermokarst subsidence and erosion produces amorphous results.

In some instances thermokarst subsidence is the dominant mechanism; Mackay (personal communication, 1990) cites several thousand thaw lakes in Alaska and Canada, and Washburn (1980, p. 276) identifies initiation of Yakutian alases. However, there appears to be considerable scope for interaction between thermokarst subsidence and erosion, as in the development of alases (Washburn, 1980, p. 276).

The dramatic impact of marine erosion on ice-rich permafrost shorelines is quite evident in such locations as the Tuktoyaktuk Peninsula of the Northwest Territories. However, the rapid rates of retreat are not associated with discrete forms which may be assigned uniquely to permafrost.

PERMAFROST-INTENSIFIED PROCESSES

A comprehensive review of all the geomorphic processes intensified and/or accelerated by the presence of permafrost is not possible here. Suffice it to say that permafrost degradation is likely to produce rapid rates of change in comparison to similar processes operating on only seasonally frozen materials. Fundamental to increased rates is the ice content of the permafrost, its temperature, and the nature of the host material.

NON-FROST PROCESSES

As noted earlier, Washburn (1980) asserted that some periglacial phenomena are not frost-derived. Such a view dismembers the conceptual unity maintained hitherto; nevertheless, it is a perfectly valid point if the objective is to determine the distinctive characteristics of cold, but unglaciated, landscapes. The critical prerequisite is to maintain the same rigorous standards of both theoretical and/or empirical development and field testing.

Mass wasting

The logical candidates for periglacial mass wasting processes of non-frost origin are those generated by seasonal snowpack and by wind. While recognizing that seasonal snowpacks are wind-dependent and that eolian processes are partially dependent upon the hardness of windblown snow, the two issues may be treated discretely.

Domination of a precipitation regime by seasonal snowpack accumulation and subsequent meltout, as opposed to domination by rainfall, may represent a climatic variable of fundamental geomorphic importance (Thorn, 1978). However, in at least some alpine periglacial regimes summertime rainfall remains an important input (Caine, 1976). Although seasonal snowpacks may have a profound impact on landscape development, they frequently fail to generate discrete landforms. Perhaps the best examples of landforms clearly derived from snowpack activity are those associated with avalanches (e.g. Luckman, 1977). Avalanches make their most pronounced impact upon taluses where they produce features ranging from debris tails (Luckman, 1977; White, 1981), through protalus ramparts (White, 1981), to avalanche boulder tongues (Rapp, 1959) and entire talus profiles (Caine, 1969; White, 1981).

Unfortunately, periglacial geomorphologists have focused most of the attention they pay to snowpacks upon the concept of nivation (Matthes, 1900; Thorn, 1988b). This morphogenetic term embraces numerous component concepts and is unlikely to be defined operationally in the foreseeable future. Its assumptions may be misleading and, consequently, it fails to provide the kind of definitive process-form relationship sought. Indeed it represents the kind of dated approach which must be abandoned in favor of unmuddled terminology and critical analysis. The issues imbedded in a term such as nivation are critical. Indeed they are so central to periglacial geomorphology that we can no longer afford to take them for granted and must subject them to the harsh light of the criteria proposed earlier in this paper.

Eolian processes are clearly important in many periglacial climates where cold temperatures preclude vegetation cover or destroy it (e.g., "turf exfoliation," Troll, 1944), and strong winds deflate bare bedrock and debris surfaces. Depending upon the partical sizes transported, the depositional phase of periglacial eolian processes may range from cover sands to loess. Although eolian lag surfaces are apparent in many periglacial environments (e.g., in Iceland, Schunke, 1975; in Scotland, Ballantyne, 1987; and in Colorado, U.S.A., Thorn and Darmody, 1985), there appear to be no definitive measures with which to differentiate between periglacial and desert lag surfaces. Moreover, it is difficult to identify any prospective diagnostic features with which to attempt this task. Again comparative scanning electron microscopy studies may provide an avenue of advancement (e.g., Dowdeswell, 1982; Dowdeswell, Osterman, and Andrews, 1985).

Identification of periglacial eolian deposition suffers many of the same problems recognized in the preceding paragraph. Niveo-eolian deposits form a link between snowpack and eolian concepts. However, the ephemeral nature of the forms produced by snow/debris mixtures (Koster, 1988) means that interpretation of such features is still extremely tenuous. Dune forms alone are rarely distinctively periglacial as such, although the extreme cold of Antarctic air masses produces a density relationship which, when combined with high windspeeds, may result in unusually large particles being incorporated in periglacial dunes (Smith, 1966; Selby, Rains, and Palmer, 1974).

The voluminous literature on loess cannot disguise the fact that although its origin is commonly in a cold environment, other conditions may prevail in some source regions. Furthermore, long distance transport of loess (e.g., Ruhe, 1969) prior to deposition means that nothing relevant to the issues at hand may be inferred directly about an environment merely from the presence of loess. Under these circumstances loess appears to be a poor diagnostic tool for present purposes.

Fluvial, lacustrine, and marine environments

While the roles of seasonal ice cover on water bodies and ground ice within land areas abutting water bodies are reasonably easy to conceptualize, identification of periglacial fluvial, lacustrine, and marine processes not associated with frost appears difficult. Although such mechanisms may exist, there appear to be none for which a rigorous operational definition can be created and, consequently, the topic will not be developed here.

Conclusions

Before attempting to answer the three direct questions posed in the title of this paper, it is appropriate to emphasize the context. Geomorphologists seek to explain landscapes past, present, and future. In undertaking this task they recognize that they are trying to unravel a web of interactions between forms and processes occurring at a multitude of scales. Within this context identification and isolation of individual forms and processes is an admission of temporary failure; artificial limits are established to simplify the matter at hand by eliminating unmanageable complexity. Such a strategy is an essential element in scientific progress. Nevertheless, if this strategy is to be successful, it places some very stringent demands upon the isolation procedure.

Identification of discrete forms (where form is used in the broad sense defined earlier) is necessary because it provides the only means of tying together the past, present, and future. Although focusing attention on form is necessary, it is also flawed. Geomorphic processes are pervasive and the ultimate source of explanation, but much of the landscape appears amorphous or poorly formed when measured against specific morphologic criteria.

For this reason form should also be seen as a means to an end and should never take center stage itself.

The only forms worthy of detailed attention are those whose explanation adds something to existing knowledge of landscape development. Conceptually, it would be preferable if these explained forms provided some unique element; however, such an attribute is unlikely if the form is drawn from a continuum. Therefore attention is most likely to be directed at forms whose explanation embraces intense development of a trait relatively weakly developed elsewhere. The traits that are likely to be of interest to geomorphologists are broad, among the more obvious are responses to dominance by a particular process or rock type (see Caine, 1983, for a good periglacial illustration of the latter).

However exotic or rare the object of the geomorphologist's attention, it is commonly drawn from a continuum. Initial investigation invariably involves isolating the object and distinguishing between noise and signal. Nevertheless, a comprehensive understanding of any process or feature must ultimately include attempts to understand its relationships within a larger framework.

The general points raised above reveal contemporary periglacial geomorphology to be a diffuse, and in many ways unsatisfactory, subdiscipline. It may be characterized as a potpourri of elements held together by a fuzzily defined framework which, in turn, is tied to a larger framework by concepts which are too infrequently examined.

The key issue in creating an operational definition of periglacial geomorphology is identification of a unique, or more probably intensely developed, geomorphic process or processes. Commonly this goal has been sought indirectly through identification of readily characterized surrogate measures, such as climatic or ecological parameters. These approaches clearly substitute ease of generalization for accuracy and, consequently, embrace uncertainties which are becoming increasingly unsatisfactory. A seemingly more accurate approach has focused upon temperature, and particularly upon freezing and thawing. However, while temperature alone may serve as an adequate surrogate for the state (but not volume) of ground moisture when monitoring established mechanisms, it cannot serve a definitive role.

The objective for a physically based definition of periglacial geomorphology should be evaluation of the geomorphic role of ground ice. It follows axiomatically from such a choice that responses of the landsurface to the aggradation and degradation of ground ice (i.e., freezing and thawing), rather than to the mere presence of ice, are probably going to constitute the dynamic components of the subdiscipline. It also follows that the behavior of the ice itself, or alone, is going to be a closely allied field.

The geomorphic role of snow, particularly seasonal snowpacks, may be logically assigned to periglacial or glacial geomorphology, depending upon personal preference. However, as contemporary definitions of periglacial geomorphology invariably exclude glacial ice, and as glacial geomorphologists have no apparent interest in the geomorphic role of snowpacks per se, there is a prima facie argument for including the geomorphic role of snowpacks in periglacial geomorphology. The existence of a blurred boundary between periglacial and glacial geomorphology must simply be recognized as having a spatial implication as well as the more commonly recognized temporal one (i.e., the earlier glaciation of many present-day periglacial areas).

A focus upon the geomorphic impact of ground ice and snowpacks will have the additional benefit of diminishing periglacial geomorphologists' preoccupation with temperature alone. Although it is self-evident that the concept of ground ice implicitly embraces temperature, the impact of ground ice also implicitly embraces water supply and the nature of the host material. A shift from a single-cause perspective to an interactive perspective should be a healthy one.

The suggested restriction of periglacial geomorphology to a focus upon ground ice and snowpack phenomena is really only a formal recognition of reality. In a text which is certainly outstanding and comprehensive, Washburn (1980) could only muster seven pages on fluvial environments, eight pages on coastal environments, and six pages on eolian processes out of a grand total of 320 pages. The remaining ca. 300 pages are overwhelmingly devoted to ground ice phenomena in one way or another. It is food for thought to juxtapose these data on Washburn's book (the "Bible" of our discipline) with Rapp's (1960) summary data for Kärkevägge. If nothing else, Rapp's relative estimates of removal by solution versus all other processes should serve to prompt the question: "What exactly is the periglacial component in the geomorphology of cold regions?". Are we perhaps devoting all of our attention to the icing on the cake and paying no attention to the cake itself?

If periglacial geomorphology is centered upon the geomorphic impact of ground ice, then there is some reasonable prospect of creating a unified body of theory. If this can be done, it is essential that periglacial geomorphologists immediately begin to investigate over what scale range the theory is applicable. At present the largest theoretically substantiated landform of exclusively periglacial origin appears to be a pingo. Larger features, such as alases, may have an exclusively periglacial origin but differing developmental pathways; yet others, such as cryoplanation terraces, are still open to serious debate. Stoddart's criticism, cited earlier, that climatic geomorphologists consider only type-landforms and not landform assemblages is valid with respect to all periglacial geomorphologists regardless of whether they approach the field from a climatic or a process perspective.

The development of periglacial geomorphology has been such that while numerous landforms have been designated periglacial, similar identification of landscapes remains elusive. A very high proportion of what we know about periglacial geomorphology is represented by small landforms confined to highly specialized micro- and meso-environments. This leaves large swaths of the periglacial landscape geomorphically unknown, a point made by Barsch and Caine (1984). It also raises the issue of what kind of coverage by periglacial elements is going to be required if an entire landscape is to be labeled periglacial. This kind of difficulty is not unique to periglacial geomorphology but must be addressed immediately. A few papers contain some very interesting approaches to this issue: for example, Caine's (1979) synthetic or statistical approach, Morris and Olyphant's (1990) facies approach, and Cook's (1989) gradient or catena-like approach.

The difficulties in assessing the spatial contribution made to a landscape by periglacial geomorphology are matched by difficulties associated with estimating the contribution made over time. Not only have many present-day periglacial regions experienced glaciation but many regions which presently exhibit periglacial landforms may not presently experience the climate to sustain them at anything like their original levels of activity. For example, movement rates exhibited by some rock glaciers in the Front Range of Colorado are extremely slow (Vitek and Giardino, 1987). The fact that high standing masses of unconsolidated debris move is hardly surprising. The real question is whether or not the measured rates merely represent degeneration or sustainable development. Questions such as this are only likely to be answerable from a sound theoretical perspective.

Given the tenor of this paper, the appropriate answer to the tripartite question in the title is a tentative operational definition of periglacial geomorphology.

> Periglacial geomorphology is that part of geomorphology which has as its primary object physically based explanations of the past, present, and future impacts of diurnal, seasonal, and perennial ground ice on landform and landscape initiation and development. Additional components of the subdiscipline include similar investigations of the geomorphic roles of snowpacks (but not glaciers) and fluvial, lacustrine, and marine ice.

This definition embraces process and form operationally and symbiotically and reflects a coherent objective. It also implies that periglacial (as now defined) geomorphologists must address the contextual question: "What was/ is (/will be?) the relative significance of ground ice processes in the development of cold region landscapes?"

Acknowledgments

I should like to thank J. Ross Mackay and Scott Morris for their helpful comments on an earlier version of this paper. I should also like to thank A.

Lincoln Washburn for a particularly penetrating discussion of the issues raised. Needless to say, any and all remaining errors and flaws are entirely my own responsibility.

Notes

1. In writing this paper I do not wish to focus upon semantics. I will adhere to the usage proposed by Washburn (1980, pp. 1–2) in which geocryology is the study of "frozen ground (seasonally frozen ground as well as permafrost) but not . . . glaciers" and the term periglacial is used (instead of geocryologic) "to designate nonglacial processes and features of cold climates regardless of age and of any proximity to glaciers." Deviations from this usage, as proposed by others, will be appropriately emphasized in the text.
2. As noted by Haines-Young and Petch (1983), it is quite clear that to a geomorphologist the concept landform embraces more than just the external morphology. It may also include internal structure and known functional attributes.
3. Frost action is used here as a synonym for any process stemming directly from the freezing and/or thawing of ground ice.

References

Andersson, J.G., Solifluction, a component of subaerial denudation, *Journal of Geology*, **14**, 91–112, 1906.

Ballantyne, C.K., The present-day periglaciation of upland Britain, in *Periglacial Processes and Landforms in Britain and Ireland*, edited by J. Boardman, pp. 113–126, Cambridge University Press, Cambridge, 1987.

Barsch, D., and N. Caine, The nature of mountain geomorphology, *Mountain Research and Development*, **4**, 287–298, 1984.

Black, R.F., Ice-wedge polygons of northern Alaska, in *Glacial Geomorphology*, edited by D.R. Coates, pp. 247–275, Publications in Geomorphology, Binghamton, N.Y., 1974.

Black, R.F., Periglacial features indicative of permafrost: ice and soil wedges, *Quaternary Research*, **6**, 3–26, 1976.

Black, R.F., Patterned-ground studies in Victoria Land, *Antarctic Journal of the United States*, **17**, 53–54, 1982.

Brunsden, D., and J.B. Thornes, Landscape sensitivity and change, *Transactions of the Institute of British Geographers, New Series,* **4**, 463–484, 1979.

Büdel, J., *Climatic geomorphology*, 443 pp., translated by L. Fischer and D. Busche, Princeton University Press, Princeton, N.J., 1982.

Caine, N., A model for alpine talus slope development by slush avalanching, *Journal of Geology*, **77**, 92–100, 1969.

Caine, N., The distribution of sorted patterned ground in the English Lake District, *Revue Géomorphologie Dynamique*, **21**, 49–56, 1972.

Caine, N., Summer rainstorms in an alpine environment and their influence on soil erosion, San Juan Mountains, Colorado, *Arctic and Alpine Research*, **8**, 183–196, 1976.

Caine, N., The problem of spatial scale in the study of contemporary geomorphic activity on mountain slopes (with special reference to the San Juan Mountains), *Studia Geomorphologica Carpatho-Balcanica*, **13**, 5–22, 1979.

Caine, N., *The Mountains of Northeastern Tasmania: A Study of Alpine Geomorphology*, 200 pp., Balkema, Rotterdam, 1983.

Cook, J.D., Active and relict sorted circles, Jotunheimen, Norway: A study of the

altitudinal zonation of periglacial processes, Ph.D. thesis, 566 pp., University of Wales, Cardiff, 1989.
Cooke, R.U., and A. Warren, *Geomorphology in Deserts*, 394 pp., University of California Press, Berkeley, Calif., 1973.
Corte, A.E., Particle sorting by repeated freezing and thawing, *Biuletyn Peryglacjalny*, **15**, 175–240, 1966.
Davis, W.M., The geographical cycle, *The Geographical Journal*, **14**, 481–504, 1899.
Davis, W.M., The geographical cycle in an arid climate, *Journal of Geology*, **13**, 381–407, 1905.
Demek, J., Cryoplanation terraces, their geographical distribution, genesis and development, *Československé Akademie Věd Rozpravy, Řada Matematických a Přírodních Věd*, **79**, 80 pp., 1969.
Derbyshire, E., Geomorphology and climate: background, in *Geomorphology and Climate*, edited by E. Derbyshire, pp. 1–24, John Wiley, London, 1976.
Dionne, J.-C., Ice action in the lacustrine environment. A review with particular reference to subarctic Quebec, Canada, *Earth-Science Reviews*, **15**, 185–212, 1979.
Dionne, J.-C., A boulder-strewn tidal flat, north shore of the Gulf of St. Lawrence, Quebec, *Géographie physique et Quaternaire*, **35**, 261–267, 1981.
Dionne, J.-C., Ploughing boulders along shorelines, with particular reference to the St. Lawrence estuary, *Geomorphology*, **1**, 297–308, 1988.
Dionne, J.-C., An estimate of shore ice action in a Spartina tidal marsh, St. Lawrence estuary, Quebec, Canada, *Journal of Coastal Research*, **5**, 281–293, 1989.
Dionne, J.-C., and D. Brodeur, Frost weathering and ice action in shore platform development with particular reference to Quebec, Canada, *Zeitschrift für Geomorphologie, Supplement-Band*, **71**, 117–130, 1988.
Dionne, J.-C., and C. Lavierdière, Ice formed beach features from Lac St. Jean, Quebec, *Canadian Journal of Earth Sciences*, **9**, 979–990, 1972.
Dowdeswell, J.A., Scanning electron micrographs of quartz sand grains from cold environments examined using Fourier shape analysis, *Journal of Sedimentary Petrology*, **52**, 1315–1323, 1982.
Dowdeswell, J.A., L.E. Osterman, and J.T. Andrews, Quartz sand grain shape and other criteria used to distinguish glacial and non-glacial events in a marine core from Frobisher Bay, Baffin Island, N.W.T., Canada, *Sedimentology*, **32**, 119–132, 1985.
Eakin, H.M., The Yukon-Koyukuk region, Alaska, *U.S. Geological Survey Bulletin 631*, 88 pp., 1916.
Evans, I.S., and N. Cox, Geomorphometry and the operational definition of cirques, *Area*, **6**, 150–153, 1974.
Fahey, B.D., An analysis of diurnal freeze-thaw and frost heave cycles in the Indian Peaks regions of the Colorado Front Range, *Arctic and Alpine Research*, **5**, 269–281, 1973.
Fahey, B.D., Seasonal frost heave and frost penetration measurements in the Indian Peaks region of the Colorado Front Range, *Arctic and Alpine Research*, **6**, 63–70, 1974.
French, H.M., *The Periglacial Environment*, 309 pp., Longman, London, 1976.
Fukuda, M., The pore water pressure profile in porous rocks during freezing, *Proceedings of the Fourth International Conference on Permafrost*, pp. 322–327, 1983.
Gilpin, R.R., A model for the prediction of ice lensing and frost heave in soils, *Water Resources Research*, **16**, 918–930, 1980.
Gleason, K.J., W.B. Krantz, N. Caine, J.H. George, and R.D. Gunn, Geometrical aspects of sorted patterned ground in recurrently frozen soil, *Science*, **232**, 216–220, 1986.

Grawe, O.R., Ice as an agent of rock weathering: a discussion, *Journal of Geology*, **44**, 173–182, 1936.

Haines-Young, R.H., and J.R. Petch, Multiple working hypotheses: equifinality and the study of landforms, *Transactions of the Institute of British Geographers, New Series*, **8**, 458–466, 1983.

Haines-Young, R.H., and J.R. Petch, *Physical Geography: its Nature and Methods*, 230 pp., Harper and Row, London, 1986.

Hall, K., Daily monitoring of a rock tablet at a maritime Antarctic site: Moisture and weathering results, *British Antarctic Survey Bulletin*, **79**, pp. 17–25, 1988.

Hallet, B., and S. Prestrud, Dynamics of periglacial sorted circles in western Spitsbergen, *Quaternary Research*, **26**, 81–99, 1986.

Harris, S.A., The alpine periglacial zone, in *Advance in Periglacial Geomorphology*, edited by M.J. Clark, pp. 369–413, John Wiley, Chichester, England, 1988.

Horton, R.E., Erosional development of streams and their drainage basins; hydrophysical approach to quantitative morphology, *Geological Society of America Bulletin*, **56**, 275–370, 1945.

Hudec, P.P., Weathering of rocks in arctic and sub-arctic environment, in *Proceedings of the Symposium of the Canadian Arctic*, edited by J.D. Aitken and D.J. Glass, pp. 313–335, Geological Association of Canada, Waterloo, 1973.

Jahn, A., Walery Łoziński's merits for the advancement of periglacial studies, *Biuletyn Peryglacjalny*, **1**, 117–124, 1954.

Johnson, P.G., Mass movement processes in Metalline Creek, southwest Yukon Territory, *Arctic*, **28**, 100–139, 1975.

Johnson, P.G., Rock glacier: glacier debris systems or high-magnitude low-frequency flows, in *Rock Glaciers*, edited by J.R. Giardino, J.F. Shroder, Jr., and J.D. Vitek, pp. 175–192, Allen and Unwin, Boston, 1987.

Kaplar, C.W., Stone migration by freezing of soil, *Science*, **149**, 1520–1521, 1965.

Koster, E.A., Ancient and modern cold-climate aeolian sand deposition: a review, *Journal of Quaternary Science*, **3**, 69–83, 1988.

Lachenbruch, A.H., Mechanics of thermal contraction cracks and ice-wedge polygons in permafrost, *Geological Society of America Special Paper 70*, 67 pp., 1962.

Lautridou, J.-P., Recent advances in cryogenic weathering, in *Advances in Periglacial Geomorphology*, edited by M.J. Clark, pp. 33–47, John Wiley, Chichester, England, 1988.

Lewkowicz, A.G., Slope processes, in *Advances in Periglacial Geomorphology*, edited by M.J. Clark, pp. 325–368, John Wiley, Chichester, England, 1988.

Loewenherz, D.S., C.J. Lawrence, and R.L. Weaver, On the development of transverse ridges on rock glaciers, *Journal of Glaciology*, **35**, 383–391, 1989.

Łoziński, W., Uber die mechanische Verwitterung der Sandsteine im gemässigten Klima, *Bulletin International de l'Académie des Sciences de Cracovie class des Sciences Mathématique et Naturelles*, **1**, 1–25, 1909.

Łoziński, W., Die periglaziale Fazies der mechanischen Vervitterung, *Proceedings of the 11th International Geological Congress*, 1039–1053, 1912.

Luckman, B.H., The geomorphic activity of snow avalanches, *Geografiska Annaler*, **59A**, 31–48, 1977.

Mackay, J.R., Disturbances to the tundra and forest tundra environment of the western Arctic, *Canadian Geotechnical Journal*, **7**, 420–432, 1970.

Mackay, J.R., The origin of massive icy beds in permafrost, western Arctic coast, Canada, *Canadian Journal of Earth Sciences*, **8**, 397–422, 1971.

Mackay, J.R., The world of underground ice, *Annals of the Association of American Geographers*, **62**, 1–22, 1972.

Mackay, J.R., Contemporary pingos: a discussion, *Biuletyn Peryglacjalny*, **27**, 133–154, 1978.
Mackay, J.R., Active layer slope movement in a continuous permafrost environment, Garry Island, Northwest Territories, Canada, *Canadian Journal of Earth Sciences*, **18**, 1666–1680, 1981.
Mackay, J.R., Downward water movement into frozen ground, western arctic coast, Canada, *Canadian Journal of Earth Sciences*, **20**, 120–134, 1983.
Mackay, J.R., The frost heave of stones in the active layer above permafrost with downward and upward freezing, *Arctic and Alpine Research*, **16**, 439–446, 1984.
Mackay, J.R., The first 7 years (1978–1985) of ice wedge growth, Illisarvik experimental drained lake site, western Arctic coast, *Canadian Journal of Earth Sciences*, **23**, 1782–1795, 1986.
Mackay, J.R., and R.V. Leslie, A simple probe for the measurement of frost heave within frozen ground in a permafrost environment, *Current Research, Part A, Geological Survey of Canada, Paper 87-1A*, 37–41, 1987.
Mackay, J.R., and W.H. Mathews, Needle ice striped ground, *Arctic and Alpine Research*, **6**, 79–84, 1974.
Mackay, J.R., and O. Slaymaker, The Horton River breakthrough and resulting geomorphic changes in a permafrost environment, western arctic coast, Canada, *Geografiska Annaler*, **71A**, 171–184, 1989.
Matthes, F.E., Glacial sculpture of the Bighorn Mountains, Wyoming, *U.S. Geological Survey 21st Annual Report 1899–1900*, pp. 167–190, 1900.
McGreevy, J.P., and W.B. Whalley, The geomorphic significance of rock temperature variations in cold environments: a discussion, *Arctic and Alpine Research*, **14**, 157–162, 1982.
McRoberts, E.C., and N.R. Morgernstern, The stability of thawing slopes, *Canadian Geotechnical Journal*, **11**, 447–469, 1974.
Medawar, P.B., *The Art of the Soluble*, 160 pp., Methuen, London, 1967.
Michaud, Y., J.-C. Dionne, and L.D. Dyke, Frost bursting: a violent expression of frost action in rock, *Canadian Journal of Earth Sciences*, **26**, 2075–2080, 1989.
Morawetz, S.O., Beobachtungen an Schutthalden, Schuttkegeln und Schuttflecken, *Zeitschrift für Geomorphologie*, **7**, 25–43, 1932.
Morris, S.E., and G.A. Olyphant, Alpine lithofacies variation: working toward a physically-based model, *Geomorphology*, **3**, 73–90, 1990.
Mortenson, H., Über die physikalische Möglichkeit der "Brodel"—Hypothese, *Centralblatt für Mineralogie, Geologie und Palaontologie Abt. B*, 417–422, 1932.
Outcalt, S.I., An algorithm for needle ice growth, *Water Resources Research*, **7**, 394–400, 1971.
Outcalt, S.I., A numerical model of ice lensing in freezing soils, *Technical Note, Corps of Engineers, U.S. Army Cold Regions Research and Engineering Laboratory, Hanover, New Hampshire*, 20 pp., 1976.
Outcalt, S.I., The effect of iteration frequency on a numerical model of near surface ice segregation, *Engineering Geology*, **13**, 111–124, 1979a.
Outcalt, S.I., The influence of the addition of water vapor diffusion on a numerical simulation of the process of ice segregation, *Frost i Jord*, **20**, 45–57, 1979b.
Outcalt, S.I., and K.M. Hinkel, Night-frost modulation of near-surface soil-water ion concentration and thermal fields, *Physical Geography*, **10**, 336–348, 1989.
Peltier, L.C., The geographic cycle in periglacial regions as it is related to climatic geomorphology, *Annals of the Association of American Geographers*, **40**, 214–236, 1950.
Péwé, T.L., The periglacial environment, in *The Periglacial Environment*, edited by T.L. Péwé, pp. 1–9, McGill-Queen's University Press, Montreal, 1969.

Pissart, A., Pingos: an overview of the present state of knowledge, in *Advances in Periglacial Geomorphology*, edited by M.J. Clark, pp. 279–297, John Wiley, Chichester, England, 1988.
Pollard, W.H., Seasonal frost mounds, in *Advances in Periglacial Geomorphology*, edited by M.J. Clark, pp. 201–229, John Wiley, Chichester, England, 1988.
Priesnitz, K., Cryoplanation, in *Advances in Periglacial Geomorphology*, edited by M.J. Clark, pp. 49–67, John Wiley, Chichester, England, 1988.
Rapp, A., Avalanche boulder tongues in Lappland, *Geografiska Annaler,* **41**, 34–48, 1959.
Rapp, A., Recent development of mountain slopes in Kärkevägge and surroundings, northern Scandinavia, *Geografiska Annaler*, **42**, 65–200, 1960.
Ray, R.J., W.B. Krantz, T.N. Caine, and R.D. Gunn, A model for sorted patterned-ground regularity, *Journal of Glaciology*, **29**, 317–337, 1983.
Reger, R.D., and T.L. Péwé, Cryoplanation terraces: indicators of a permafrost environment, *Quaternary Research*, **6**, 99–109, 1976.
Ruhe, R.V., *Quaternary Landscapes in Iowa*, 255 pp., Iowa State University Press, Ames, Iowa, 1969.
Sauchyn, D.J., and J.S. Gardner, Morphometry of open rock basins, Kananaskis area, Canadian Rocky Mountains, *Canadian Journal of Earth Sciences*, **20**, 409–419, 1983.
Schunke, E., Formungsvorgange an Schneeflecken im islandische Hochland, in Geomorphologische Prozesse und Prozesskombinationen in der Gegenwart unter verschiedenen Klimabedingungen, edited by H. Poser, pp. 274–286, *Akademie Wissenschaften Gottingen Abhandlungen Mathematisch-Physikalische Klasse Folge 3*, **29**, 1974.
Schunke, E., Die Periglazialerscheinungen Islands in Abhängigkeit von Klima und Substrat, *Akademie Wissenschaften Gottingen Abhandlungen Mathematisch-Physikalische Klasse Folge 3*, **30**, 273 pp., 1975.
Selby, M.J., R.B. Rains, and W.P. Palmer, Eolian deposits of the ice-free Victoria valley, southern Victoria Land, Antarctica, *New Zealand Journal of Geology and Geophysics,* **16**, 543–562, 1974.
Seppälä, M., An experimental study of the formation of palsas, *Proceedings Fourth Canadian Permafrost Conference*, pp. 36–42, 1982.
Seppälä, M., Palsas and related forms, in *Advances in Periglacial Geomorphology*, edited by M.J. Clark, pp. 247–278, John Wiley, Chichester, England, 1988.
Smith, D.G., River ice processes: thresholds and geomorphological effects in northern and mountain rivers, in *Thresholds in Geomorphology*, edited by D.R. Coates and J.D. Vitek, pp. 323–343, Allen and Unwin, London, 1980.
Smith, H.T.U., Wind-formed pebble ripples in Antarctica, *Geological Society of America Special Paper 87*, 160 pp., 1966.
Soons, J.M., and D.E. Greenland, Observations on the growth of needle ice, *Water Resources Research*, **6**, 579–593, 1970.
Stoddart, D.R., Climatic geomorphology: review and re-assessment, *Progress in Geography*, **1**, 159–222, 1969.
Strahler, A.N., Dynamic basis of geomorphology, *Geological Society of America Bulletin*, **63**, 923–938, 1952.
Taber, S., Frost heaving, *Journal of Geology*, **37**, 428–461, 1929.
Taber, S., The mechanics of frost heaving, *Journal of Geology*, **38**, 303–317, 1930.
Taber, S., Perenially frozen ground in Alaska: its origin and history, *Geological Society of America Bulletin,* **54**, 1433–1548, 1943.
Thorn, C.E., The geomorphic role of snow, *Annals of the Association of American Geographers*, **68**, 414–425, 1978.

Thorn, C.E., Bedrock freeze-thaw weathering regime in an alpine environment, Colorado Front Range, *Earth Surface Processes*, **4**, 211–228, 1979.

Thorn, C.E., *Introduction to Theoretical Geomorphology*, 247 pp., Unwin Hyman, Boston, 1988a.

Thorn, C.E., Nivation: a geomorphic chimera, in *Advances in Periglacial Geomorphology*, edited by M.J. Clark, pp. 3–31, John Wiley, Chichester, England, 1988b.

Thorn, C.E., and R.G. Darmody, Grain-size sampling and characterization of eolian lag surfaces with alpine tundra, Niwot Ridge, Front Range, Colorado, U.S.A., *Arctic and Alpine Research*, **17**, 443–450, 1985.

Trenhaile, A.S., The development of shore platforms in high latitudes, in *Shorelines and Isostasy*, edited by D.E. Smith and A.G. Dawson, pp. 77–93, Academic Press, London, 1983.

Tricart, J., Étude expérimentale du problème de la gélivation, *Biuletyn Peryglacjalny*, **4**, 285–318, 1956.

Tricart, J., *Geomorphology of Cold Environments*, translated by E. Watson, 320 pp., Macmillan, London, 1970.

Tricart, J., and A. Cailleux, *Introduction to Climatic Geomorphology,* translated by C.J. Kiewet de Jonge, 295 pp., Longman, London, 1972.

Troll, C., Strukturböden, Solifluktion und Frostklimate der Erde, *Geologische Rundschau*, **34**, 545–694, 1944.

Vitek, J.D., and J.R. Giardino, Rock glaciers: a review of the knowledge base, in *Rock Glaciers*, edited by J.R. Giardino, J.F. Shroder, Jr., and J.D. Vitek, pp. 1–26, Allen and Unwin, Boston, 1987.

Walder, J., and B. Hallet, A theoretical model of the fracture of rock during freezing, *Geological Society of America Bulletin*, **96**, 336–346, 1985.

Washburn, A.L., Classification of patterned ground and review of suggested origins, *Geological Society of America Bulletin*, **67**, 823–865, 1956.

Washburn, A.L., Instrumental observations of mass-wasting in the Mesters Vig district, northeast Greenland, *Meddelelser om Gronland*, **166**, pp. 318, 1967.

Washburn, A.L., *Geocryology: A Survey of Periglacial Processes and Environments*, 406 pp., John Wiley, New York, 1980.

Washburn, A.L., Near-surface soil displacement in sorted circles, Resolute area, Cornwallis Island, Canadian High Arctic, *Canadian Journal of Earth Science*, **26**, 941–955, 1989.

Whalley, W.B., Rock glaciers and their formation as part of a glacier debris transport system, *Geographical Paper No. 24*, 60 pp., Department of Geography, University of Reading, Reading, England, 1974.

Whalley, W.B., Rockfalls, in *Slope Instability*, edited by D. Brunsden and D.B. Prior, pp. 217–256, John Wiley, Chichester, England, 1984.

White, S.E., Alpine mass movement forms (noncatastrophic): classification, description, and significance, *Arctic and Alpine Research*, **13**, 127–137, 1981.

Woo, M.-K., Permafrost hydrology in North America, *Atmosphere-Ocean*, **24**, 201–234, 1986.

2 Origin of Certain High-Elevation Local Broad Uplands in the Central Appalachians South of the Glacial Border, U.S.A.—A Paleoperiglacial Hypothesis

G. Michael Clark
Department of Geological Sciences, University of Tennessee

James Hedges
Big Cove Tannery, Pennsylvania, U.S.A.

Abstract

Form, relief, and coincidence in elevation of broad, flat ridge crests developed on Central Appalachian summits are vital characteristics that bear on the origins of Appalachian topography and drainage. Investigated sites include summit flats and bordering risers and their subjacent terraces that truncate both lithology and bedrock structure. In order of descending elevation are summit tors, summit flats, one to several terrace scarps or risers, and subjacent treads or terraces that slope gently valleyward. Summit tors are free-standing, tower-like masses of bedrock or shattered rock. Where bedrock exposures exist, topographic flats truncate both lithology and structure and may have examples of large-scale sorted patterned ground. Terrace scarps may be rubble covered, mantled by shattered but essentially in-place rock, or exposed bedrock cliffs. Treads may be blanketed by contiguous block rubble (block slopes), block streams, or stony soils that in some localities exhibit sorted patterned ground. Described features have weathering and soil

Periglacial Geomorphology. Edited by J. C. Dixon and A. D. Abrahams

development properties that are evidence of inactive or fossil states. A tentative hypothesis of cryoplanation explains known field relationships, may help to account for the origin of certain colluvium deposits by providing a source area, and obviates the necessity for a complicated erosional history based on long-term cyclical events to explain the origin of summit morphology. Problems of subsurface geometry, age and correlation, mechanisms of origin, and environmental conditions of development are unsolved and must be addressed to test this hypothesis and to develop a thorough understanding of summit-level morphogenesis and history.

Introduction

The origin and evolution of Appalachian topography and drainage are classic and monumental problems in historical and regional geomorphology. Mills, Brakenridge, Jacobson, Newell, Pavich, and Pomeroy (1987) summarized the state of Appalachian geomorphic research and noted some pressing landscape origin problems in the region. Gardner and Sevon (1989) brought together papers on geomorphic evolution of the Appalachians that focused attention on fundamental questions about the geomorphic history of this mountain system. Among the many unsolved geomorphic problems is the existence of accordant ridge tops that locally truncate lithology and structure in the upland areas (Bryan, Cleaves, and Smith, 1932/33). For many decades, the most common hypothesis to explain Appalachian upland surfaces was the peneplain concept (Sevon, Potter, and Crowl, 1983). Davis (1889) elaborated details of his peneplain concept using the Ridge and Valley province in Pennsylvania and interpreted the highest levels as remnants of a once-continuous peneplain. Davis's promulgation of the geographical cycle of erosion was so effective that for decades many workers concentrated on definition, description, and correlation of numerous summit level accordances (Sevon, Potter, and Crowl, 1983) within the Central Appalachians south of the glacial border (Figure 2.1). Since the peneplain concept fell out of favor (Flemal, 1971), a common practice has been to deny, ignore, or explain away the fact that ridge top relationships do exist. The ascendance in importance of process geomorphology in the United States might have discouraged researchers from investigation of field relationships cited by earlier workers as evidence for various regional denudational chronologies. Mills et al. (1987) remarked on the overall shortage of geomorphologists working in the Appalachians and noted many pressing applied geomorphic problems in the region. These trends in American geomorphology, and perhaps other factors, have resulted in a nearly complete stasis of geomorphic research on upland surfaces in the region.

Monmonier (1967, 1968, 1971), however, employed trend surface analysis to study ridge crest topography and elevation in an objective manner and to determine quantitatively whether or not accordance exists. Data were taken from large-scale topographic maps that cover the Ridge and Valley area of Pennsylvania. Moreover, Monmonier divided ridge samples into a population of narrow ridge crests and one of local broad uplands. In effect, this subdivision separates summits controlled by different aspects of structure and rock character. Monmonier found that much local variation in elevation can

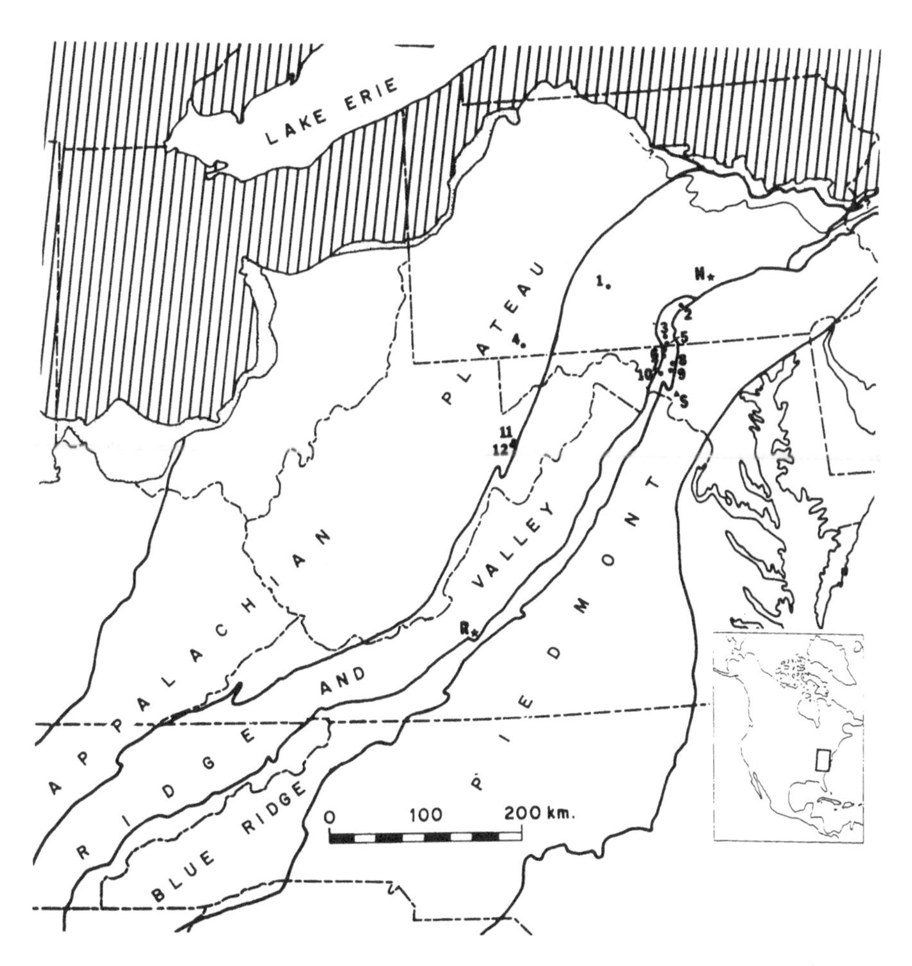

Figure 2.1. Location map with geomorphic provinces in central portion of Appalachian Highlands Major Division (Thornbury, 1965). Site numbers refer to Table 2.2. R = Roanoke, H = Harrisburg, S = Sugarloaf Mountain. Striped pattern = Late Wisconsinan drift; dotted pattern = Pre-Wisconsinan drift (undifferentiated). In northeastern Pennsylvania only, heavy line with question mark = Illinoian drift; light line with question mark = Pre-Illinoian drift (Braun, 1989a, and personal communication, 1990)

be explained by the presence of only one formation underlying local broad uplands.

Hack (1975), by contrast, studied the overall summit relationships among rock types, outcrop width, and structural attitude at a scale of 1:250,000 for an area north of Harrisburg, Pennsylvania. He showed that, at this scale, the apparent accordance of summits is due to fold regularity, thinness of the resistant sandstones, and the spacing of the sandstone units in the geologic column. At a comparable scale, Monmonier (1971) demonstrated that local broad uplands in Pennsylvania also have a regional relationship to the major rivers that tend to flow down the dip of the trend surface and in broad troughs in it. Hack (1989) concluded that in most regions the general accordance of summits within a landscape can be related to uniform properties of resistant rock units, and that peneplanation is not necessary. At the scales required to generalize on a regional basis (cf. Table 2.1), we concur with these two authors. Left unanswered, however, are the problems cited by Bryan, Cleaves, and Smith (1932/33): the upland surface problem, the origin of the unconformity beneath the Atlantic Coastal Plain, and the origin of the transverse drainage (Clark, 1989b). With respect to the upland surface problem, the elements in summit topography that, in the aggregate, give rise to the remarkable evenness of skyline so admired by adherents to the geographical cycle of erosion have yet to be determined. Thus, despite efforts to ignore, refute, or explain away the existence of enigmatic upland surfaces, the problem of the origin of the local broad uplands still remains.

Another problem in the Appalachians at least spatially related to the uplands is the presence of complex, thick, and extensive diamicton deposits along many mountain flanks that border the upland surfaces, especially in the

Table 2.1. Ranking of Landscape Units Used in this Paper[a]

Rank	Area (km^2)	Basis (dominant entity)	Examples
Realm	10^7	Largest plate-tectonic units	North American Plate
Major division	10^6	Sub-continental entities	Appalachian Highlands
Province	10^5	Regional similarity	Ridge and Valley
Section	10^4	One tectonic-landscape style	Northern Blue Ridge
Subsection	10^3	Structure-landform similarity	South Mountain
District	10^2	Form-material relationships	Ridge Road Upland Area
Subdistrict	10^1	Direct material-form linkage	Local broad upland
Zone	10^0	Few form-relief parameters	Single summit cryoplain
Locale	10^{-1}	Individual landforms	Cryoplanation terrace
Compartment	10^{-2}	Single form/relief units	Lobe or terrace slope break
Feature	10^{-3}	Specific microform	Opferkessel; expanded joint

[a]Modified from Godfrey and Cleaves (1991).

Ridge and Valley and Blue Ridge provinces (King, 1950; Pierce, 1966; Moss, 1976; Clark and Ciolkosz, 1988; Braun, 1989b; Ciolkosz, Carter, Hoover, Cronce, Waltman, and Dobos, 1990). What were the specific source areas for the debris? What were the mechanisms of entrainment, transport, and deposition of the sediments? And when and under what conditions did these events occur? Except in several areas (cf. Braun, 1989b) little is known about the geometry and chronology of diamicton emplacement, so that questions of erosion magnitude and rates, geomorphic effectiveness of the responsible process groups, and Quaternary landscape history have been difficult to address in a quantitative manner.

The object of this paper is to examine local broad upland ridge top form and relief on the zonal landform level (Table 2.1). Implications for interpretation of Appalachian summits on local to broader scales are then drawn and their logical consequences tested. We consider the results of this reconnaissance study to be preliminary only. The conclusions, therefore, are presented in the form of a hypothesis that will require detailed testing and evaluation. Caution is especially necessary because of the dearth of subsurface information and lack of numerical age data. For example, Büdel (1977, pp. 57–59) interprets certain bedrock erosional surfaces as pre-Quaternary features simply on the grounds that they have been overprinted by periglacial processes.

Methodology

An area in the Central Appalachians south of the glacial border (Braun, 1989a) that included both the southern portion of the region studied by Monmonier (1967) and parts of Maryland and West Virginia was chosen for investigation (Figure 2.1). Our selection objective was to determine field relationships at significant distances (ca. 200 km) from Quaternary ice margins. General agreement exists today that periglacial effects near the glacial border were severe and that extremely harsh environmental conditions also extended to low elevations (Clark and Ciolkosz, 1988, and references contained therein). For example, Watts (1979) reported clear evidence of tundra vegetation in association with colluvium in areas within 35 km of the Late Wisconsinan border at the Longswamp, Pennsylvania site (40° 29′ N; 75° 40′ W) at an elevation of only 192 m. Marsh (personal communication, 1991) remarked on the great number of flat uplands in central Pennsylvania, and noted that contour maps from digital elevation tapes display flat, abrupt-edged uplands better than standard U.S. Geological Survey 7.5-minute quadrangle maps, even with the same contour interval. In one specific case, part of a digital elevation map covering the Hartleton, Pennsylvania, 7.5-minute quadrangle shows excellent examples of these highest flats that Marsh reports break abruptly into blocky slopes.

Topographic maps at 1:24,000 scale were examined to identify local broad upland sites on the zonal scale (Table 2.1). Many of the identified uplands display prominent topographic benches and some show chimney-like summit protuberances. Localities for field study were chosen where comparable-scale geologic maps were available or where it was known from other sources that bedrock exposures were present. Topographic profiles were constructed from theodolite and rod traverses orthogonal to the contours. These slope profiles and surficial materials were described utilizing standard geomorphological procedures (Gardiner and Dackombe, 1983). Terminology for landscape size classification follows Godfrey and Cleaves (1991) with minor modification (Table 2.1). Although most locations have road access, several sites do not, including locations along the Appalachian Trail or near power transmission lines.

Our rationale was to select a number of representative sites (Table 2.2) sufficiently distant from the glacial border and determine ridge top geomor-

Table 2.2. Locations of Selected Central Appalachian Local Broad Upland Sites

Map Site[a]	U.S.G.S. 7.5′ Map (Locality Name)	Latitude Longitude	Elevation (ft) Elevation (m)	Feature name Aspect
1	Saxton, Pa.	40° 14′ 46″ N	2280–2300	Subsummit terraces
	(Broad Top)	78° 07′ 40″ W	695–701	N 71.5° E
2	Mount Holly Springs, Pa.	40° 04′ 05–20″ N	1405–1520	Summit with tors
	(Hammonds Rocks area)	77° 14′ 30–55″ W	428–463	Crestal area
3	Iron Springs, Pa.	39° 50′ 03″ N	2020–2040	Summit w/tor remnant
	(Snowy Mountain)	77° 29′ 30″ W	616–622	Crestal area
4	Markleton, Pa.	39° 47′ 09″ N	3180–3213	Summit plain
	(Mt.Davis)	79° 10′ 38″ W	969–979	Summit plain
5	Blue Ridge Summit, Md./Pa.	39° 43′ 44″ N	1680–1694	Summit area
	(Mount Dunlop)	77° 29′ 23″ W	512–516	Crest
6	Smithsburg, Md./Pa.	39° 41′ 44″ N	2120–2140	Riser, terrace
	(Quirauk Mtn.)	77° 30′ 48″ W	646–652	S 10° W
7	Smithsburg, Md./Pa.	39° 39′ 38″ N	1560–1580	Summit tor
	(Buzzard Knob)	77° 32′ 20″ W	475–481	Crest
8	Catoctin Furnace, Md.	39° 36′ 58″ N	1520–1560	Tors, subsummit flat
	(Cat Rock)	77° 26′ 54″ W	463–476	N 86° W–S 41° E
9	Catoctin Furnace, Md.	39° 30′ 36″ N	1600–1620	Tor, risers
	(Hamburg Lookout Tower)	77° 28′ 31″ W	488–494	Crest
10	Keedysville, Md.	39° 25′ 00–52″ N	1320–1400	Risers, terraces
	(South Mountain)	77° 38′ 19–27″ W	402–433	N 62–82° W
11	Blackbird Knob, W.Va.	39° 03′ 53″ N	3880–3900	Tors, riser, terrace
	(Bear Rocks area)	79° 18′ 01″ W	1183–1189	S 50° E
12	Hopeville, W.Va.	38° 58′ 40″ N	3880–3920	Risers, terraces
	(Allegheny Front)	79° 20′ 08″ W	1183–1195	S 43° E

[a]Keyed to Figure 2.1.

phology. Another factor in site selection was to pick a number of sites outside the Ridge and Valley province (Figure 2.1) to see whether local broad uplands also exist in neighbouring geomorphic provinces. Locations in the following section on description of sites refer to Figure 2.1 and Table 2.2.

Description of sites

The Mt. Davis area, Pennsylvania, is an essentially horizontal to gently sloping local broad upland of topography underlain by quartz sandstone and conglomerate assigned to the Pottsville Group (Flint, 1965). The exposed bedrock (Figure 2.2A) crests in an anticlinal pattern and then dips beneath stony colluvium that mantles the upland surface of the mountain. Large-scale, sorted stone nets (Figure 2.2B) on the surface are either inactive or relict sorted patterned ground features as evidenced by soil-geomorphic relationships that require thousands of years for development (G. M. Clark and E. J. Ciolkosz, unpublished data). A seismic refraction traverse run with a Geospace GT-2B portable seismograph indicates that in the area profiled the bedrock surface is gently dipping and only several meters deep (Figure 2.2A).

The Broad Top synclinorium in Pennsylvania exposes clastic sedimentary rocks of Pennsylvanian age surrounded by much older Paleozoic rocks in the Ridge and Valley province (Berg, Edmunds, Geyer et al., 1980). Along the east-southeast edge of the broad upland underlain by the synclinorium, resistant rocks of the Lower Pennsylvanian Pottsville Group overlie less resistant clastic rocks of the Mississippian Mauch Chunk Formation. There are topographic relationships in the northeastern corner of the Saxton quadrangle that cut across the bedding in shattered exposures of Pottsville Group rocks. These landforms consist of a shattered low cliff (riser) of quartz-rich sandstone and conglomerate and downslope, a bordering, gentle slope (tread) which has large-scale, sorted stripes whose trend is generally normal to the contours and which can be traced to the shattered cliff (Figure 2.3A). Riser slope direction is essentially 180° to the bedding dip direction and discordant with the amount of dip by approximately 5° to 15° (amount of slope plus dip); the uncertainty is due to the completely shattered nature of the bedrock exposed in the cliff. Minor breaks in slope continuity, or steps, are also present in the slope profile (Figure 2.3B). Some of these irregularities represent large-scale anastomosing sorted stripes crossing the profile; others occur in areas of relatively stone-free soil.

In the Hammonds Rocks area on South Mountain, Pennsylvania, in the northern section of the Blue Ridge province, Freedman (1967, Figure 21 and Plate 1) mapped structural elements that detail the trend of the Hammonds Rocks anticline (Figure 2.4A). Dips of bedding, jointing, and cleavage are discordant with the topography but are also nearly perpendicular to the overall topographic trend of tor-like features on the upland (Figure 2.4B).

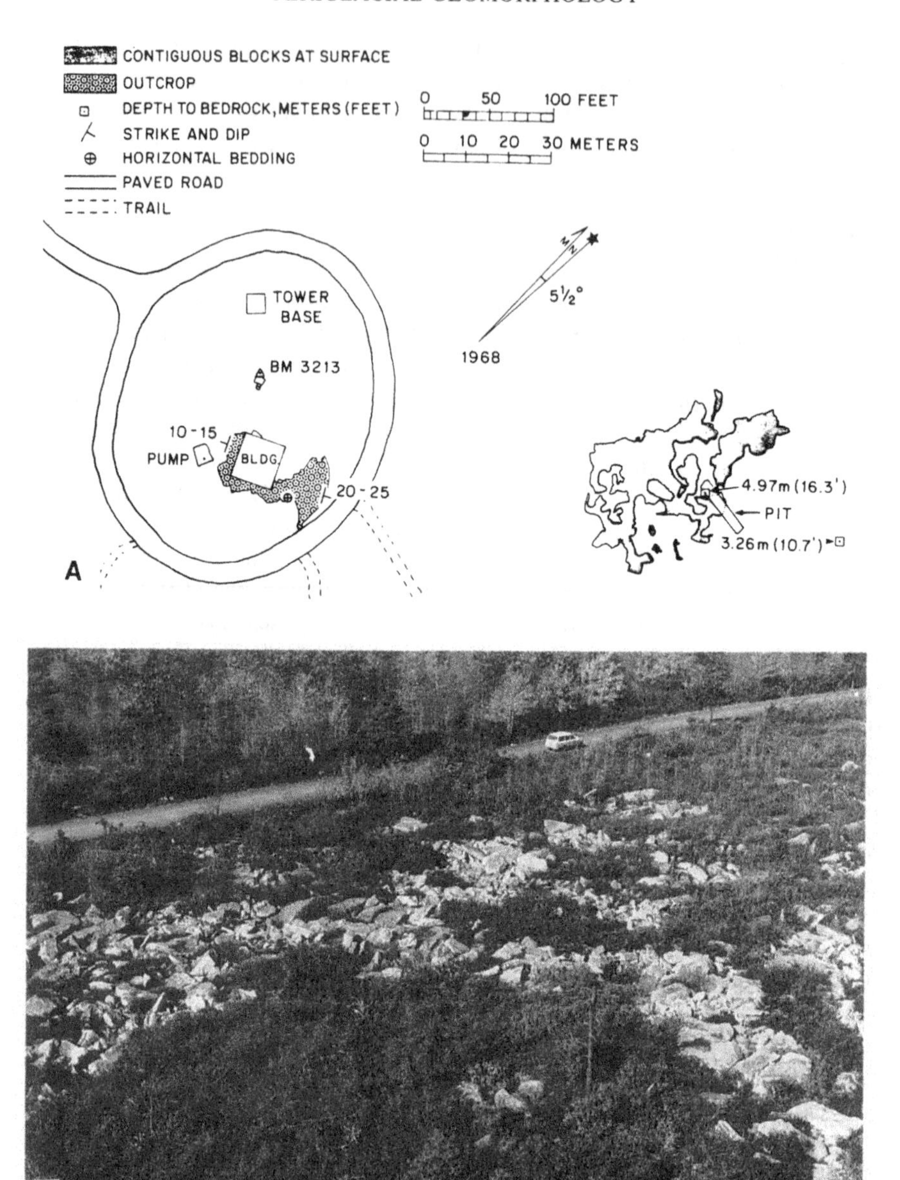

Figure 2.2. (A) Map showing relationships among outcrop, sorted patterned ground, and computed depths to bedrock in an area perpendicular to dip direction as determined by seismic refraction on the upland surface, Mt. Davis, Pennsylvania. Computed depths to bedrock shown by open squares with central dots flanking soil pit on east side of map. (B) Sorted nets on the upland surface northwest of the Mt. Davis, Pennsylvania, observation tower

Farther northeast, along the Hammonds Rocks Ridge Road, flat upland surfaces exist that are underlain by the Montalto Member of the Harpers Formation as mapped by Freedman (1967). However, there are few outcrops on the upland surface, and structural, lithologic, and topographic relationships are obscure.

Snowy Mountain, Pennsylvania, also on South Mountain, provides an example of a summit tor rising above bordering treads that constitute the local broad upland surface at this site (Figure 2.3B). Bedrock exposures in the area comprise thick-bedded, coarse-grained sandstone of the Weverton Formation (Fauth, 1978). Dip of rock cleavage and bedding are strongly discordant with the local tread topography. Large-scale, sorted patterned ground on the tread can be traced upslope to where talus breaks down below the outcrop.

On Mount Dunlop, Pennsylvania, the summit topography is characterized by gentle slopes over distances of hundreds of meters. The Weverton Formation bedrock (Fauth, 1978, Plate 1) dips from 10° northwest to 75° overturned to the southeast. Outcrop-level structures display even more spectacular discordances between topography and bedrock attitudes. Farlekas (1961) mapped folds in a thin but locally continuous "quartz" unit over a portion of the summit area and showed the striking discordance between the bedrock and summit topography.

Along South Mountain in Washington County, Maryland, bedrock attitudes in the Waverton Formation (Edwards, 1978) are discordant with the trend and slope of local upland topographic surfaces. Summit topography on Quirauk Mountain, Maryland, follows in general the major structures and outcrop belt in the Weverton Formation, as do most of the summits and knolls on South Mountain (Godfrey, 1975, p. 5). At the outcrop level, however, there are discordances between topography and the sense of bedding in shattered outcrops. An excellent example of cross-cutting relationships occurs south of the U.S. Army Information Systems Command Site C installation, where most of the riser and tread topography is still preserved. Large-scale sorted stripes on the tread can be traced upslope to contiguous blocks that comprise the riser. Also present are funnel-shaped depressions and micro-hollows visually similar to features that Demek (1969, Photos 15 and 16) termed respectively "solifluction forms" and "nivation funnels." Farther south on South Mountain at the Buzzard Knob site, tor microtopography and a bordering terrace, both discordant with structure, are well displayed. At the South Mountain site Hedges (1975) noted excellent riser and terrace development that cuts across bedding in the Waverton Formation along the Appalachian Trail between the Townsend Memorial and Lambs Knoll.

Similar antipathy between bedrock attitudes and local summit topography also obtains at Cat Rock and east of Hamburg Lookout Tower on Catoctin

Mountain, Maryland, east of South Mountain. Whitaker (1955) mapped and cross-sectioned an area east of Hamburg Fire Tower in detail (Figure 2.5, p. 454) and showed the contrast between asymmetrically overturned shear folds in two members of the Weverton Formation and the plateau-like bench and scree summit topography. Mapping by Fauth (1977) also revealed discordance between structure and topography at these two locations. Taken together, the Maryland Blue Ridge sites demonstrate that local summit topography is discordant with the dip of bedding structure and lithology on both limbs of the South Mountain Anticlinorium where a variety of structural attitudes exist.

The eastern edge of the Appalachian Plateau in northern West Virginia has localities where sickle-shaped topographic terraces convex to slope cut across bedding on the eastern edge of the plateau escarpment. At two sites investigated, orthoquartzite sandstones and conglomerates of Lower Pennsylvanian Pottsville Group overlie the Mauch Chunk Formation (Cardwell, Erwin, and Woodward, 1968). Bedding dips to the northwest. The local broad upland summit topography is punctuated by occasional shattered bedrock tors that rise above the plateau surface (Figure 2.5A, B). The topographic crest of the plateau edge in the Bear Rocks area (Figure 2.6A) varies from a bedrock cliff through shattered but essentially in-place bedrock to a jumble of blocks (Figure 2.6B). A topographic profile east-southeast from the Allegheny Front site illustrates that bedding dip and topographic slope on the terrace are antipathetic (Figure 2.7).

Rock weathering and surface soil horizon characteristics at all the sites provide qualitative evidence of relatively prolonged slope stability (Figures 2.5B, 2.8, 2.9). When broken open, both bedrock ledges and float blocks show different degrees of weathering (e.g., pitting, bleaching, and staining) on their top as opposed to their bottom surfaces. Large blocks are weathered and broken up in place, with little separation of the constituent fragments Gently inclined surfaces of some large quartzite blocks show well-developed Opferkessel (weathering pits in quartzite) that show no morphological evidence of block disturbance during the time they have developed. Hedges

---

Figure 2.3. (A) Tread (foreground) and riser (background, to skyline) topography bordering the east-southeast side of the upland surface underlain by the Broad Top synclinorium, Pennsylvania. Note large-scale sorted stripes on tread that can be traced as contiguous blocks to shattered bedrock of riser in background. Near individual is at junction between tread and talus breakdown; far individual is at base of shattered cliff. (B) Topographic profiles across edge of upland surfaces, risers, and treads; no vertical exaggeration. (a) Broad Top site, Pennsylvania; downslope trend of profile is N71.5°E. (b) Snowy Mountain site, Pennsylvania; profile line is due South (left)–North (road). (c) Allegheny Front site, West Virginia; downslope trend of profile is S43°E. The profile shows from left to right the riser (cliff is outcrop sloping 51 to 90°), slumped and jumbled blocks comprising talus breakdown, and tread surface (slopes 1 to 7°) (compare with Figure 2.7)

A

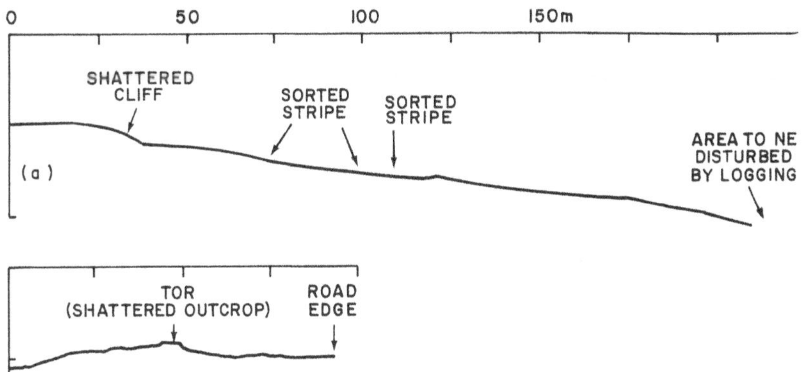
0
50
100
150m
SHATTERED CLIFF
SORTED STRIPE
SORTED STRIPE
AREA TO NE DISTURBED BY LOGGING
(a)
TOR (SHATTERED OUTCROP)
ROAD EDGE
(b)

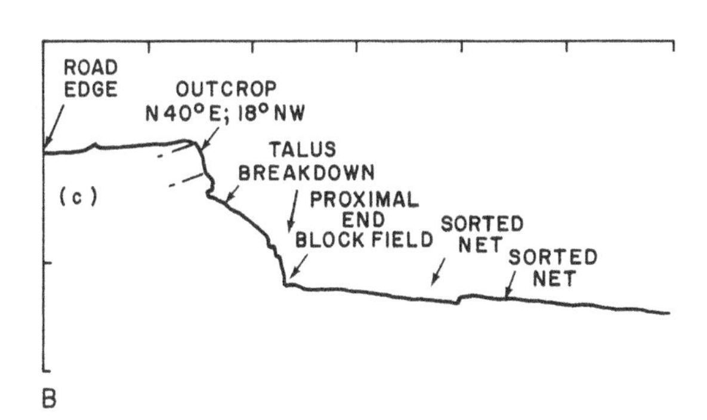
ROAD EDGE
OUTCROP N 40° E; 18° NW
TALUS BREAKDOWN
PROXIMAL END BLOCK FIELD
SORTED NET
SORTED NET
(c)
B

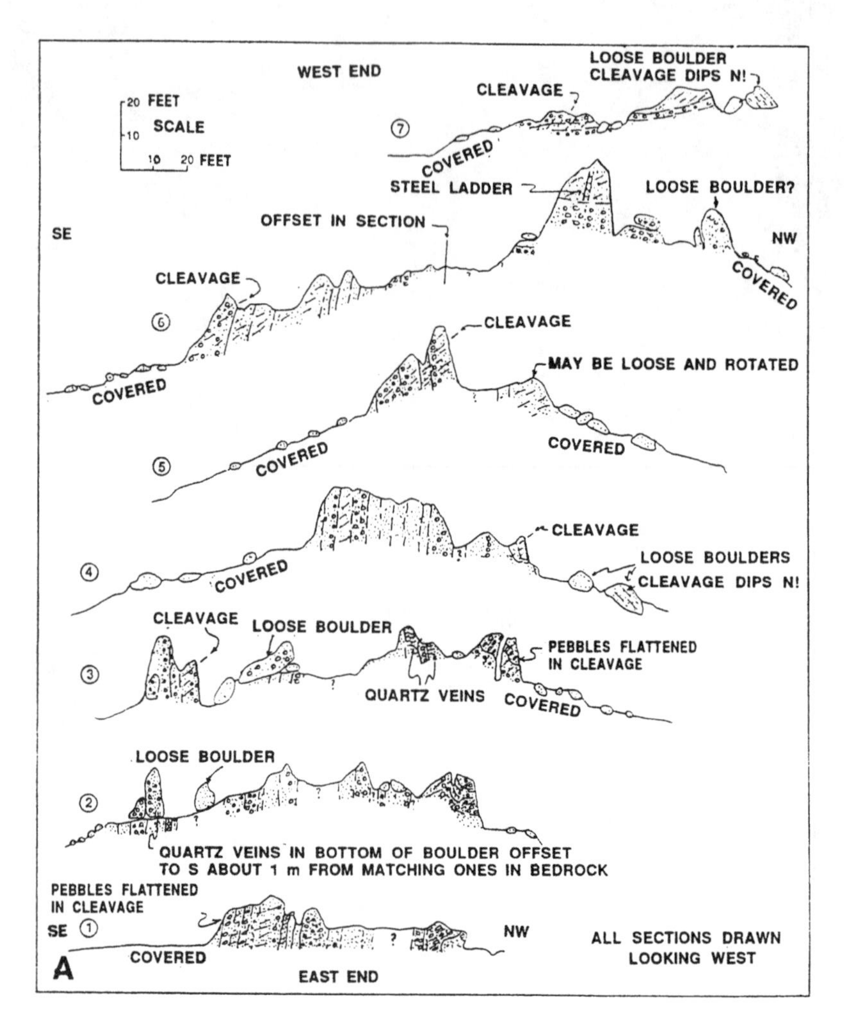

Figure 2.4. (A) Structural cross-sections across Hammonds Rocks, Pennsylvania. Lithology consists of sandstones and conglomerates of the Weverton Formation, generally considered to be Early Cambrian. Structure drawn by N. Potter, Jr., Dickinson College. Profile lines from microtopographic mappings by H. W. A. Hanson, III, Dickinson College. See Figure 2.4B for profile line locations. Reproduced, with permission, from Sevon and Potter (1991)

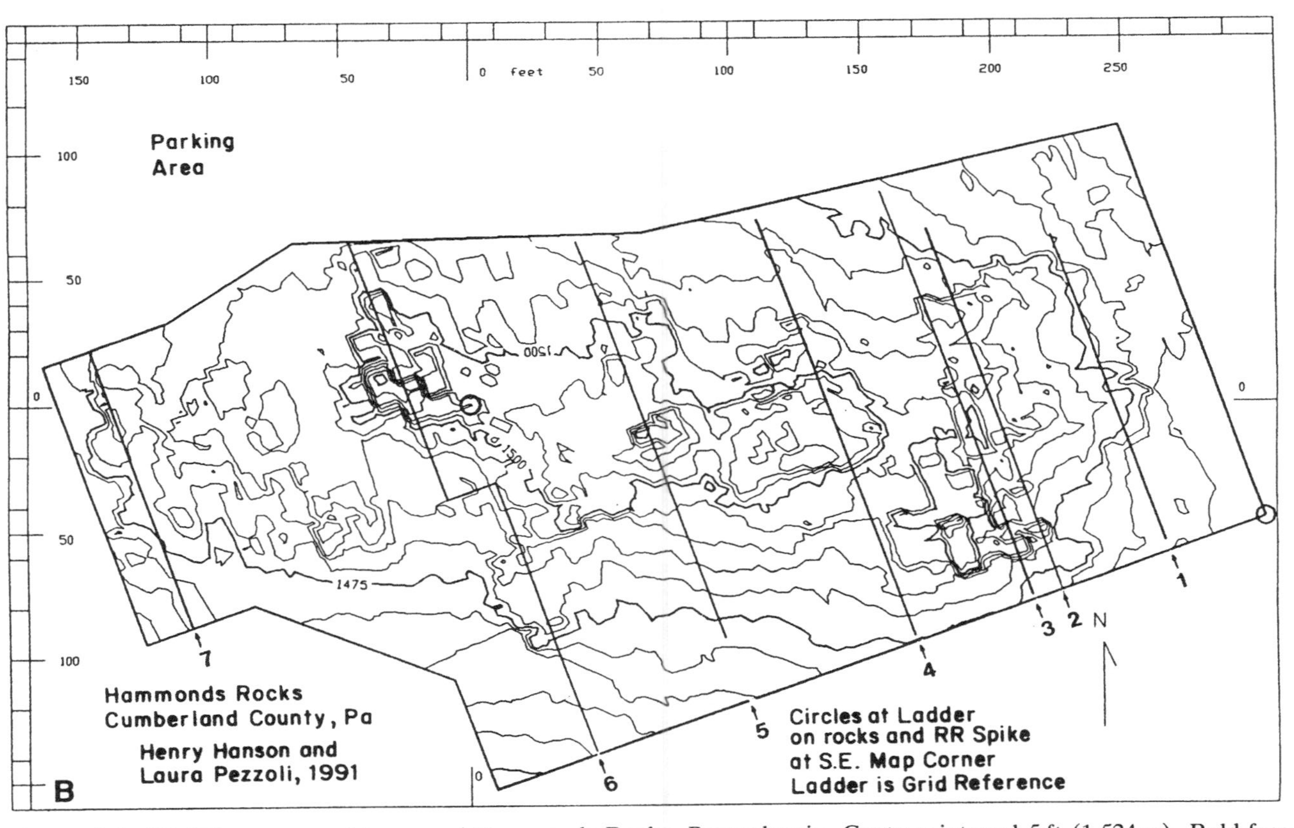

Figure 2.4. (B) Microtopographic map of Hammonds Rocks, Pennsylvania. Contour interval 5 ft (1.524 m). Bold-face numbers and arrows refer to cross-sections in Figure 2.4A. Computer-generated from field measurements both by H. W. A. Hanson, III, Dickinson College. Reproduced, with permission, from Sevon and Potter (1991)

Figure 2.5. (A) Tors on the upland surface in the Bear Rocks area, West Virginia. Notebook case dimensions are 23 × 30 cm. (B) Tor in gently dipping orthoquartzite that has been shattered along joints at Bear Rocks area. Shovel is 1.1 m long

---

Figure 2.6. (A) Terrace (east-southeast of BM 3954) near crest of Allegheny Front south of Bear Rocks (from U.S. Geological Survey Blackbird Knob, WV 7.5-minute quadrangle). Map portion covers 1.2 × 1.7 km; Contour interval = 20 ft (6.1 m). Arrow shows view direction for Figure 2.6B. (B) Terrace and riser of Figure 2.6A; view is to south. Bedding dips 16° to right against slope of terrace. There is a large-scale sorted stripe on tread from just right of lower center to left edge, parallel with bottom border of photograph

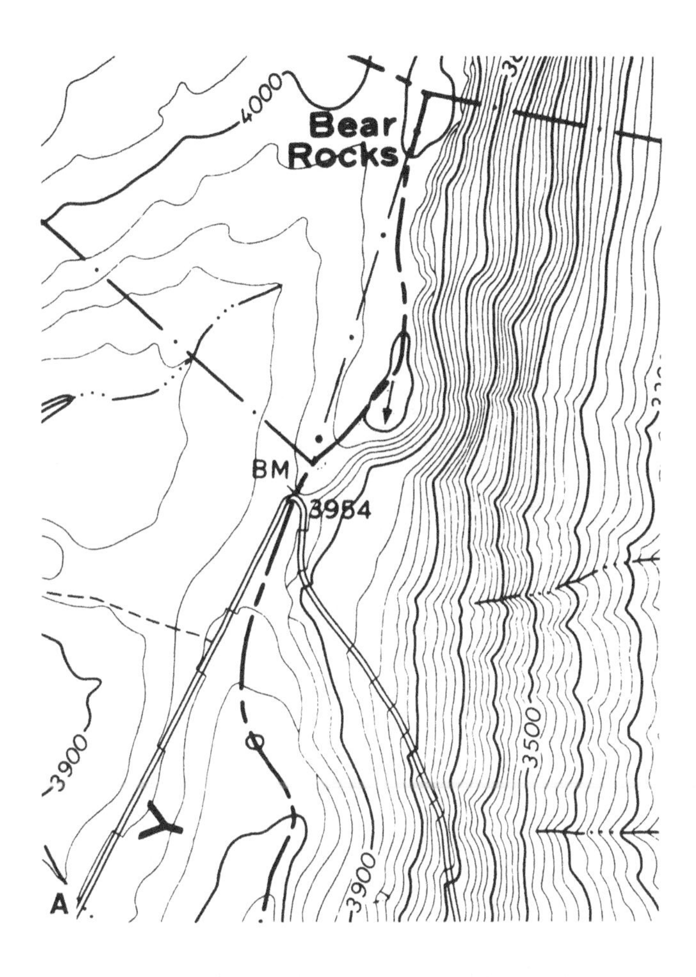
4000
Bear
Rocks
BM
3954
3900
3900
3500
A

B

Figure 2.7. Tread, talus breakdown, riser, and upland surface at Allegheny Front site; view is to southwest. The left person is standing on bedrock dipping 18° to the northwest. Line of section of Figure 2.3B(c) is from right to left across center of photograph

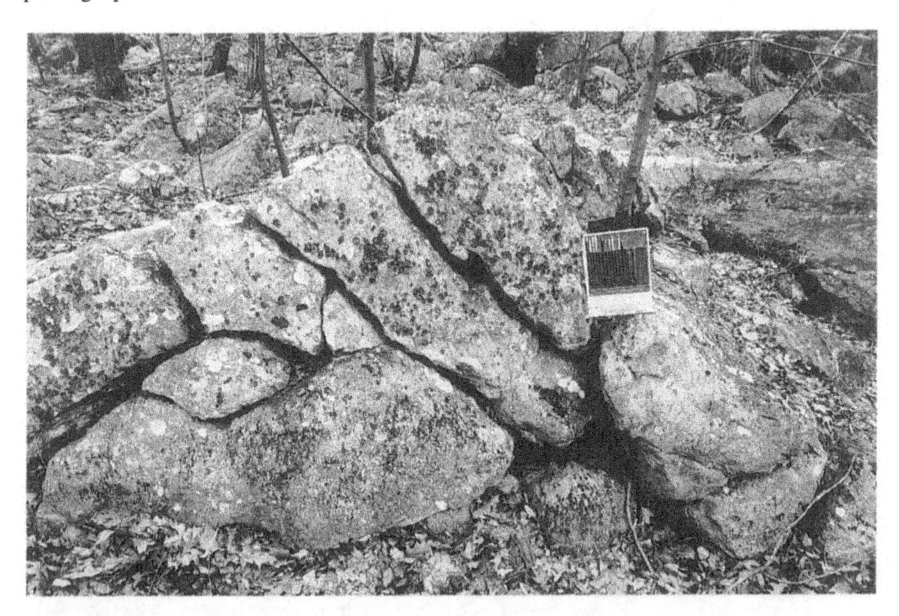

Figure 2.8. Block weathering and breakup in place on terrace tread with riser in background, Quirauk Mountain, Maryland site. Notebook case dimensions are 23 × 30 cm

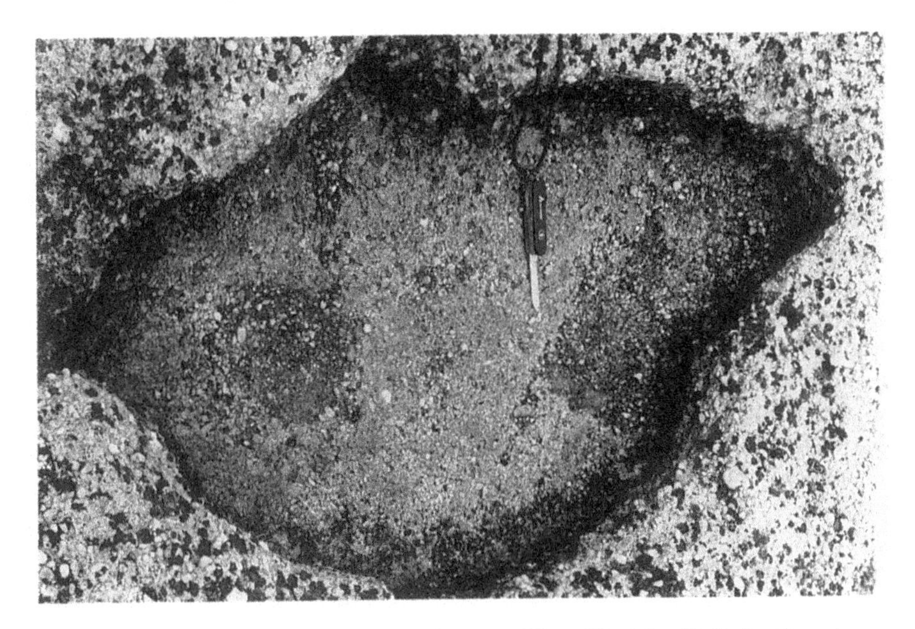

Figure 2.9. Opferkessel in Bear Rocks area, West Virginia. Knife is 16 cm long

(1969) assembled convincing evidence that Opferkessel are contemporary features that develop slowly under modern environmental conditions, although the rates of formation are unknown.

Discussion

INTRODUCTION

The American foci on planation surfaces (Adams, 1975) have dealt primarily with peneplains, pediplains, and etchplains, although Adams (1975, p. 12) mentioned altiplanation. Major emphasis on landscape-forming processes has traditionally centered on the hardy triumvirate of structure, process, and stage, with process being generally neglected in many studies. The European literature, by contrast (Baulig, 1952), has treated a wider variety of processes and environments that can produce planation surfaces which truncate bedrock structures and rocks of differing relative resistances to weathering and erosion. Until recently, effects of climates on landform morphogenesis have never been accorded the attention by geomorphologists in the United States that they have in Europe (Büdel, 1977), although Peltier (1950) was an exception. Sevon (1985) has called climate "the unknown factor" in development of the landscapes in Pennsylvania.

PENEPLAIN REMNANTS

An extensive literature exists on the subject of Central Appalachian summits as peneplain remnants (Sevon, Potter, and Crowl, 1983). Nor are the Central Appalachians the only old eroded "middle mountain" range in the United States where the highest flat areas have been interpreted as peneplain remnants. Péwé (1970) reported that the highest levels at about 650–900 m elevation in the Fairbanks, Alaska, area were related to peneplain surfaces by early workers. Many modern writers prefer alternative interpretations to the peneplain remnant concept. Our investigation unearthed no evidence that requires peneplanation as an explanation. Neither did our observations disprove peneplanation, but our data certainly render it unnecessary. With respect to the study sites in the Maryland Blue Ridge, Godfrey (1975) noted that the overall form of ridge crests closely follows the trend of the Weverton Formation, and that large knolls and summits are underlain by crests of folds in the Weverton Formation. Thus the major form and relief elements in this part of the present study area can be explained by lithology and structure in the same way as Hack (1975) was able to interpret regional relationships north of Harrisburg, Pennsylvania, and in the region as a whole (Hack, 1989).

STEADY STATE

Hack (1960) applied the concept of equilibrium to landscape development and indicated that the downwasting rates of adjacent areas underlain by differentially resistant rocks should be in dynamic equilibrium when slopes are adjusted to the work to be done. He later applied the dynamic equilibrium concept to the origin of topography and ore deposits in the Shenandoah Valley, Virginia, about 200 km south-southwest of the Maryland Blue Ridge (Hack, 1965), including areas with steeply sloping screes. Hupp (1983) has also presented convincing evidence for contemporary scree movement on steep slopes at relatively low elevations in Virginia.

Some confusion therefore exists as to what landforms in the region are evolving today versus features that may be inactive or fossil. Acceptance of the steady state concept is not difficult for steeply sloping landforms with underlying soil and rock materials that can equilibrate rapidly to changing environmental conditions. However, the local broad uplands in this study display weathering features and soil development that suggest long-term stability. An interpretation of inactive or fossil state is not unreasonable considering the low slope gradients, blocky armor, resistant lithologies, and distance from active stream channels that characterize these features.

DEEP WEATHERING STRIPPING

Büdel (1977) advanced the hypothesis that certain planation surfaces predate Late Cenozoic glacial ages and have simply been veneered by periglacial features and debris. There is ample evidence of older unconsolidated sediments in the Appalachians that probably formed under conditions that fostered deep and prolonged chemical and biochemical weathering (Pierce, 1965, 1966; Sevon, 1985). In places deposits are found even at relatively high elevations. For example, Sears (1957) described lignite buried by colluvium at an elevation of 960 m in southwest Virginia.

Have the processes which formed these older deposits had a lasting effect on present local broad upland form and relief, and can subsurface evidence of these events be found? Braun (1989b) calculated that ridge crests have been lowered tens of meters by periglacial processes during Pleistocene time. This amount of summit reduction could have effectively removed most, if not all, vestiges of pre-Quaternary surfaces and their underlying soil and weathered rock. Such stripping not only has the effect of making analysis of deep weathering hypotheses difficult on uplands but also masks direct effects of early events. With the present information, it is not possible to either accept or reject deep weathering arguments for local broad upland morphogenesis. Although no surface evidence of deep weathering phenomena was observed during this study, deep exposures are lacking, and our geophysical data are limited to one site.

REVERSAL OF RELIEF

Mills (1981) revived and refurbished the concept of gully gravure (or lateral migration of rills along topographic contours) to explain topographic reversals of noses and hollows on side slopes where shales are capped and protected by colluvial sandstone clasts derived from upslope rock units. In this modified form of gully gravure, termed gully planation, erosion is most effective at the junction between boulder-covered hollow floors and shale side slopes on mountain flanks. Could an analogous process operate on ridge tops, involve colluvium and bedrock, and develop surfaces that truncate different lithologies? The surfaces of local broad uplands investigated in this study are predominantly veneered with stony colluvium (Ciolkosz et al., 1990) that mantles and largely obscures the underlying bedrock. Slopes are much gentler on these ridge tops than on the mountain flanks that Mills (1981) studied, and the lack of fluvial incision is striking. If a variation of gully planation operated in the past to produce local broad uplands, it has left no visible surface evidence of the process. Also, the areas where Mills (1981, 1988) demonstrated excellent evidence for the operation of gully gravure exhibit much greater relief than the local broad uplands described in this paper.

Visual evidence of active surficial processes on local broad uplands is largely confined to weathering, and the estimated rates of denudation are low. Ciolkosz et al. (1990) reviewed estimates of modern rates of soil formation from sandstones on ridge crests that range from 0.026 to 1 cm ka^{-1}. Godfrey (1975) calculated chemical erosion rates for metaquartzites of 1.8 m my^{-1} for a drainage basin; ridge top rates might be lower or higher. Unless estimates of current rates of weathering and denudation on resistant rocks that underlie these uplands are radically in error, topographic development during Holocene time has been negligible, requiring that summit landscape development occurred in Pleistocene or earlier time.

PROPOSED HYPOTHESIS: PALEOPERIGLACIATION

Within a framework of cyclic-time, cold-climate denudation, Troll (1948) enumerated six general periglacial process groups. These were congelifraction, cryoturbation, solifluction, river gravel deposition, gelideflation, and cryoplanation. We understand the term cryoplanation (Bryan, 1946) to mean cold-climate land reduction with concomitant development of conspicuous upland flats, risers, terraces, and other features (Demek, 1969, pp. 5–8; Washburn, 1980, p. 237). Although periglacial environments have also been considered as regions of extremely active valley incision and destruction of plains (Büdel, 1977), modern workers increasingly recognize the presence of planar upland landforms (Priesnitz, 1988).

Extensive documentation exists on cryoplanation forms from a wide variety of localities (Demek, 1969). Given suitable bedrock structure, lithology, and climate, one overall impact of a periglacial environment on summit and near-summit landforms is the production of upland flats, risers, and treads, although all of the responsible processes and required ground frost environments have yet to be elucidated (Washburn, 1985). Mention of cryoplanation in the Central Appalachians is not new. Peltier (1949, pp. 30, 67–69) invoked cryoplanation as a process to explain the form and relief of mountain tops in the Susquehanna River Valley. North of the glacial border Berg (1975, p. 32) indicated that the surface morphology of Wisconsinan-age till had been modified by periglacial cryoplanation. Godfrey (1975, p. 7) noted that areas with "flat outcrops" along South Mountain, Maryland, closely resemble the "cryoplanation terraces with frostriven scarps" of Demek (1972). Olson (1989) also noted the presence of steplike landforms in the Blue Ridge Province in Maryland. Péwé (1983, Figure 9-11, p. 169 and Table 9-7, p. 177) reported an unpublished observation of cryoplanation for the Mt. Davis site we studied.

That periglacial processes might be responsible for major modification of summit topographic forms over wide areas to the extent that they could be mistaken for surfaces of other origins is, however, a more sweeping proposal. Russell (1933) studied geomorphic form in western United States and stated this concept clearly:

> Herein lies the explanation of numerous forms, ranging in size from minute, steplike benches to slopes covering whole mountain sides and broad surfaces across highlands which may readily be mistaken for parts of peneplains. (p. 939)

There are few references to cryoplanation summits and terraces as major elements of Appalachian ridge tops over wide or disparate areas. Hedges (1975) proposed that cryoplanation had produced the truncation of bedrock structure and lithology on Sugarloaf Mountain on the Maryland piedmont and also noted well-developed terraces at the South Mountain, Maryland, site. Clark (1989a) offered the option that Central Appalachian ridge crests might be incipient cryoplanation features rather than remnants of peneplains, but he did not publish site data in support of this speculation.

Present climates in the study region are humid temperate. Soil temperature regimes are mesic on the lower ridge crests and frigid at the higher elevations (Carter and Ciolkosz, 1980). Cryic soils are unknown and frost pockets are unlikely in the exposures reported here. Under natural forest conditions with snow cover, Central Appalachian mountain soils are frozen in winter to depths of less than 25 cm (Carter and Ciolkosz, 1980). Moreover, Leffler (1981) demonstrated that Appalachian summit temperatures can be estimated reliably and in the study region are characteristic of middle-latitude forested mountains. The predominant natural vegetation is deciduous forest cover with spruce-fir forests only at the highest elevations (Braun, 1950).

Set against the modern climatic environment are reports refining the Pleistocene glaciation sequence (Braun, 1989a) and studies of paleoclimates and paleovegetation beyond the ice sheets (Delcourt and Delcourt, 1981, 1984), notably during the last deglaciation (Jacobson, Webb, and Grimm, 1987). Larabee (1986) extracted a 2.3-m core from Big Run Bog, West Virginia (39° 07′; 79° 35′ W; 980 m), which is on the Allegheny Plateau ca. 25–35 km northwest of upland flats we found along the eastern edge of the Appalachian Plateau, including the Bear Rocks and Allegheny Front sites. He found that between 17,040 and 13,860 years B.P. plant communities surrounding the site were a mosaic of alpine tundra dominated by sedges and grasses. By 13,860 years B.P., spruce and fir had invaded the area, and wet meadow and disturbed ground conditions were prevalent until 11,760 years B.P., indicating the dominance of colluvial activity in the drainage basin. Thus there is increasing independent evidence that severe cold climates were accompanied by landscape disturbance on the district to lower landform scales (Table 2.1) at the higher elevations, even in the southern part of the study area.

The morphological features reported in this study are similar or identical to forms described as cryoplanation features by previous workers. Tors on the local broad upland summits resemble tors that have been interpreted as periglacial in origin. For example, Ehlen (1990) studied topographically similar summit landforms developed on granites in Dartmoor. She found that the summit tors are composed of rocks that are highly resistant to erosion and are parted by widely spaced vertical joints and medium-spaced horizontal secondary joints. Ehlen (personal communication, 1990) indicated that there is a consensus among researchers that the Dartmoor tors are of periglacial origin. In the Appalachians, Braun and Inners (1990) and Braun (personal communication, 1990) interpreted tors morphologically identical to those we found as periglacial in origin and not as relict Tertiary features.

Cryoplanation summit flats described by Demek (1969, p. 7) resemble those of the Mt. Davis area as well as the uplands surmounting the Broad Top, Bear Rocks, and Allegheny Front sites. At the outcrop scale, trend and dip of bedding are discordant with local upland surfaces at the localities we examined, although over broad areas (such as major fold) there is a general positive relationship between major bedrock structures, lithologies, and mountain summit topography.

The form, relief, and materials of risers we describe (Table 2.3) conform remarkedly well with descriptions of cryoplanation risers reported in the

Table 2.3. Form and Material Attributes of Central Appalachian Upland Sites[a]

Map Site[b]	Width[c] Length[d] (m)	Tread Gradient Riser Gradient (degrees)	Tread Material Riser Material	Tread Features Riser Height (m)
1	60–80	3–11.5	Stony soil	Sorted stripes
	190	22–24	Shattered outcrop, blocks	2
2	30–40	4	Soil w/blocks 1–8 m a-axis	Sparse stones
	65	20–38	Outcrop, loose blocks	4–12
3	43–105	2–4	15–25% blocks	Sorted stripes
	170	8.5–15	Outcrop; shattered outcrop	2–3
4	400	1–7	Stony soil, outcrop	Sorted nets
	600	N/A	N/A	N/A
5	40	0–11	Outcrop, stony soil	Shallow
	150	summit	Summit	N/A
6	190	3.5–6	Stony soil	Sorted stripes
	190	>10	Blocks	5
7	50	2–4	5–35% blocks	Float
	62	27–30	Outcrop over scree apron	15
8	120	1.5–16.5	0–30% blocks	Shallow to rock
	215	52.5–113	Bedrock	4.5–5.8
9	330	1–5	Outcrop, stony soil	Two treads
	275	30–85	Outcrop, shattered outcrop	1–4
10	40–115	2–10	<5% stones	Few blocks
	235–445	44–53	Blocks, outcrop	3–11
11	100	1–7	Blocky colluvium	Sorted nets
	450	30–80	Outcrop, blocks	15–30
12	75	0–10	Blocky colluvium	Sorted stripes
	780	51–90	Outcrop, blocks	0–30

[a]Numerical values given in the table are for representative features.
[b]Keyed to Figure 2.1.
[c]Measured orthogonal to contours.
[d]Measured parallel to contours.

literature (Table 2.4). Features composed of firm bedrock, shattered bedrock, and slumped to jumbled blocks singly and in various combinations are present in the study sample.

The sizes, shapes, surficial geomorphic materials, and site factors of treads we studied are also similar to features of cryoplanation terraces described in the literature (Tables 2.3 and 2.4). For example, Demek (1969, pp. 55–56) noted that although cryoplanation terraces are rare to absent on steep slopes, their development is not excessively dependent on leveled surfaces. The same topographic situations are present at the study sites. In another case, the treads we studied along the eastern edge of the Appalachian Plateau in West Virginia mimic in both form and aspect treads reported by Demek (1969, p. 40) in that they are sickle-shaped, wedging out and starting again, a little higher, then a little lower, along crestal areas. One shortcoming of this study is that, except for the Mt. Davis site, subsurface data are lacking. Many investigators consider cryoplanation terraces as transport slopes where 3 m or less of debris mantles the bedrock. On the other hand, Demek (1969, pp. 6–8) also recognized less common cryoplanation terraces with external parts composed of loose material. He termed these complex features compound cryoplanation terraces.

If cryoplanation terraces formed primarily as bedrock erosional features, where did the volumes of coarse to fine clastic sediment go? There are numerous reports on the widespread nature and great thickness of sideslope, footslope, hollow, and toeslope colluvium in the Central Appalachians south

Table 2.4. Geomorphic Form and Material Attributes of Reported Cryoplanation Terraces[a]

Characteristic	Reported Data		
Slope and slope position			
Summits	Summit cryoplains, with or without tors		
Slopes	Interrupt middle and upper slopes; generally several levels		
Width[b]	30 m (low)	400 m (common)	>10 km (high)
Length[c]	5 m (low)	100 m (common)	> 1 km (high)
Riser or scarp height	1 m (low)	6 m (common)	>50 m (high)
Gradient			
Tread or flat	1° (low)	6° (common)	14° (high)
Riser or scarp	9° (low)	30° (common)	90° (high)
Tread material	<1–3 m debris depth, rarely more; sorted patterned ground		
Riser material	Bedrock, shattered bedrock, block rubble 1–3 m thick		
Rock type	Resistant lithologies that produce coarse debris are favored		
Lithologic control	Commonly initiate at or near breaks in lithologic resistance		

[a]Compiled from Demek (1969), Priesnitz (1988), Péwé (1975), and Reger and Péwé (1976).
[b]Measured parallel to contours.
[c]Measured orthogonal to contours.

Figure 2.10. Clearcut area showing terrace wraparound of outcrop area, South Mountain, Pennsylvania. (Dickinson, Pa.7.5-minute quadrangle; elevation 442 to 463 m)

of the glacial border (King, 1950; Pierce, 1966; Moss, 1976; Clark and Ciolkosz, 1988; Clark, Ciolkosz, Kite, and Lietzke, 1989; Ciolkosz et al., 1990). Until recently, few references were made to the sources of these great volumes of sediment. Carter and Ciolkosz (1986) conducted a systematic study of bedrock depth on a ridge crest and found that the soil is 3 to 4 m deep. Based on seismic and rock fragment orientation data and the presence of sorted and stratified material in the upper 1 to 2 m of the profiles, they concluded that the upper 1 to 2 m of the soil is colluvium and the underlying material is residuum. Braun (1989b, p. 249) calculated that eight to ten periglacial episodes of the last 850,000 years were capable of lowering ridge tops by 8 to 10 m. Ciolkosz et al. (1990) showed that the upper parts of ridge-crest residual soils in the Ridge and Valley province of central Pennsylvania were truncated during Late Wisconsinan time and then either buried with local colluvium or cryoturbated. These nearly in situ parent materials have been stable since Late Wisconsinan time as evidenced by the nature and properties of the soils developed in them.

We therefore hypothesize that the individually small zone or locale forms (Table 2.1) and materials described in this paper are relict cryoplanation features that formed under former rigorous periglacial environments (with or without permafrost). We suggest that the aggregate visual effect of these features, developed to common elevational ranges in local areas, is that of a

Table 2.5. Environmental (site) Associations on Cryoplanation Terraces[a]

Influencing Factor	Favorable Conditions	Unfavorable Conditions
Wind effects on snow drifting	High	Low
Exposure to prevailing winds	Lee slopes	Windward slopes
Proportion of snow in precipitation	High proportion	Low proportion
Sublimation rates	Low	High
Relation to (local) snowline	300–500 m below to very little above snowline	Rarely down to 1000 m below snowline
Summer temperatures	Cold	Warm
Freezing index	>1000 degree days (°C)	–
Permafrost required (continental areas)	Yes	–
Permafrost required (maritime areas)	No	–
Formation time (estimated)	ca. 10,000 years	>10,000 years

[a]Compiled from Priesnitz (1988), Péwé (1975), and Reger and Péwé (1976).

much larger summit "plain" when seen from a distance. This interpretation argues that large areas of Central Appalachian local broad uplands can be viewed as relict incipient-to-essentially-complete assemblages of much smaller individual surfaces of cryoplanation (Figure 2.10), as opposed to remnant fragments of a peneplain. Further, we conjecture that, in the aggregate, the visual "evenness of skyline" effect that these summit and bordering summit landforms have on our visual perception is that very impression which led early workers to adopt the peneplain remnant hypothesis.

Important problems remain, however, and some new ones are raised. The volume of debris remaining on terraces cannot be computed until the soil-bedrock interface is mapped with tomographic geophysical techniques. Rates of removal cannot be computed and geomorphic effectiveness of the proposed periglacial processes cannot be assessed until age(s) of the terrace deposits and the downslope diamictons into which they grade are determined. Numerical ages will, in most cases, have to be measured with new methods of dating of geomorphic surfaces and materials. The time span required for cryoplanation terrace development has only been estimated (Table 2.5). Further complication is indicated by evidence that individual terraces may record more than one episode of development (Lauriol, 1990).

Conclusions

We have presented a hypothesis that relates the origin of local broad uplands to cryoplanation during cold phases of Late Cenozoic time, and we have speculated that the accordance of their summit levels may be due to the effects of periglacial processes that worked to produce a common elevation range. The intent of this paper is not to promulgate yet another grand scheme of landscape origin for ridges in the Central Appalachians. Rather, our goal is to stimulate detailed geomorphic research on upland rocks, soils, and landforms by providing a viable alternative hypothesis that can be tested with newly developing field and laboratory techniques. These techniques will also be useful in searching for and evaluating evidence for or against deep weathering stripping (Büdel, 1977) as a group of processes that may help to explain the formation of local broad uplands and perhaps summit accordance. The same methodologies should also help unravel geometries and depositional histories of the complex diamicton deposits that border these Appalachian upland surfaces.

It is premature to make paleoclimatic interpretations until definitive studies have been completed of both the Appalachian features and comparable active analogs. However, some thoughts on the potential of cryoplanation terraces for providing paleoenvironmental information may be in order (Table 2.5). Karte (1982) noted that, in addition to forms found in treeless, cold continental permafrost environments, cryoplanation terraces are forming above the forest limit in maritime "Icelandic-type" areas without permafrost where mean annual air temperatures are between −1° and −2° C. When definitive work on active forms becomes available, moreover, terraces may provide opportunity for other types of paleoenvironmental inferences in a region where such data are difficult to obtain. In Alaska, for example, the altitudes of cryoplanation terraces plot ca. 100–300 m below the Wisconsinan snow line (Péwé, 1975, Figure 8, p. 22). The proportion of rain to snow in total precipitation is another possible paleointerpretation. Péwé (1975, Figure 8, p. 22) showed that the elevation of cryoplanation terraces in central Alaska is lower in areas where the proportion of snow to rain is higher. Paleowind interpretations and other aspect inferences may also be possible in the future (Table 2.5).

The genesis of accordant ridge crests remains unaddressed. Tarr (1898) suggested that rock above the tree line undergoes more rapid weathering and erosion than below it. Daly (1905) discussed the accordance of summit elevations above the forest limit in mountains and argued for the effectiveness of alpine processes in both rock disintegration and removal of rock waste (pp. 120–123). In the Appalachians much work must be done in order to define montane paleoclimatic altitudinal zones, the landcape-forming processes that operated there, and their geomorphic effectiveness.

Could erosional magnitudes sufficient to form the features described in this paper have accounted for at least some of the thick diamicton deposits that underlie simple side slope and complex fan-shaped landforms in the Ridge and Valley and Blue Ridge provinces? Cryoplanation terrace development requires the production of large volumes of debris that are transported valleyward across the terraces, although the rates of terrace development are presumably slow and not well known (Washburn, 1980, p. 240). Braun (1989b) enumerated numerous major cold-phase events that would have affected the study area. One might speculate that if cryoplanation terrace development accompanied a number of these events, the cumulative production of debris may have been quite large. However, much better knowledge of the origin and age of these diamicton deposits will be required to evaluate this scenario.

Assessment of the quantitative effects of long-term erosional activity on overall relief between mountain summits and bordering valleys is beyond the scope of this paper. Godfrey (1975) measured and compared present chemical and physical rates of erosion in the South Mountain area, Maryland. He speculated that cold-climate processes would have reduced overall relief between ridge crests and the Middletown Valley, while temperate climatic weathering and erosional processes would have increased relief. During times of temperate climate Godfrey (1975) estimated that the Catoctin Metabasalt beneath the valley would be eroded at a rate three to four times faster than the Loudon and Weverton Formations on the ridge crests. Under Pleistocene cold-phase conditions he concluded that the ridge crests would have been lowered faster than the valley, causing an overall reduction in landscape relief. Braun (1989b) hypothesized that Pleistocene periglacial erosion should be the dominant Pleistocene erosional process in the Appalachian Highlands, and that present processes are shaping the landscape to provide slope forms necessary to transport the periglacial debris. To address the hypotheses and speculations presented by Godfrey (1975) and Braun (1989b), the form, volumes, stratigraphy, and ages of summit erosional landforms and diamicton deposits will need to be determined. New methods for the numerical age dating of geomorphic surfaces hold great promise, as do evolving geophysical tomographic techniques that image subsurface soil and rock conditions. With the aid of these new technologies, we may begin to address questions of landscape equilibrium and disequilibrium in a quantitative fashion in areas where surficial evidence indicates that effects of Late-Cenozoic environmental change have been severe.

Acknowledgments

Field research in Pennsylvania was supported by the Pennsylvania Academy of Science and in West Virginia by the West Virginia Geological and Economic Survey. D. D. Braun provided updated map information on glacial borders in eastern Pennsylvania (Figure 2.1). N. Potter, Jr., permitted publication of a manuscript cross-section of Hammonds Rocks (Figure 2.4A), and H. W. A. Hanson, III, permitted publication of a manuscript map of Hammonds Rocks (Figure 2.4B). We thank D. D. Braun, E. T. Cleaves, B. Marsh, N. Potter, Jr., and W. D. Sevon for valuable suggestions that led to manuscript improvement.

References

Adams, G.F. (ed.), *Planation Surfaces, Peneplains, Pediplains, and Etchplains*, 476 pp., Dowden, Hutchinson & Ross, Harrisburg, 1975.

Baulig, H., Surfaces d'Aplanissement, *Annales de Géographie*, **61**, 161–183, 245–262, 1952.

Berg, T.M., Geology and mineral resources of the Brodheadsville Quadrangle, Monroe and Carbon Counties, Pennsylvania, *Pennsylvania Geological Survey, 4th Series, Atlas 205a*, 60 pp., 1975.

Berg, T.M., W.E. Edmunds, A.R. Geyer, and others (Compilers), Geologic Map of Pennsylvania, scale 1:250,000, *Pennsylvania Geological Survey, Fourth Series*, 1980.

Braun, D.D., A revised Pleistocene glaciation sequence in eastern Pennsylvania: Support for limited Early Wisconsin ice and a single Late Illinoian advance beyond the Late Wisconsin border, *28th International Geological Congress, Absracts, 1*, 1-196–1-197, 1989a.

Braun, D.D., Glacial and periglacial erosion of the Appalachians, in *Appalachian Geomorphology*, edited by T.W. Gardner and W.D. Sevon, pp. 233–256, Elsevier, Amsterdam, 1989b.

Braun, D.D., and J.D. Inners, Weathering of conglomerate ledges and tors within the glacial limit in northeastern Pennsylvania: Evidence for single stage tor development under Pleistocene periglacial conditions, *Geological Society of America, Abstracts with Programs, 22*(2), **6**, 1990.

Braun, E.L., *Deciduous Forests of Eastern North America*, Blackiston, Philadelphia, 596 pp., 1950.

Bryan, K., Cryopedology—the study of frozen ground and intensive frost-action with suggestions on nomenclature, *American Journal of Science*, **244**, 622–642, 1946.

Bryan, K., A.B. Cleaves, and H.T.U. Smith, The present status of the Appalachian problem. *Zeitschrift für Geomorphologie*, **7**, 312–320, 1932/33.

Büdel, J., *Klima-Geomorphologie*, 304 pp., Gebrüder Borntrager, Berlin, 1977.

Cardwell, D.H., R.B. Erwin, and H.P. Woodward (Compilers), Geologic Map of West Virginia, scale 1:250,000, *West Virginia Geological and Economic Survey*, 1968.

Carter, B.J., and E.J. Ciolkosz, Soil temperature regimes of the Central Appalachians, *Soil Science Society of America Journal*, **44**, 1052–1058, 1980.

Carter, B.J., and E.J. Coilkosz, Sorting and thickness of waste mantle material on a sandstone spur in central Pennsylvania, *Catena*, **13**, 241–256, 1986.

Ciolkosz, E.J., B.J. Carter, M.T. Hoover, R.C. Cronce, W.J. Waltman, and R.R. Dobos, Genesis of soils and landscapes in the Ridge and Valley province of central Pennsylvania, in *Soils and Landscape Evolution*, edited by P.L.K. Knuepfer and L.D. McFadden, pp. 245–261, Elsevier, Amsterdam, 1990.

Clark, G.M., Central and southern Appalachian accordant ridge crest elevations south of glacial border: Regional erosional remnants or incipient periglacial surfaces? *28th International Geological Congress, Abstracts, 1*, 1-299–1-300, 1989a.

Clark, G.M., Central and southern Appalachian water and wind gap origins: Review and new data, in *Appalachian Geomorphology*, edited by T.W. Gardner and W.D. Sevon, pp. 209–232, Elsevier, Amsterdam, 1989b.

Clark, G.M., and E.J. Ciolkosz, Periglacial geomorphology of the Appalachian Highlands and Interior Highlands south of the glacial border—a review, *Geomorphology*, **1**, 191–220, 1988.

Clark, G.M., E.J. Ciolkosz, J.S. Kite, and D.A. Lietzke, Central and Southern Appalachian Geomorphology—Tennessee, Virginia, and West Virginia, *28th International Geological Congress, Field Trip Guidebook T150*, 105 pp., 1989.

Daly, R.A., The accordance of summit levels among Alpine mountains: the fact and its significance, *Journal of Geology*, **13**, 105–125, 1905.
Davis, W.M., The rivers and valleys of Pennsylvania, *National Geographic Magazine*, **1**, 183–253, 1889.
Delcourt, P.A., and H.R. Delcourt, Vegetation maps for eastern North America; 40,000 yr B.P. to the present, in *Geobotany II*, edited by R.C. Romans, pp. 123–165, Plenum, New York, 1981.
Delcourt, P.A., and H.R. Delcourt, Late Quaternary paleoclimates and biotic responses in eastern North America and the western North Atlantic Ocean, *Palaeogeography, Palaeoclimatology, Palaeoecology*, **48**, 263–284, 1984.
Demek, J., Cryoplanation terraces, their geographical distribution, genesis and development, Rozpravy Československé Akademie Věd, Řada Matematických A Přírodních Věd, *79*(4), 80 pp., 1969.
Demek, J., *Manual of Detailed Geomorphological Mapping*, Academia, Prague, 344 pp., 1972.
Edwards, J., Jr. (Compiler), Geologic Map of Washington County, scale 1:62,500, *Maryland Geological Survey*, 1978.
Ehlen, J., Geomorphic, petrographic and structural classification of granite landforms using spatial patterns, *Geological Society of America, Abstracts With Programs, 22*(7), A21, 1990.
Farlekas, G., The geology of part of South Mountain of the Blue Ridge province north of the Pennsylvania-Maryland border, M.S. Thesis, 64 pp., The Pennsylvania State University, University Park, 1961.
Fauth, J.L., Geologic map of the Catoctin Furnace and Blue Ridge Summit Quadrangles, Maryland, scale 1:24,000, *Maryland Geological Survey*, 1977.
Fauth, J.L., Geology and mineral resources of the Iron Springs area, Adams and Franklin Counties, Pennsylvania, *Pennsylvania Geological Survey, Fourth Series, Atlas 129c*, 72 pp., 1978.
Flemal, R.C., The attack on the Davisian system of geomorphology: a symposium, *Journal of Geologic Education*, **19**, 3–13, 1971.
Flint, N.K., Geology and mineral resources of southern Somerset County, Pennsylvania, *Pennsylvania Geological Survey, 4th Series, County Report C56A*, 267 pp., 1965.
Freedman, J., Geology of a portion of the Mount Holly Springs quadrangle, Adams and Cumberland Counties, Pennsylvania, *Pennsylvania Geological Survey, 4th Series, Progress Report 169*, 66 pp., 1967.
Gardiner, V., and R. Dackombe, *Geomorphological Field Manual*, 254 pp., George Allen & Unwin, London, 1983.
Gardner, T.W., and W.D. Sevon, (eds), *Appalachian Geomorphology*, 318 pp., Elsevier, Amsterdam, 1989.
Godfrey, A.E., Chemical and physical erosion in the South Mountain anticlinorium, Maryland, *Maryland Geological Survey, Information Circular 19*, 35 pp., 1975.
Godfrey, A.E., and E.T. Cleaves, Landscape analysis: Theoretical considerations and practical needs, *Environmental Geology and Water Sciences*, **17**, 141–155, 1991.
Hack, J.T., Interpretation of erosional topography in humid temperate regions, *American Journal of Science*, **258-A**, 80–97, 1960.
Hack, J.T., Geomorphology of the Shenandoah Valley Virginia and West Virginia and origin of the residual ore deposits, *U.S. Geological Survey Professional Paper 484*, 84 pp., 1965.
Hack, J.T., Dynamic equilibrium and landscape evolution, in *Theories of Landform Development*, edited by W.N. Melhorn and R.C. Flemal, pp. 87–102, George Allen & Unwin, London, 1975.

Hack, J.T., Geomorphology of the Appalachian Highlands, in *The Appalachian-Ouachita Orogen in the United States*, edited by R.D. Hatcher, Jr., W.A. Thomas, and G.W. Viele, pp. 459–470, Geological Society of America, Boulder, Colo., 1989.

Hedges, J., Opferkessel, *Zeitschrift für Geomorphologie*, **13**, 22–55, 1969.

Hedges, J., Multiple cycles of cryoplanation on Sugarloaf Mountain, Maryland, *Biuletyn Peryglacjalny*, **24**, 233–243, 1975.

Hupp, C.R., Geo-botanical evidence of Late Quaternary mass wasting in block field areas of Virginia, *Earth Surface Processes and Landforms*, **8**, 439–450, 1983.

Jacobson, G.L., Jr., T. Webb, III, and E.C. Grimm, Patterns and rates of vegetation change during the deglaciation of eastern North America, in *North America and Adjacent Oceans During the Last Deglaciation*, edited by W.F. Ruddiman and H.E. Wright, Jr., pp. 277–288, Geological Society of America, Boulder, Colo., 1987.

Karte, J., Development and present state of German periglacial research in arctic and alpine environments, *Biuletyn Peryglacjalny*, **29**, 183–201, 1982.

King, P.B., Geology of the Elkton area, Virginia, *U.S. Geological Survey Professional Paper 230*, 82 pp., 1950.

Larabee, P.A., Late-Quaternary vegetational and geomorphic history of the Allegheny Plateau at Big Run Bog, Tucker County, West Virginia, M.S. Thesis, 115 pp., University of Tennessee, Knoxville, 1986.

Lauriol, B., Canadian landform examples—18: Cryoplanation terraces, northern Yukon, *The Canadian Geographer*, **34**, 347–351, 1990.

Leffler, R.J., Estimating average temperatures on Appalachian summits, *Journal of Applied Meteorology*, **20**, 637–642, 1981.

Mills, H.H., Boulder deposits and the retreat of mountain slopes, or "gully gravure" revisited, *Journal of Geology*, **89**, 649–660, 1981.

Mills, H.H., Surficial geology and geomorphology of the Mountain Lake area, Giles County, Virginia, including sedimentological studies of colluvium and boulder streams, *U.S. Geological Survey Professional Paper 1469*, 57 pp., 1988.

Mills, H.H., G.R. Brakenridge, R.B. Jacobson, W.L. Newell, M.J. Pavich, and J.S. Pomeroy, Appalachian Mountains and Plateaus, *Geological Society of America, Centennial Volume 2*, 5–50, 1987.

Monmonier, M.S., Upland accordance in the Ridge and Valley section of Pennsylvania. M.S. Thesis, 58 pp., Pennsylvania State University, University Park, 1967.

Monmonier, M.S., Trends in upland accordance in Pennsylvania's Ridge and Valley section, *Proceedings of the Pennsylvania Academy of Science*, **42**, 157–162, 1968.

Monmonier, M.S., Upland adjustment to regional drainage in central Pennsylvania: an application of trend surface analysis, *Journal of Geography*, **70**, 360–370, 1971.

Moss, J.H., Periglacial origin of extensive lobate colluvial deposits on the south flank of Blue Mountain near Shartlesville and Strausstown, Berks County, Pennsylvania, *Proceedings of the Pennsylvania Academy of Science*, **50**, 42–44, 1976.

Olson, C.G., Mountain soils of eastern Appalachians, *28th International Geological Congress, Abstracts, 2*, 2-548, 1989.

Peltier, L.C., Pleistocene terraces of the Susquehanna River, Pennsylvania, *Pennsylvania Geological Survey, 4th Series, Bulletin G 23*, 158 pp., 1949.

Peltier, L.C., The geographic cycle in periglacial regions as it is related to climatic geomorphology, *Association of American Geographers Annals*, **40**, 214–236, 1950.

Péwé, T.L., Altiplanation terraces of early Quaternary age near Fairbanks, Alaska, *Acta Geographica Lodziensia*, **24**, 357–363, 1970.

Péwé, T.L., Quaternary geology of Alaska, *U.S. Geological Survey Professional Paper 835*, 145 pp., 1975.

Péwé, T.L., The periglacial environment in North America during Wisconsin time, in *Late-Quaternary Environments of the United States, Volume 1, The Late Pleistocene*, edited by S.C. Porter, pp. 157–189, University of Minnesota Press, Minneapolis, 1983.

Pierce, K.L., Geomorphic significance of a Cretaceous deposit in the Great Valley of southern Pennsylvania, *U.S. Geological Survey Professional Paper 525-C*, pp. C152–C156, 1965.

Pierce, K.L., Bedrock and surficial geology of the McConnellsburg quadrangle, Pennsylvania, *Pennsylvania Geological Survey, 4th Series, Atlas 109a*, 111 pp., 1966.

Priesnitz, K., Cryoplanation, in *Advances in Periglacial Geomorphology*, edited by M.J. Clark, pp. 49–67, John Wiley, Chichester, England, 1988.

Reger, R.D., and T.L. Péwé, Cryoplanation terraces: indicators of a permafrost environment, *Quaternary Research*, **6**, 99–109, 1976.

Russell, R.J., Alpine land forms in western United States, *Geological Society of America Bulletin*, **44**, 927–950, 1933.

Sears, C.E., Late Cretaceous erosion surface in southwest Virginia (abstract), *Geological Society of America Bulletin*, **68**, 1883, 1957.

Sevon, W.D., Pennsylvania's Polygenetic Landscape, 4th Annual Field Trip Guidebook, *Harrisburg Area Geological Society*, 55 pp., 1985.

Sevon, W.D., and N. Potter, Jr. (eds), Geology in the South Mountain Area, Pennsylvania, *56th Annual Field Conference of Pennsylvania Geologists*, Field Conference of Pennsylvania Geologists, Harrisburg, Pa., 236 pp., 1991.

Sevon, W.D., N. Potter, Jr., and G.H. Crowl, Appalachian peneplains: an historical review, *Earth Sciences History*, **2**, 156–164, 1983.

Tarr, W.S., The peneplain, *American Geologist*, **21**, 351–370, 1898.

Thornbury, W.D., *Regional geomorphology of the United States*, 609 pp., John Wiley, New York, 1965.

Troll, C., Der subnivale oder periglaziale Zyklus der Denudation, *Erdkunde*, **2**, 1–21, 1948.

Washburn, A.L., *Geocryology: A Survey of Periglacial Processes and Environments*, 406 pp., John Wiley, New York, 1980.

Washburn, A.L., Periglacial problems, in *Field and Theory: Lectures in Geocryology*, edited by M. Church and O. Slaymaker, pp. 166–202, University of British Columbia Press, Vancouver, 1985.

Watts, W.A., Late Quaternary vegetation of central Appalachia and the New Jersey Coastal Plain, *Ecological Monographs*, **49**, 427–469, 1979.

Whitaker, J.C., Geology of Catoctin Mountain, Maryland and Virginia, *Geological Society of America Bulletin*, **66**, 435–462, 1955.

3 Spatial Patterns of Geochemical Denudation in a Colorado Alpine Environment

Nelson Caine
Institute of Arctic & Alpine Research and Department of Geography, University of Colorado

Abstract

Up to ten years of discharge and water quality records from Green Lakes Valley, a nonglacierized alpine catchment in the Colorado Front Range, define a consistent spatial pattern of geochemical denudation. The solute yields suggested by these records are generally low (between 5 and $20\,g\ m^{-2}\ yr^{-1}$) but exceed by an order of magnitude rates of suspended sediment transport in the basin. Geochemical denudation rates vary spatially, being highest in subcatchments where winter snow accumulation is greatest. Within these subcatchments, spatial variability of at least the same magnitude may be predicted from the local water budget and is supported by rock weathering studies (e.g., Thorn, 1975). These results suggest that solute removal from sites of greatest snow accumulation is important in maintaining the hollows associated with nivation and similar forms in mountain terrain.

Introduction

In his classical study of geomorphic activity in Northern Sweden, Rapp (1960) found solute removal to be the most important transport mechanism acting on the slopes of Karkevagge today. This conclusion has been corroborated by other studies in mountain drainage basins during the past 30 years (e.g., Barsch and Caine, 1984). In the Southern Rocky Mountains the importance of geochemical denudation was indicated by the early studies of Miller (1961), Caine (1976), and Thorn (1976), and these have been extended by work

Periglacial Geomorphology. Edited by J. C. Dixon and A. D. Abrahams

published in the last decade (Vitek, Deutch, and Parson, 1981; Baron, 1983; Baron and Bricker, 1987; Rochette, Drever, and Sanders, 1988; Caine and Thurman, 1990; Mast, Drever, and Baron, 1990). The importance of geochemical processes in alpine areas is also suggested by an extensive body of research on soil processes (Dixon, 1986; Litaor, 1987) and the use of rock weathering in relative age dating (Benedict, 1970; Birkeland, 1984; Ballantyne, Black, and Finlay, 1989).

However, many of these studies are based on only short periods of record, and few have been used in the context of landscape development. Here the results of a consistent, decade-long research program are used to evaluate further the effectiveness of geochemical denudation in the Colorado Front Range and to estimate its spatial variability. This spatial variability is significant for local landform development. Two problems are encountered in this study of geochemical denudation.

First, there is the estimation of the magnitude of solute removal from alpine drainage basins over a decadal time scale, which includes the annual variability of climatic conditions. Second is the evaluation of the variability in geochemical activity within the drainage basin for which integrated estimates are made. This variability is expected to be high because of the local variability in water inputs from snowmelt and rainfall and in the pathways by which this water moves through the system (Trudgill, 1988).

Previous work

In the Colorado Front Range the early observations of Thorn (1975, 1976) of weathering rinds and solute removal indicated the significance of solution processes as a component in nivation. These studies were extended during the past decade by work associated with a continuing program of Long-Term Ecological Research in Green Lakes Valley (e.g. Litaor and Thurman, 1988; Caine, 1989a,b; Caine and Thurman, 1990). This work has extended the empirical data base from the Martinelli Snowpatch of Thorn's studies to a hierarchy of catchments along North Boulder Creek above 3250 m, the approximate elevation of the forest limit in the Front Range. This work has also provided estimates of atmospheric inputs to the alpine system (Greenland, 1986; Reddy and Caine, 1990).

Elsewhere in the Colorado Front Range, research in a variety of field areas suggests the significance of solute removal in contemporary erosion (Lewis and Grant, 1979; Baron, 1983; Baron and Bricker, 1987; Rochette, Drever, and Sanders, 1988; Stednick, 1989; Mast, Drever, and Baron, 1990). Given the low rates of clastic sediment removal from high elevation basins in the range (e.g., Caine, 1986), these studies all point to solute removal as a dominant influence in contemporary landscape modification.

Field area

Above 3250 m, Green Lakes Valley (Figure 3.1) is a 7.1-km^2 catchment which is essentially alpine in nature, although 40 ha of forest occurs on the north valley wall at the lowest elevations, and stunted spruces may be found in sheltered sites up to 3600 m. The catchment appears typical of the high-elevation environment of the Colorado Front Range and includes the south slope of Niwot Ridge, where work on alpine environments has been conducted for more than 30 years (Ives, 1980).

Green Lakes Valley includes two contrasting sections: the upper basin (2.1 km^2 in area) comprising the area draining through Green Lake 4, and the lower valley (5.0 km^2 in area) between Green Lake 4 and the former settlement of Albion. The upper valley is typical of high alpine environments. It has relatively little vegetation (about 20% areal cover: Caine, 1986), steep rock walls and talus slopes, and a valley floor of glaciated bedrock. It includes numerous semipermanent snowpatches and the small, almost stagnant Arikaree Glacier at the valley head. The lower valley has less exposed bedrock, fewer cliff-talus slopes, more debris-mantled valley sides with lower gradients, rounded interfluves, and a more extensive soil and vegetation cover on the valley floor. A glaciated valley step about 75 m high between Green Lakes 3 and 4 separates the two sections of the valley.

The bedrock of the lower Green Lakes Valley is dominated by granites and quartz monzonites of two different intrusions: the Silver Plume monzonite of Precambrian age and the Audubon-Albion stock of Miocene age. The Silver Plume monzonite extends into the upper valley where the metasediments of Precambrian age, into which it was intruded, are exposed on the northern slopes of Kiowa Peak and on Niwot Ridge above Green Lake 5.

The Green Lakes Valley has a continental, high-mountain climate (Table 3.1). The 40-year record for the D-1 station on Niwot Ridge, which forms the northern drainage divide of the basin, has been summarized by Barry (1973), Losleben (1983) and Greenland (1989). Shorter periods of observations for stations on the valley floor suggest a less exposed and windy environment with greater snow accumulation than on the surrounding ridges. Within the valley, south-facing slopes are warmer than north-facing ones, which are underlain by permafrost (Ives, 1973).

In this environment, about 80% of the annual precipitation occurs as snow, which is especially prone to wind transport during and after deposition. The relocation of snow on the landscape gives a patchy accumulation pattern that tends to be reproduced each winter. Berg (1986) recorded 69% of mean hourly wind speeds in winter at 7.5 m s^{-1} or more and suggested that wind drifting of snow occurs on 75% of winter days. This extreme windiness in combination with generally low vapor pressures suggests a high potential for water loss by sublimation (Schmidt, 1982) in addition to the relocation of

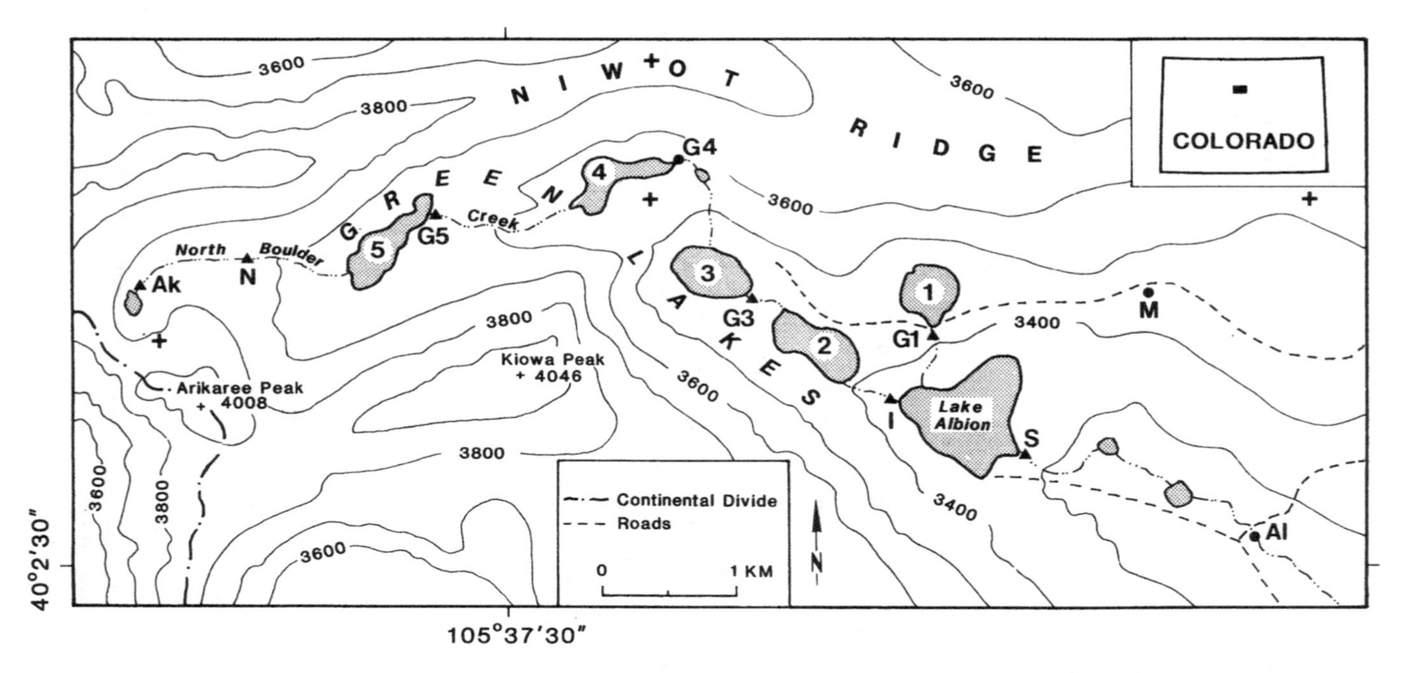

Figure 3.1. Location map of Green Lakes Valley. Study sites with a continuous discharge record (●), other sampling sites (▲), and climate stations (+) are shown.
Site identification: Al = Albion, S = Spillway of Lake Albion, I = Inlet to Lake Albion, G4 = Green Lake 4, G5 = Green Lake 5, N = Navajo, Ak = Arikaree, G1 = Green Lake 1, and M = Martinelli Snowpatch

Table 3.1. Hydroclimate: Green Lakes Valley and Niwot Ridge[a]

Characteristic	Jan.	Feb.	March	April	May	June	July	Aug.	Sept.	Oct.	Nov.	Dec.	Year
Temperatures (°C)[b]	−13.2	−12.8	−11.2	−7.0	−0.9	4.6	8.2	7.1	3.4	−2.1	−8.9	−11.8	−3.7
Precipitation (mm)[b]	102.5	80.1	127.8	100.1	87.5	59.5	50.5	62.1	49.1	40.3	80.3	90.1	930
Runoff (mm)[c]					39	250	286	177	94	26			872

[a]Sources: Greenland (1989) and LTER streamflow files for Green Lake 4.
[b]Temperature and precipitation values are based on a 36-year record at the D-1 station on Niwot Ridge.
[c]Runoff is based on the 1981–1990 record from the outlet of Green Lake 4.

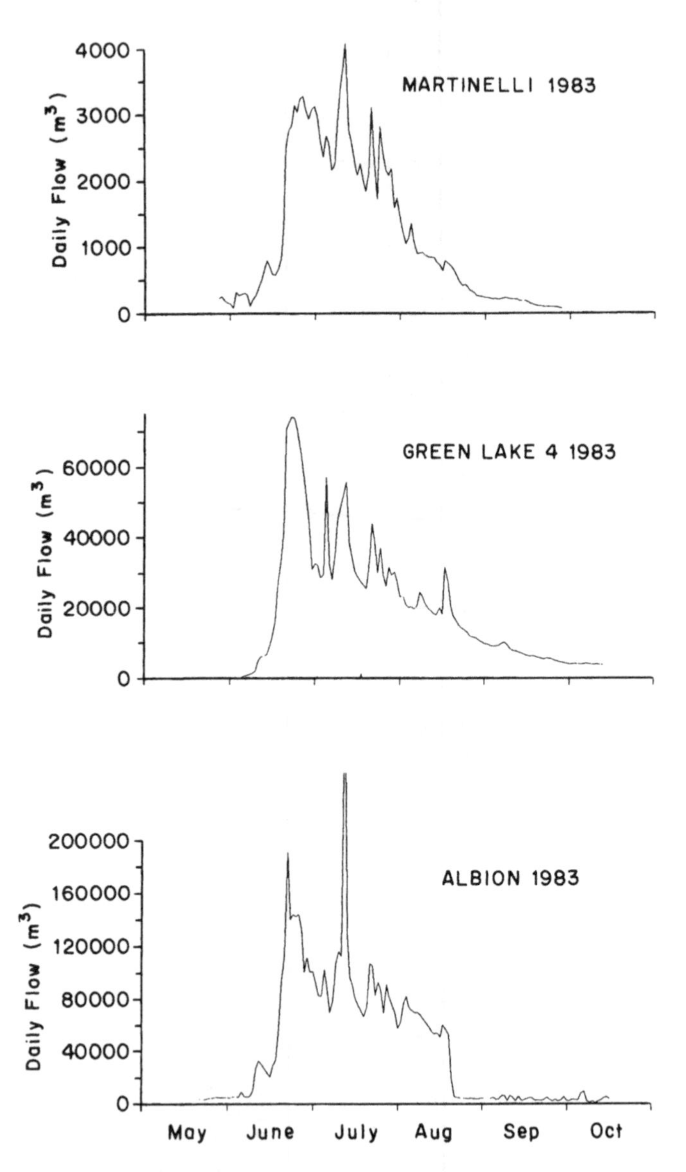

Figure 3.2. Flow hydrographs for Martinelli, Green Lake 4, and Albion in 1983. The Albion record shows the influence of reservoir manipulations in the lower Green Lakes Valley during August

water mass by drifting. On the basis of two winters of record, Berg (1986) suggested that sublimation losses could exceed 250 mm yr^{-1}—that is, more than 30% of the snowfall.

Hydrographically, the catchment is a linear cascade of 5 lakes with only Green Lake 1, on the north valley wall, tributary to this sequence (Figure 3.1). The channels between and above the lakes are invariably steep and rocky. They are sufficiently well armored to prevent sediment movement even at peak flows, except when such flows are forced out of the channels by ice blockages and superimposed from the snow cover onto the valley floor in a manner similar to that reported from the high arctic by Woo (1982).

Streamflows are markedly seasonal, varying from less than 0.1 m^3 s^{-1} during the winter months to 1.5 m^3 s^{-1} at the peak of snowmelt in June at the Albion site (Figure 3.2). High flows are usually associated with snowmelt rather than summer rainstorms (Jarrett, 1989), though the latter may be locally important in sediment movement (Caine and Swanson, 1989). A 24-hour periodicity in stream discharges, associated with the diurnal pattern of snowmelt, is normal during high flows (e.g., Caine, 1989a) but attenuates in late-summer as groundwater contributions become proportionately greater. Both periodicities in streamflow volumes are reflected in the temporal pattern of solute concentrations, which vary by a factor of 2 or more seasonally and diurnally (Caine, 1989a; Caine and Thurman, 1990).

The alpine tundra vegetation of Niwot Ridge has been mapped and described by Komarkova and Webber (1978). Above treeline at 3400 m elevation, a complex mosaic of turf and shrub communities reflects local contrasts in soil moisture, wind exposure, and snow cover. In Green Lakes Valley, wetter communities are more extensive than on Niwot Ridge and reflect the same controls and the effects of drainage on the valley floor. Komarkova and Webber (1978) show the upper Green Lakes Valley as part of a subnival belt of bare rock, talus and boulder slopes, with little vegetation other than mosses and lichens. Soils and surficial materials match the vegetation. They consist of a heterogeneous cover of cryic inceptisols and entisols with histosols in the wetter parts of the valley floor (Burns, 1980). The valley floor appears to have been deglaciated about 12,000 years ago (Harbor, 1984). The soils and superficial deposits are all of Holocene age. The interfluves, like Niwot Ridge, were not glaciated in the late-Pleistocene (Madole, 1982) and are mantled with older deposits in which relict patterned ground and mass wasting features are extensively developed (Benedict, 1970).

Procedures

Stream discharges have been continuously monitored during the May-to-October flow season at Albion, the outlet from Green Lake 4 and the

Martinelli Snowpatch (Figure 3.1). At these three sites, water-level records digitized with a 1-hour interval have been converted to volumetric discharges by empirical ratings and integrated to daily, or longer period, totals. The stage-discharge ratings have been verified by direct measurement at approximate monthly intervals during summers. At the other six sites in the basin, water levels have been measured on a weekly or shorter interval (i.e., whenever an observer is at the site) and converted to discharges by empirical ratings or weir formulas. These estimates have then been correlated with those at the nearest downstream station with a continuous record and that relationship used to estimate flows over intervals between visits.

In summer, water sampling has been conducted, usually once a week, at the seven sites along the channel of North Boulder Creek and at the streams draining the Martinelli Snowpatch and Green Lake 1 on the north side of the valley (Figure 3.1). During October to May, sampling has been limited to once a month at the Albion and Green Lake 4 sites. Sampling on a systematic time interval leads to an underestimation of sediment yields because of the probability that short periods of high sediment concentrations are missed (Walling and Webb, 1981). This effect should be recognized, but its magnitude has not yet been estimated, as the empirical relationships between water discharge and sediment concentration in the catchment streams are usually weak ($r < 0.2$), except over short time intervals (Caine, 1989a). Sample treatment and analytical procedures are described in Caine and Thurman (1990).

Surface water quality

The stream waters of Green Lakes Valley have low concentrations of dissolved and particulate material (Caine and Thurman, 1990). Except at the outlet from Green Lake 1, dissolved solids concentrations rarely exceed $30\,mg\,l^{-1}$ and fall to less than $3\,mg\,l^{-1}$ on occasion in the meltwaters from Arikaree Glacier. Suspended sediment concentrations are generally less than $1.0\,mg\,l^{-1}$ at the three sites where they have been estimated over an eight-year period (Albion, Green Lake 4, and Martinelli). Low sediment yields reflect low rates of geomorphic activity in the catchment (Caine, 1986) and low sedimentation rates in its lakes (Harbor, 1984).

The concentrations of major ions and silica in the streams of Green Lakes Valley are summarized in Table 3.2. The same constituents, in similar relative concentrations, are found at all nine sites: calcium is the dominant cation, and bicarbonate and sulfate are usually the most important anions. At sites with the greatest snow accumulations, nitrate concentrations may exceed those of bicarbonate and sulfate early in the flow season when meltwater flushes the snowpack (Brimblecombe, Tranter, Abrahams, Blackwood, Davies, and Vincent, 1985). Except at the highest elevations, the pH of the stream water is

Table 3.2. Surface Water Quality Characteristics[a]

Site[b]	Ca	Mg	Na	K	HCO_3	SO_4	NO_3	Cl	SiO_2	Cond.	*p*H
Arikaree	0.42	0.04	0.10	0.13	2.00	0.60	0.70	0.09	0.48	5.68	5.48
Navajo	0.73	0.07	0.21	0.19	1.90	1.02	1.28	0.11	0.77	7.28	5.69
Green L 5	1.08	0.11	0.22	0.22	2.98	1.42	0.78	0.10	0.94	8.85	6.21
Green L 4	1.27	0.17	0.29	0.28	4.43	1.86	0.58	0.18	1.17	10.7	6.35
Inlet	2.30	0.26	0.49	0.42	5.69	2.93	0.66	0.30	1.14	13.5	6.41
Spillway	2.05	0.20	0.43	0.27	5.59	2.31	0.21	0.12	0.80	16.7	6.55
Albion	2.45	0.28	0.64	0.31	7.30	2.78	0.30	0.29	1.72	20.9	6.51
Green L1	6.83	0.47	1.17	0.55	20.8	7.59	1.12	0.21	2.42	37.9	6.60
Martinelli	1.26	0.11	0.43	0.22	5.16	0.92	0.71	0.11	1.39	8.79	6.20

[a]Values are discharge-weighted averages with concentrations in mg l^{-1} and conductances in $\mu S\ cm^{-1}$. Record lengths vary between 5 and 9 years.
[b]Site locations are shown in Figure 3.1.

generally above 6.0 but tends to increase, along with alkalinity, in the downstream direction (Table 3.2).

Factor analysis of the concentration data, by identifying surrogate variables, might allow the identification of sources and flowpaths of the water (e.g., Baron, 1983). Postulating two sources for the dissolved material in the stream water (an atmospheric one and a soil and bedrock one) suggests that two factors be sought, and these are usually sufficient to account for 70% of the variability in the correlation matrix. The factors show no consistent differences between years of study or sites within the catchment, despite changes in absolute concentrations that amount to an order of magnitude according to site and time of year.

The first factor, usually identified by the major cations, bicarbonate, sulfate, and specific conductance, often accounts for more than 50% of the total variance and seems to reflect catchment sources. Chloride and nitrate are important constituents associated with the second factor and may be identified with atmospheric sources. When a third factor is extracted, it is frequently aligned with silica and so may be associated with a groundwater influence, although it is usually weak (eigenvalue of less than 1.0). These results are consistent with those derived from other studies of alpine waters in the Colorado Front Range (e.g. Baron, 1983; Baron and Bricker, 1987; Mast, Drever, and Baron, 1990).

Two water quality characteristics are used here in evaluating chemical denudation: specific conductance and SiO_2 (Table 3.3). Conductance has been measured in the field and is closely associated with the total dissolved solids concentration, as estimated by the sum of major ions and SiO_2 (e.g., in 1985, $r = 0.979$ with $N = 219$). SiO_2 concentrations in bulk precipitation samples are low (ca. 1 $\mu g\ l^{-1}$: Reddy and Caine, 1990; Mast, Drever, and

Table 3.3. Annual Specific Yields: Green Lakes Valley[a]

Site[b]	Area (ha)	Water (mm)	Conductance ($\mu S\ cm^{-1}\ m^{-2}$)	Silica ($g\ m^{-2}$)
Arikaree	9.2	4086	22.68	1.94
Navajo	41.8	1626	11.72	1.25
Green L 5	135.2	991	8.73	0.88
Green L 4	220.9	872	9.34	0.94
Inlet	354.7	546	7.28	0.63
Spillway	546.2	668	11.10	0.53
Albion	709.7	612	12.73	0.81
Green L1	47.3	78	2.94	0.19
Martinelli	8.0	964	8.45	1.40

[a]Values are mean annual rates based on records for 1982–1990 (water and conductance) or 1984–1990 (silica).
[b]Site locations are shown in Figure 3.1.

Baron, 1990). So the catchment solute yield can be assumed to derive from its bedrock and debris cover. SiO_2 makes up less than 10% of the solutes in the drainage waters (Table 3.2) and is not strongly correlated with conductance ($r = 0.377$ with $N = 195$ in the 1985 records).

The mean values of specific conductance and SiO_2 tend to increase with drainage area along the main drainage of Green Lakes Valley (Figure 3.3). However, there are two notable departures from these trends. (1) The values for the Green Lake 1 and Martinelli sites are higher than those for sites of equivalent area on the main channel, probably reflecting a high groundwater contribution to their flows (Caine and Thurman, 1990). (2) Reduced SiO_2 concentrations at the Inlet and Spillway of Lake Albion seem to be explained by increased rates of biological extraction of silica in the surface waters of the lower valley lakes.

Snow distribution and geochemical denudation

Even at the end of the winter accumulation season, large areas of the alpine carry only a slight snow cover, whereas snow accumulation sites (e.g., the Martinelli basin (Figure 3.4)) may have a depth of more than 12 m of snow (up to 6 m water equivalent). Johnson (1980) estimated average accumulations for the entire surface of Arikaree Glacier in the early 1970s at about 4 m (water equivalent), 3 to 4 times the average precipitation.

These contrasts are shown in the results of snow surveys in the upper valley reported by Carroll (1974) who found that at the end of the accumulation season, 35% of the snow water equivalent in the basin was stored on only 15% of its area (Figure 3.5). In contrast, the 32% of the basin area with the

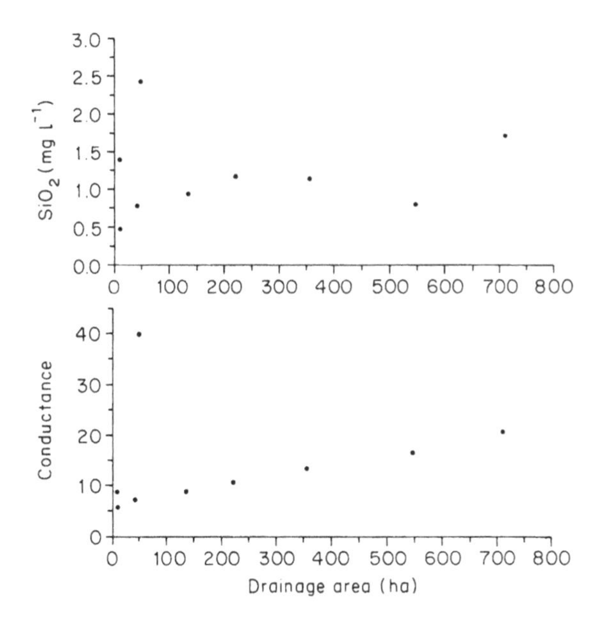

Figure 3.3. Variations in silica concentrations and specific conductance with drainage area in Green Lakes Valley. Points shown are discharge-weighted mean values for the 1982–1990 period

lowest snow accumulations carried only 5% of the snow water equivalent and was less efficient in converting that snow to streamflow. A similar pattern was found in the Martinelli basin in 1983, the year of greatest water yield from this basin during the 1980s. Figure 3.6 shows that snow depths in early May exceed 10 m on 9% of the area, representing 33% of the snow stored on the catchment. Thirty-six per cent of the basin area carried less than 1 m depth of snow, that is 4% of the total volume of snow. The zone of least snow cover was around the catchment edge, especially at its head (Figure 3.6), where it is least likely to contribute to streamflow after melting (Caine and Swanson, 1989).

In 1985, surveys of the snow-covered area in the entire Green Lakes basin during the ablation season showed that the patchy nature of the distribution becomes more marked as the season progresses (Figure 3.5; Buchanan, 1986). Approximately 100% snow cover in the lower part of Green Lakes Valley in late March 1985 was reduced to 15% by late June. In contrast, the valley above Green Lake 5 changed from 80% to 50% snow cover in the same period. Such variability in the water equivalent on the ground and snowcover duration implies a corresponding variability in local water budgets which should then be reflected in local weathering rates and solute yields.

Summer rainfall in Green Lakes Valley is also spatially variable and does lead to flashy flow responses in the streams. However, it contributes less than

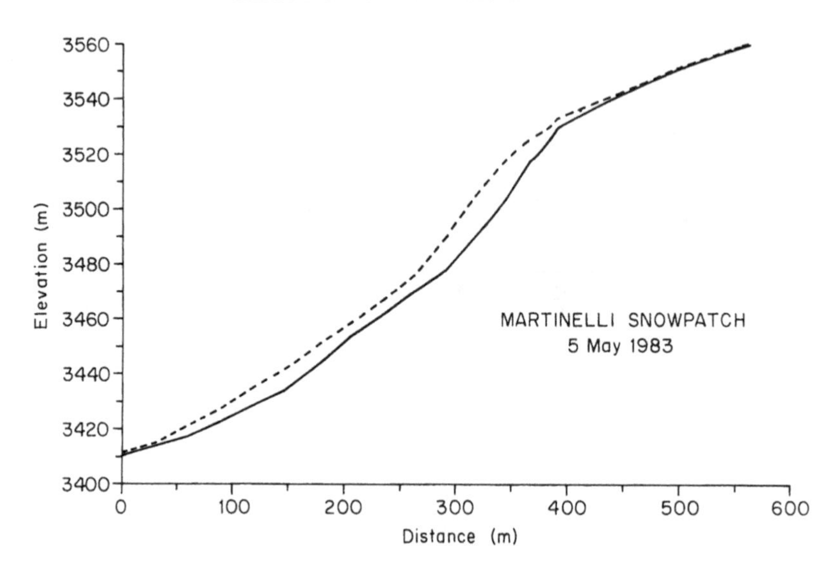

Figure 3.4. Profile of Martinelli Snowpatch. The survey approximately follows the axis of the basin from its outlet. The solid line represents the ground surface and the broken line the snow surface on May 5, 1983

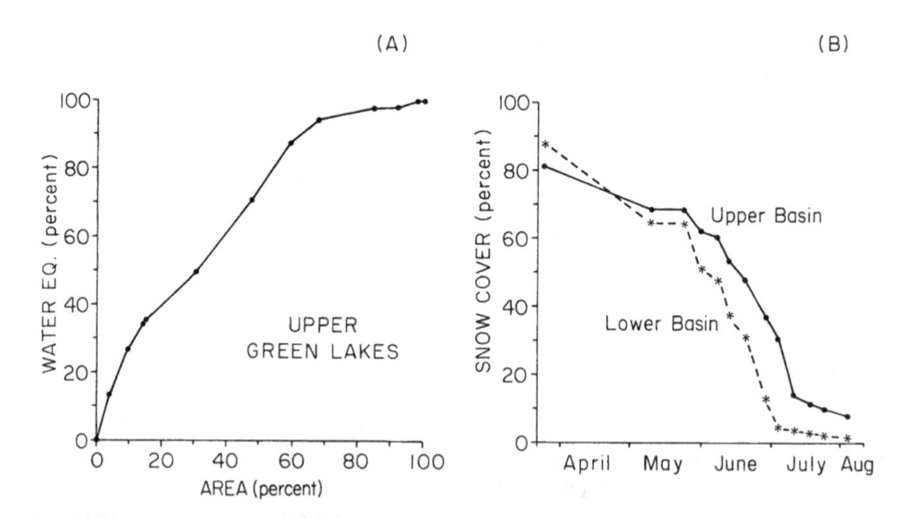

Figure 3.5. Snow surveys in Green Lakes Valley. (A) Snow water storage by area in the Upper Green Lakes Valley on May 18, 1973. Surveys were made by Federal snow sampler at 89 points in the basin (after Carroll, 1974). (B) Changes in the area of snow cover in the Green Lakes Valley during 1985 (after Buchanan, 1986)

5% to annual stream discharges and even less to solute transport and so is not treated here.

In general, water yields per unit area in Green Lakes Valley increase with decreasing basin size (Table 3.4, Figure 3.7). For sites along the main drainage, the relation between average water yield WY (mm) and basin area A (ha) is estimated by

$$WY = 9558\,A^{-0.444} \qquad (3.1)$$
$$(r = -0.974,\ N = 7)$$

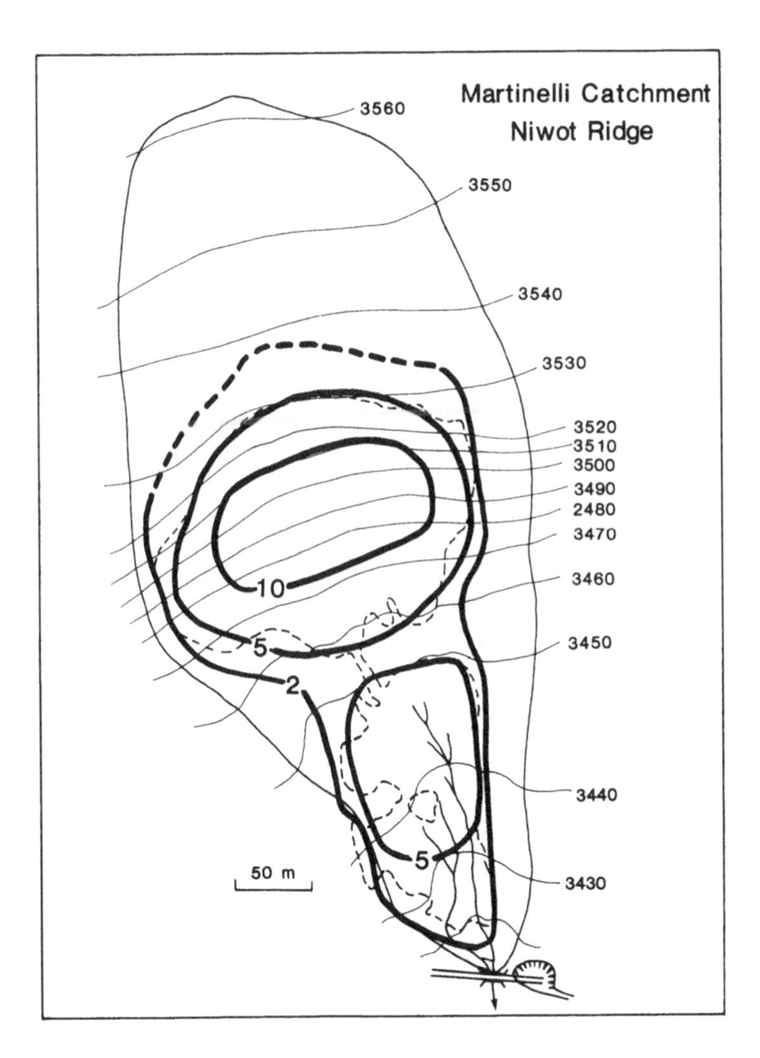

Figure 3.6. Snow depths on the Martinelli Catchment on May 5, 1983. Contours of snow depth from probe surveys, including those used in Figure 3.4

Table 3.4. Discharge–Conductance Relationships[a]

Site	Sample Size	Intercept a	Exponent b	Correlation Coefficient
Hourly records:	Martinelli, June 1987[b]			
	47	14.54	−0.308	−0.979
	18	11.09	−0.222	−0.913
Weekly records:				
Albion	214	52.54	−0.097	−0.522
Green Lake 4	145	35.15	−0.128	−0.460
Martinelli	120	16.33	−0.100	−0.602
Annual records:				
Arikaree	8	166.0	−0.396	−0.689
Navajo	8	35.7	−0.216	−0.591
Green Lake 5	8	21.4	−0.131	−0.306
Green Lake 4	8	8.7	0.026	0.053
Inlet	8	22.2	−0.081	−0.605
Spillway	8	13.0	0.042	0.197
Albion	8	27.3	−0.043	−0.162
Green Lake 1	8	75.1	−0.149	−0.966
Martinelli	8	9.2	−0.010	−0.073

[a]The relationships are of the form $y = a\,x^b$, where x = discharge and y = conductance.
[b]Source: Caine (1989a).

The inverse relation, indicating a loss of basin efficiency (Carroll, 1976), reflects greater evaporative and consumptive losses in large basins and at lower elevations.

The Martinelli and Green Lake 1 sites are, again, not part of this general trend (Figure 3.7) but contribute to the wide between-site variability evident in basins smaller than 50 ha. In part, this variability is an artefact of uncertainties in estimating runoff totals from these small catchments, only one of which (Martinelli) is continuously gauged. However, this is probably not sufficient to account for deviations of more than an order of magnitude from the trend. A second factor contributing to this variability is the corresponding variability in winter snow accumulation in the smaller basins, some of which are effective accumulators of snow while others appear to lose large volumes of snow to wind drifting. Arikaree Glacier and Martinelli snowpatch exemplify the former, although the topographic catchment at Martinelli includes a large proportion which accumulates little snow (Figure 3.6). Green Lake 1 illustrates the latter.

The importance of contrasts in local water flows on geochemical denudation is defined by the empirical relationships between specific conductance (a surrogate for dissolved solids concentration) and water discharge (Table 3.4,

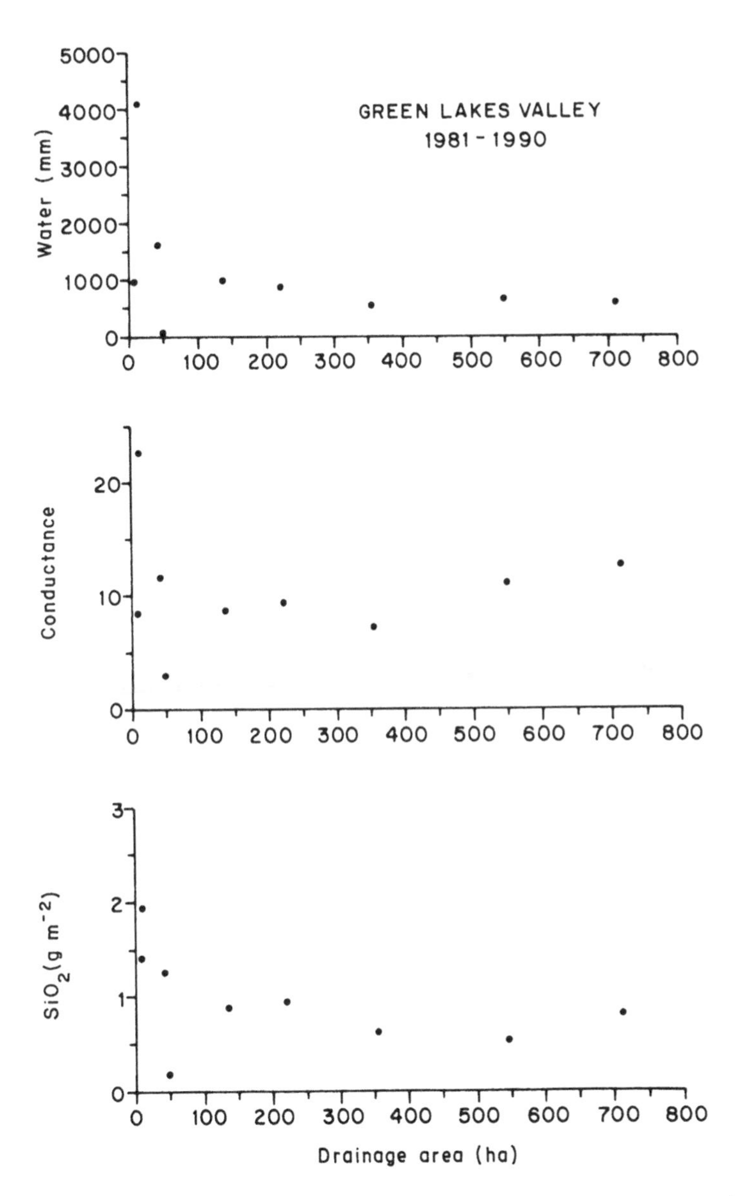

Figure 3.7. Catchment yields plotted against drainage area. Yields of silica (g m^{-2}) are annual averages for a period of 7 to 9 years at each site. Specific conductance is expressed as (μS cm^{-1}) · m (water) and is approximately equivalent to 1.15 · total dissolved solids concentration. Conductance is based on a 10-year record at each site, except Green Lake 1, where it is based on the 1985–1990 period

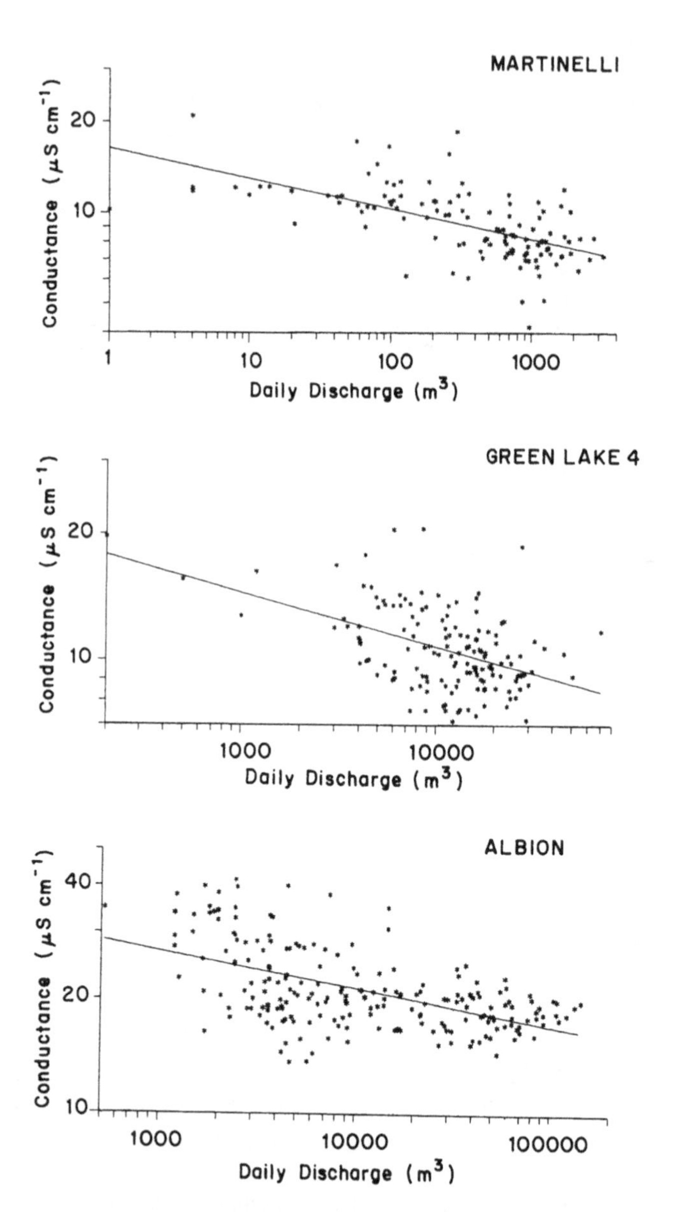

Figure 3.8. Specific conductance–discharge relations for three sites in Green Lakes Valley. Data are derived from weekly observations during the May–October season. The least-squares regressions are summarized in Table 3.4

Figure 3.8). For all sites and a variety of time scales (hourly to annual), the power relationship between conductance and flow has an exponent (*b*) in the range −0.1 to −0.3. These values (which are less negative than −1.0) indicate that variations in discharge volumes are more important in accounting for changes in solute yield (the product of concentration and water volume) than variations in concentration (Gunnerson, 1967). Thus the between-basin variations in water yield should be translated into roughly equivalent variations in solute yield (Table 3.3).

More detailed patterns are defined by a survey of surface water quality conducted between June 3 and 5, 1985, at the time of peak flow for that season (Hoffman, Caine, and Caine, 1985). Spatial variations in conductance of SiO_2 (Figure 3.9) reflect two influences: (1) elevation along the valley, with a pattern of increasing concentrations down valley; and (2) aspect, with higher conductances and SiO_2 concentrations on the south-facing valley wall. The first of these matches the average pattern along the main drainage (data from the main stream have not been used in Figure 3.9), while the second may be a response to the date of the survey. In early June, snowmelt and the thawing of exposed ground have usually progressed much further on southern exposures than on northern ones. Figure 3.9 also shows locations of high SiO_2 concentrations near the foot of the south-facing valley wall, at sites where winter icings suggest groundwater drainage. Unfortunately, the data of Figure 3.9 cannot be transformed to local yields because the discharges of the surface flows sampled, their duration, and their catchment areas cannot be estimated.

More locally, the effect of water movement has been inferred by Thorn (1975) and Caine (1979) from an upslope-downslope contrast in weathering indicators. The effect of drainage along the slope profile has also been suggested by studies of alpine soil catenas, although in these studies the effect of water movement may be confounded with topographic influences on snow distribution (Dixon, 1986). Direct observations of the volumes and chemistry of soil interstitial waters on dry tundra sites in Green Lakes Valley show that material transfers through the hillslope are not simple responses to gravity drainage (Litaor, 1987).

The weathering rates of samples of crushed rhyodacite fragments in the alpine of the San Juan Mountains (Caine, 1979) provide further empirical evidence of the link between the snow cover and solution processes on a very small scale. The results of up to nine years of record are shown in Figure 3.10, which defines the relationship between snow cover duration, a surrogate for snow water equivalent, and the rate of mass loss as

$$WR = -0.03 - 0.0009SD \quad (r = -0.659,\ N = 69) \tag{3.2}$$

where *WR* is the weathering rate, percent yr^{-1}, and *SD* is the 5-year average duration of snowcover, days after March 31. The intercept in this relation is

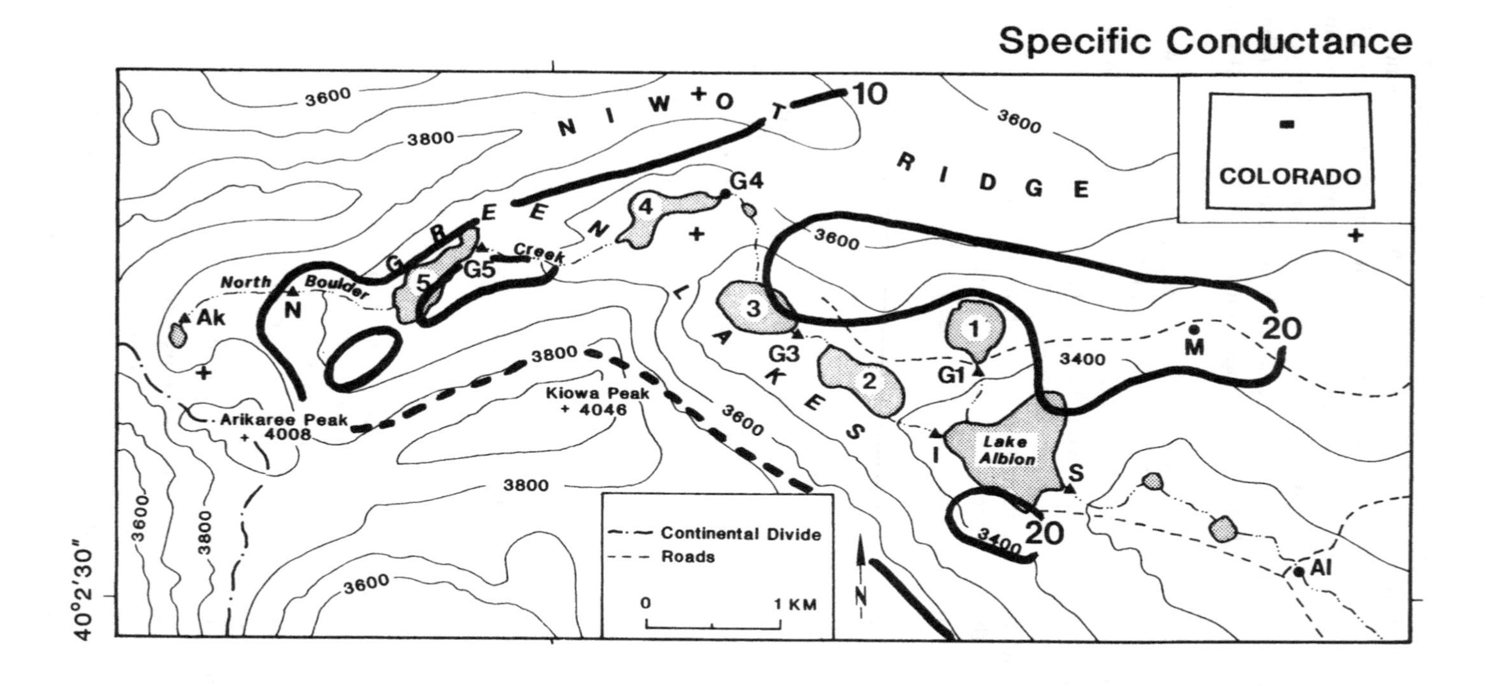
Specific Conductance
COLORADO
N I W O T
R I D G E
G R E E N
L A K E S
North Boulder Creek
Ak
N
G5
5
4
G4
3
G3
2
1
G1
M
I
S
Lake Albion
Al
Arikaree Peak
+ 4008
Kiowa Peak
+ 4046
3400
3600
3800
10
20
Continental Divide
Roads
0
1 KM
N
40°2'30"

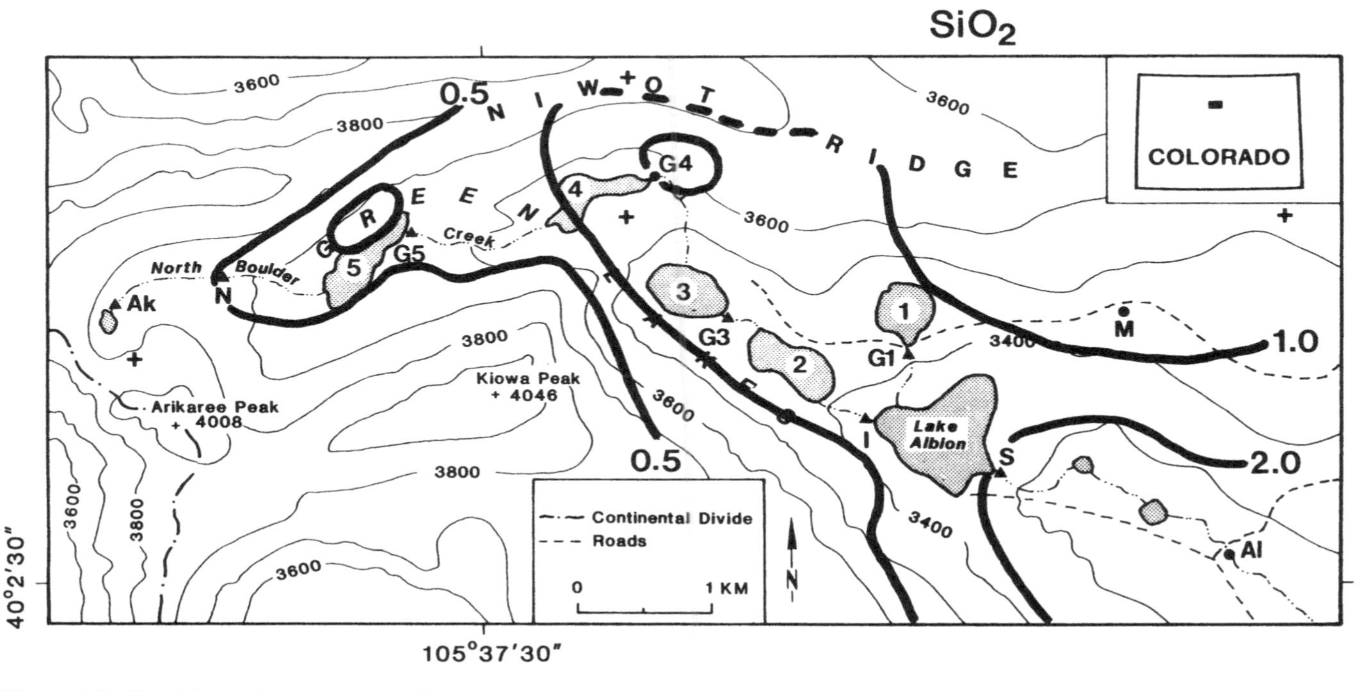

Figure 3.9. Specific conductance and silica concentrations of surface water, June 3–5, 1985. Isolines are based on 71 sampling sites of surface water flows and exclude observations made along the main stream. Conductance is in $\mu S\ cm^{-1}$ and silica in $mg\ l^{-1}$ (after Hoffman, Caine, and Caine, 1985)

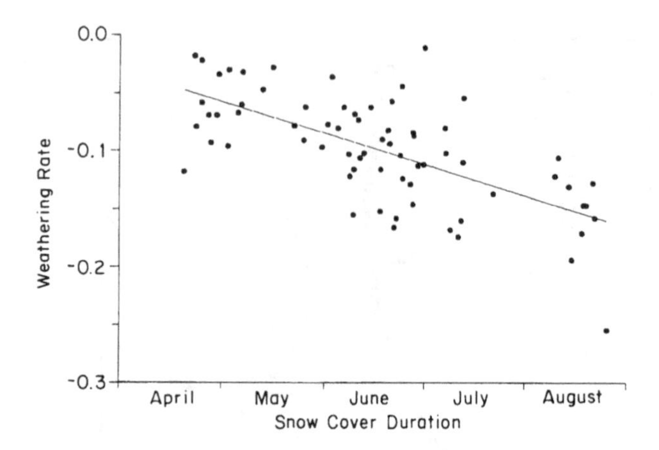

Figure 3.10. Graph of rock weathering rate *WR* versus snow cover duration *D*. The mean rates of mass loss (% yr^{-1}) are based on up to 9 years of observations in the San Juan Mountains, southwestern Colorado (Caine, 1979). The least-squares regression $WR = -0.03 - 0.0009\ D$ is based on 69 plot values ($r = -0.659$). Snow cover duration (days after March 31) is the average for 1971–1975

not different from 0, and the coefficient suggests an acceleration of rock weathering by approximately 0.001% per year for each 2 to 3 cm of accumulated snow depth (water equivalent). Within a typical alpine basin this acceleration could lead to local differences of an order of magnitude in rates of solute removal, depending on snow accumulation patterns.

Discussion

The relative significance of geochemical denudation in the contemporary geomorphic activity of alpine environments like those of the Front Range has been defined most recently by Caine and Thurman (1990), and the results presented here, based on a longer record, support those found earlier. The comparison of solute yields and suspended sediment yields is best defined for the Martinelli catchment (Caine and Swanson, 1989), where the latter are usually more than an order of magnitude lower than the former. Over the 1984–1990 period, the average solute yield of 10 g m^{-2} yr^{-1} greatly exceeded the average sediment yield of 0.6 g m^{-2} yr^{-1}. Even when corrected for solute and dust inputs from the atmosphere (Reddy and Caine, 1990) and the likely underestimation of sediment yields, a ten-fold difference remains.

Similar differences between solute and sediment concentrations are found at Green Lake 4 and Albion, the other sites where samples of suspended sediment are regularly taken. At Green Lake 4, the discharge-weighted mean solute concentration is 10.30 mg l^{-1}, and the suspended sediment concentra-

Table 3.5. Estimates of Local Geochemical Denudation Rates, Martinelli Snowpatch

Snow Survey, May 5, 1983[a]			DR_{83}[b]	DR_{LT}[c]
Area (m^2)	Area (%)	Snow Depth (m)		
29 000	36.3	0.33	2.0	1.0
15 000	18.8	1.0	5.2	3.0
14 500	18.1	3.0	17.7	9.0
14 100	17.6	8.0	47.2	23.0
7400	9.2	12.0	70.8	35.0
Area-weighted mean		3.36	19.9	10.2

[a]Snow survey data are from Figure 3.6.
[b]DR_{83}: Local denudation rates (g m^{-2} yr^{-1}) estimated for 1983 record.
[c]DR_{LT}: Local denudation rates (g m^{-2} yr^{-1}) estimated for 1982–1990 records (long-term).

tion 0.76 mg l^{-1}. At Albion, the equivalent values are 16.21 mg l^{-1} and 1.20 mg l^{-1}, respectively.

Geochemical denudation rates in Green Lakes Valley are consistently low and average only 4 gm m^{-2} yr^{-1} (equivalent to about 0.002 mm yr^{-1} of surface lowering). This estimate is slightly less than that made by Caine and Thurman (1990) because of the reduced water yields in 1988–1990, years not included in the earlier study. The estimate is comparable to estimated geochemical denudation rates for other mountain basins in the Colorado Front Range (e.g., Baron and Bricker, 1987; Stednick, 1989). However, it is appreciably lower than estimated rates for glacierized mountain catchments (e.g., Reynolds and Johnson, 1972; Eyles, 1982; Souchez and Lemmens, 1987), which supports the conclusion that the extent of glacierization has a dominant influence on alpine erosion (Slaymaker, 1977).

On the local (landform) scale, the correlation between snow depths (water equivalent) and weathering rates suggests that estimates of snow depth, like those of Figure 3.6, may be transformed to rates of solute removal. Partitioning the Martinelli solute yield of 19.9 g m^{-2} yr^{-1} for 1983 in proportion to the snow distribution on the basin surface suggests the spatial distribution of geochemical denudation shown in Table 3.5. Extending this to the 1982–1990 period gives an average solute yield of 35 g m^{-2} yr^{-1} for the 10% of the basin with greatest snow depths and less than 1 g m^{-2} yr^{-1} for the 35% of the basin periphery with least snow accumulation. Although this analysis neglects the influence of meltwater drainage and residence times, it does suggest an internal contrast in rates of solute removal of an order of magnitude. Estimates from the empirical relationship between snow cover duration and rock weathering (Figure 3.10) would be more conservative: a six- or seven-fold increase in weathering rates across the range of snow cover durations in the Martinelli basin.

Within small drainage basins, such a level of variability in solute removal should be significant in landscape evolution. For the Martinelli catchment, Caine and Swanson (1989) suggested a decay mode of landscape evolution on the basis of the inequality between sediment transport to the basin floor and sediment removal from it. As a result, the topographic hollow would tend to infill, with aggradation on the floor occurring at the rate of about 330 kg yr^{-1}. However, this conclusion was based upon an assumption of spatial uniformity in solution processes within the basin. A spatially differentiated model based on snow depths suggests differences in rates of solute removal between the basin floor and its peripheries by a factor of 35, with a solute removal rate of 267 kg yr^{-1} from the 9% of the basin with the deepest snow cover (Table 3.5). This may be sufficient to balance the rate of clastic sediment accumulation on the basin floor.

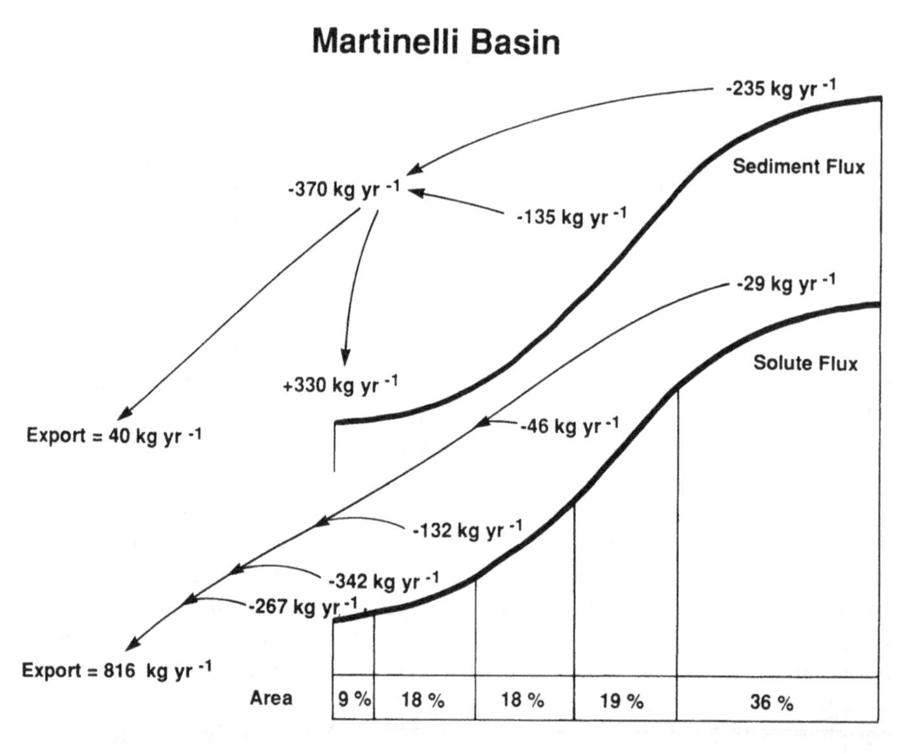

Figure 3.11. Materials budgets for the Martinelli Basin. Estimates of mass wasting and clastic sediment transport rates are shown in the upper profile (after Caine and Swanson, 1989). The solute flux estimates shown in the lower profile are the 9-year basin average weighted for the differences in snow depths in the basin (Table 3.5)

These results can be combined with those of Caine and Swanson (1989) to give the landscape model presented in Figure 3.11. This model suggests that in sites like the Martinelli catchment, which carry a long-lasting snow cover, geochemical denudation is capable of at least maintaining if not accentuating the form of the hollow. Solute removal associated with areas of concentrated snow accumulation could therefore contribute to the development of nivation hollows.

On a smaller scale, gelifluction terraces and lobes are other topographic features of the alpine that modify snow accumulation. The nature and movement of the lobes and terraces of Niwot Ridge have been documented by Benedict (1970), but the form of the hollows upslope of them, from which the moving debris is derived, has received less attention. As they accumulate drifting snow, such hollows should also be affected by locally accelerated solutional weathering. With detailed surveys (e.g., those of Benedict, 1970), the volumes of lobe and hollow can be estimated as the difference between a low-order polynomial surface fitted to the neighboring slopes and the actual topography. In the case of Lobe 45 (Benedict, 1970, p. 186), the hollow has a volume of $-335\ m^3$ and the lobe of one of $175\ m^3$, suggesting a loss from the system of $160\ m^3$. Distributing this over the area of the hollow gives an average lowering of 250 mm. This corresponds to a denudation rate of 12.5 or $50\ g\ m^{-2}\ yr^{-1}$, depending on whether the lobe is ascribed an age of 2500 yr (Benedict, 1970, p. 217) or 10 000 yr (Mahaney and Fahey, 1980). Such rates are equivalent to those estimated for the parts of the Martinelli basin with more than 2 m of snow accumulation. This analysis suggests that solutional denudation in response to local snow accumulation may be as important as wind and water erosion in removing mass from the lobe-hollow system.

Conclusion

This paper has defined two effects of geochemical denudation on alpine geomorphic processes. The first confirms the observational evidence, açcumulated over the last 25 years from non-glacierized mid-latitude mountains, that geochemical processes dominate contemporary denudation, exceeding rates of sediment export by an order of magnitude. The second suggests that spatial variations in solute processes, reflecting the uneven distribution of alpine snow covers, could have a large role in modifying, or at least preserving, topographic variability in alpine landscapes. This, in turn, implies that solute removal could be a significant landforming influence even in nonkarstic mountain terrain.

Acknowledgements

Research in Green Lakes Valley has been supported by the National Science Foundation through grants BSR 8012095 and BSR 8514329 as part of the

University of Colorado Long-Term Ecological Research Program. Colleagues at the Niwot Ridge/Green Lakes Valley LTER site have contributed greatly to the work reported here through assistance in the field and laboratory and through discussions over the 10-year life of this project. I have particularly benefited from the assistance, advice, and comments of Cathy Baganz, Jennifer Caine, Dave Furbish, Jim Halfpenny, Damian Kraus, Iggy Litaor, Mark Losleben, Mike Reddy, Mike Thurman, and Pat Webber at various times in the last decade. They all have my sincere thanks. This manuscript has benefited from the comments of Jim Gardner and Olav Slaymaker who read an earlier, less satisfactory version of it.

References

Ballantyne, C. K., N. M. Black, and D. P. Finlay, Enhanced boulder weathering under late-lying snowpatches, *Earth Surface Processes and Lanforms*, **14**, 745–750, 1989.

Baron, J., Comparative water chemistry of four lakes in Rocky Mountain National Park, *Water Resources Bulletin*, **19**, 897–902, 1983.

Baron, J., and O. P. Bricker, Hydrologic and chemical flux in Loch Vale watershed, Rocky Mountain National Park, in *Chemical Quality of Water and the Hydrologic Cycle*, edited by R. C. Averett and D. M. McKnight, pp. 141–155, Lewis Publishers, Chelsea, Mich., 1987.

Barry, R. G., A climatological transect on the east slope of the Front Range, Colorado, *Arctic and Alpine Research*, **5**, 89–110, 1973.

Barsch, D., and N. Caine, The nature of mountain geomorphology, *Mountain Research and Development*, **4**, 287–298, 1984.

Benedict, J. B., Downslope soil movement in a Colorado alpine region: rates, processes and climatic significance, *Arctic and Alpine Research*, **2**, 165–226, 1970.

Berg, N. H., Blowing snow at a Colorado alpine site: measurements and implications, *Arctic and Alpine Research*, **18**, 147–161, 1986.

Birkeland, P. W., *Soils and Geomorphology*, 372 pp., Oxford University Press, New York, 1984.

Brimblecombe, P., M. Tranter, P. W. Abrahams, I. Blackwood, T. D. Davies, and C. E. Vincent, Relocation and preferential elution of acidic solutes through the snowpack of a small, high-altitude Scottish catchment, *Annals of Glaciology*, **7**, 141–147, 1985.

Buchanan, L. R., A snow depletion curve for modelling runoff in an alpine basin, M.Sc. thesis, 143 pp., University of Colorado, Boulder, 1986.

Burns, S. F., Alpine soil distribution and development, Indian Peaks, Colorado Front Range, Ph.D. thesis, 360 pp., University of Colorado, Boulder, 1980.

Caine, N., A uniform measure of sub-aerial erosion, *Bulletin of the Geological Society of America*, **87**, 137–140, 1976.

Caine, N., Rock weathering rates at the soil surface in an alpine environment, *Catena*, **6**, 131–144, 1979.

Caine, N., Sediment movement and storage on alpine hillslopes in the Colorado Rocky Mountains, in *Hillslope Processes*, edited by A. D. Abrahams, pp. 115–137, Allen and Unwin, Boston, 1986.

Caine, N., Diurnal variations in the inorganic solute content of water draining from an alpine snowpatch, *Catena*, **16**, 153–162, 1989a.

Caine, N., Hydrograph separation in a small alpine basin based on inorganic solute concentrations, *Journal of Hydrology*, **112**, 89–101, 1989b.

Caine, N., and F. J. Swanson, Geomorphic coupling of hillslope and channel systems in two small mountain basins, *Zeitschrift für Geomorphologie*, **33**, 189–203, 1989.

Caine, N., and E. M. Thurman, Temporal and spatial variations in the solute content of an alpine stream, Colorado Front Range, *Geomorphology*, **4**, 55–72, 1990.

Carroll, T. R., The water budget of an alpine catchment in central Colorado. M.A. thesis, 124 pp., University of Colorado, Boulder, Colorado, 1974.

Carroll, T., An estimate of watershed efficiency for a Colorado alpine basin, *Proceedings of the 44th Western Snow Conference*, 69–77, 1976.

Dixon, J. C., Solute movement on hillslopes in the alpine environment of the Colorado Front Range, in *Hillslope Processes*, edited by A. D. Abrahams, pp. 139–159, Allen and Unwin, Boston, 1986.

Eyles, N., D. R. Sasseville, R. M. Slatt, and R. J. Rogerson, Geochemical denudation rates and solute transport mechanisms in a maritime temperate glacier basin, *Canadian Journal of Earth Sciences*, **18**, 1570–1581, 1982.

Grant, M. C., and M. W. Lewis, Jr., Chemical loading rates from precipitation in the Colorado Rockies, *Tellus*, **34**, 74–88, 1982.

Greenland, D., Preliminary results of NADP sampling on Niwot Ridge, Colorado, in *Proceedings of Symposium on Acid Deposition in Colorado—A Potential or Current Problem; Local versus Long-Distance Transport into the State*, edited by R. A. Pielke, pp. 75–102, Colorado State University, Cooperative Institute for Research in Atmospheric Sciences, Fort Collins, 1986.

Greenland, D., The climate of Niwot Ridge, Front Range, Colorado, USA, *Arctic and Alpine Research*, **21**, 380–391, 1989.

Gunnerson, C. G., Streamflow and quality in the Columbia River basin, *Proceedings of the American Society of Civil Engineers, Journal of Sanitary Engineering Division*, **93** (SA6), 1–16, 1967.

Harbor, J. M., Terrestrial and lacustrine evidence for Holocene climatic/geomorphic change in the Blue Lake and Green Lakes valleys of the Colorado Front Range, M.A. thesis, 205 pp., University of Colorado, Boulder, 1984.

Hoffman, K., J. Caine, and N. Caine, Surface water quality in the Green Lakes Valley prior to peak flow 1985, *University of Colorado Long-Term Ecological Research Data Report*, **85/7**, 18 pp., 1985.

Ives, J. D., Permafrost and its relationship to other environmental parameters in a mid-latitude, high-altitude setting, Front Range, Colorado Rocky Mountains, *North American Contribution to the Second International Conference on Permafrost*, 13–28, 1973.

Ives, J. D. (ed.), *Geoecology of the Colorado Front Range*, 484 pp., Westview Press, Boulder, 1980.

Jarrett, R. D. Paleohydrologic techniques used to define the spatial occurrence of floods, *Geomorphology*, **3**, 181–195, 1989.

Johnson, J. B., Mass balance studies on the Arikaree Glacier, in *Geoecology of the Colorado Front Range*, edited by J. D. Ives, pp. 209–213, Westview Press, Boulder, 1980.

Komarkova, V., and P. J. Webber, An alpine vegetation map of Niwot Ridge, Colorado, *Arctic and Alpine Research*, **10**, 1–29, 1978.

Lewis, W. M. Jr., and M. C. Grant, Relationships between stream discharge and yield of dissolved substances from a Colorado mountain watershed, *Soil Science*, **128**, 353–363, 1979.

Litaor, M. I., Aluminum chemistry: fractionation, speciation, and mineral equilibria of soil interstitial waters of an alpine watershed, Front Range, Colorado, *Geochemica et Cosmochemica Acta*, **51**, 1285–1295, 1987.

Litaor, M. I., and E. M. Thurman, Acid neutralizing capacity in an alpine watershed, Front Range, Colorado I. The chemistry of dissolved organic carbon. *Applied Geochemistry*, **3**, 645–652, 1988.

Losleben, M., Climatological data from Niwot Ridge, East Slope, Front Range, Colorado 1970–1982, *University of Colorado Long-Term Ecological Research Data Report, 83/10*, 193 pp., 1983.

Madole, R. F., Possible origins of till-like deposits near the summits of the Front Range in north-central Colorado, *U.S. Geological Survey Professional Paper 1243*, 31 pp., 1982.

Mahaney, W. C., and B. D. Fahey, Morphology, composition and age of a buried paleosol, Front Range, Colorado, USA, *Geoderma*, **23**, 209–218, 1980.

Mast, M. A., J. I. Drever, and J. Baron, Chemical weathering in the Loch Vale Watershed, Rocky Mountain National Park, Colorado, *Water Resources Research*, **26**, 2971–2978, 1990.

Miller, J. P., Solutes in small streams draining single rock types, Sangre de Cristo Range, New Mexico, *U.S. Geological Survey Water Supply Paper, 1535F*, 23 pp., 1961.

Rapp, A., Recent developments of mountain slopes in Karkevagge and surroundings, northern Scandinavia, *Geografiska Annaler*, **42**, 71–200, 1960.

Reddy, M. M., and N. Caine, Dissolved solutes budget of a small alpine basin, Colorado, in *Proceedings of the International Mountain Watersheds Symposium*, edited by I. G. Poppoff, C. R. Goldman, S. L. Loeb, and L. B. Leopold, pp. 370–385, Tahoe Resource Conservation District, South Lake Tahoe, Nev., 1990.

Reynolds, R. C., and N. M. Johnson, Chemical weathering in the temperate glacial environment of the northern Cascade mountains. *Geochimica et Cosmochimica Acta*, **36**, 537–554, 1972.

Rochette, E., J. I. Drever, and F. Sanders, Chemical weathering in the West Glacier Lake drainage basin, Snowy Range, Wyoming: Implication for future acid deposition, *Contributions to Geology*, **26**, 29–44, 1988.

Schmidt, R. A., Vertical profiles of wind speed, snow concentration and humidity in blowing snow, *Boundary Layer Meteorology*, **23**, 223–246, 1982.

Slaymaker, O., Estimation of sediment yield in temperate alpine environments. *International Association Scientific Hydrology Publication*, **122**, 109–117, 1977.

Souchez, R. A., and M. M. Lemmens, Solutes, in *Glacio-fluvial Sediment Transfer*, A. M. Gurnall and M. J. Clark (eds), pp. 285–303, John Wiley, Chichester, 1987.

Stednick, J. D., Hydrochemical characterization of alpine and alpine-subalpine stream waters, Colorado Rocky Mountains, USA, *Arctic and Alpine Research*, **21**, 276–282, 1989.

Thorn, C. E., Influence of late-lying snow on rock-weathering rinds, *Arctic and Alpine Research*, **7**, 373–378, 1975.

Thorn, C. E., Quantitative evaluation of nivation in the Colorado Front Range, *Bulletin of the Geological Society of America*, **87**, 1169–1178, 1976.

Trudgill, S., Hillslope solute modelling, in *Modelling Geomorphological Systems*, M. G. Anderson (ed.), John Wiley, Chichester, pp. 309–339, 1988.

Vitek, J. D., A. L. Deutch, and C. G. Parson, Summer measurements of dissolved ion concentrations in alpine streams, Blanca Peak Region, Colorado, *Professional Geographer*, **33**, 436–444, 1981.

Walling, D. E., and B. W. Webb, The reliability of suspended sediment load data, *International Association Scientific Hydrology Publication,* **133**, 177–194, 1981.

Woo, M.-K., Snow hydrology of the high Arctic, *Proceedings of the 50th Western Snow Conference*, 63–74, 1982.

4 The Zonation of Freeze-thaw Temperatures at a Glacier Headwall, Dome Glacier, Canadian Rockies

James S. Gardner
President's Office, University of Manitoba

Abstract

Theories about cirque development have postulated accelerated freeze-thaw weathering and erosion around and beneath semipermanent snow patches and small glaciers. Field observations and measurements over the past half-century indicate that requisite temperature conditions are found in only a very restricted zone at the glacier margin and on the headwall of a cirque-type glacier. The research reported here contributes to knowledge of the glacier-margin weathering environment by presenting temperature data incorporating spring, summer, and autumn conditions at the Dome Glacier in the Canadian Rocky Mountains. Results indicate freeze-thaw temperature conditions in the presence of moisture on rock surfaces above the glacier randkluft throughout the spring or early ablation season. Freeze-thaw regimes are restricted to the randkluft lip area in early to mid-summer. With the widening of the randkluft in late summer and early autumn a freeze-thaw regime is driven by local atmospheric frost alternations. The randkluft environment, while cool and damp, appears not to provide a special freeze-thaw environment under these conditions. Nonetheless, over the course of the observation period, significant parts of the glacier margin were exposed to accentuated freeze-thaw temperature regimes which may be a factor in rock weathering and cirque development.

Periglacial Geomorphology. Edited by J. C. Dixon and A. D. Abrahams

Introduction

This paper describes temperature conditions in a glacier-margin environment, specifically that at the headwall in the vicinity of the bergschrund and randkluft. The glacier margin is a periglacial environment in the most literal sense. This location was identified as a focus for intense physical weathering due to freeze-thaw leading to erosion and cirque development (Johnson, 1904). Further, Lewis (1938) and Nussbaum (1938) postulated that snowmelt would provide a moisture source for the rock-shattering process. However, subsequent measurements of temperature conditions in bergschrunds revealed a cold, stable thermal environment (Battle and Lewis, 1951; Battle, 1960). Following these results, Battle proposed that the vicinity of the randkluft or "moat" at the glacier headwall may provide conditions more

Figure 4.1 The Dome Glacier, Canadian Rocky Mountains. "S" marks the location of the study site

suitable for freeze-thaw. Although he was unable to test this hypothesis, it was substantiated in later research (Gardner, 1987). This paper offers further data on zonation and controls of temperature and moisture conditions from the headwall of the Dome Glacier in the Canadian Rocky Mountains (52°12′N, 117°12′W) (Figure 4.1). The Dome Glacier is one of several valley glaciers issuing from the Columbia Icefield and serving as one of the source glaciers for the Athabasca–MacKenzie River system.

Related research

A growing corpus of experimental and field observational research bears on physical and chemical weathering in periglacial environments. Reviews of this research (e.g., McGreevy, 1981; McGreevy and Whalley, 1982; Whalley and McGreevy, 1985) indicate very complex interactions between temperature, moisture, and rock conditions in the control of physical and/or chemical weathering of in situ rock masses. These reviews also emphasize the need for data collection, especially under field conditions.

The research of most direct relevance to this paper deals with rock temperatures at high altitude (Mante, 1985; Kuhle, 1988; Francou, 1988; Coutard and Francou, 1989; Matsuoka, 1990). These studies, and others (Whalley, 1984), point to the important role played by direct solar radiation in driving the rock temperature regime. Further, they allude to the importance of aspect or exposure in influencing moisture availability and, therefore, disintegration and decomposition.

Matsuoka (1990) describes empirical relationships between temperature, moisture, and rock conditions, noting that shattering and rock release (rockfall) are prominent where there is frequent freeze-thaw, a presence of moisture, and closely spaced joints. Similar results were found at lower elevations along the Niagara Escarpment by Fahey and Lefebvre (1988). The role of moisture in rock disintegration is extremely important but notoriously difficult to measure and monitor in field conditions (Hall, 1986, 1988). As a result, some of the most useful research on this aspect has been done under laboratory conditions (Lautridou and Ozouf, 1982).

None of the related research has been conducted in a glacier-margin environment. The closest physical analogue is found in the vicinity of snowpatches, where prior research has been conducted. Earlier researchers, such as Matthes (1900), Lewis (1939), and McCabe (1939), described freeze-thaw in the presence of moisture around snowpatches, and these observations became the basis for understanding the process of nivation. Later empirical research by Gardner (1969), Fukuda (1971), Thorn (1980), and Hall (1980) identified the snowpatch margin as being particularly susceptible to freeze-thaw.

The glacier margin is somewhat analogous to a snowpatch margin. This recognition and Battle's (1960) hypothesis led to the investigation of temperature conditions at a glacier headwall on Mt. Athabasca in the Canadian Rockies (Gardner, 1987). Results demonstrated the importance of direct solar radiation in quickly raising rock-surface temperatures, usually on a diurnal basis, in an otherwise cold environment. Further, in a zone greater than about 2 m beneath the lip of the randkluft it was found that cold, stable thermal condition persisted throughout the ablation season, regardless of meteorological and external energy balance conditions. Freeze-thaw occurred through the ablation season in a 2-m-wide band centered on the randkluft lip. This freeze-thaw zone migrated up and down the adjacent rockface with the seasonal migration of the randkluft lip as snow accumulated and ablated. Indeed, this migration of freeze-thaw was found to include as much as 10 m vertical of the adjacent headwall. Finally, meltwater from snow- and ice-patches on the headwall flowed and dripped into the subfreezing zone where it froze onto or into cold rock surfaces exposed there.

This paper builds on the previous research and leads to further definition of the glacier-margin, periglacial weathering environment.

Study site and data collection

The Dome Glacier is one of the major distributary glaciers of the Columbia Icefield (Figure 4.1). The study site is situated at the randkluft of the glacier on the northeast face of the Snowdome (Figures 4.2 and 4.3). The altitude of the site is 2600 m. The 800-m-high northeast face of the Snowdome in this location is composed of the Devonian limestone and dolomite "sandwich" that characterizes many of the freefaces of scarp slopes of the Canadian Rockies. Individual members of the Palliser, Banff, and Rundle formations are exposed in cross-section. Bedrock at the measurement site consists of a massive dark grey limestone unit of the Palliser formation. In addition to detailed monitoring of temperatures at this site, other objectives of the measurement program included extension of the temperature data base from late winter through the autumn seasons, and evaluation of moisture conditions on a qualitative basis at least. The former was in answer to the often-cited criticism that empirical observation in mountain environments restricted to "favorable" summer months.

Instrumentation for temperature monitoring consisted of an array of thermistors attached to Grant data loggers, including electronic storage and paperchart models. The thermistors were emplaced and secured on the surface of the rockface or inserted flush with the surface in shallow cracks in the rocks at various levels above and below the randkluft lip. Window putty was used to secure the thermistors and leads to the rockwall. In addition, several thermistors were suspended against the rockwall at depth (up to 1 m)

Figure 4.2 The study site in mid-April, 1988

within the randkluft. As the randkluft lip migrated downslope with summer ablation, more thermistors were added to the array to take account of the changing position of the randkluft lip. The greatest impediment to data collection was the severing of thermistor leads by rockfalls.

The following analysis is based on data collected during the periods of July 23 to September 13, 1987; April 13 to June 3, 1988; and July 20 to October 9, 1988 (Tables 4.1–4.3).

Results

The results support conclusions from previous studies of temperature regimes in a glacier-margin environment. The most general of these is that a freeze-thaw temperature regime on the headwall rock surface is most probable

Figure 4.3 The study site in late September, 1987. Note the width of the randkluft moat. The randkluft lip has receded approximately 4 m from its level relative to the headwall at the beginning of the ablation season (see Figure 4.2 for comparison)

within a 2-m band centered on the randkluft lip (Gardner, 1987), at least during mid-summer (mid-June to mid-August). The previous research established that within the randkluft temperatures remained stable at or below the freezing point,whereas temperatures on the rock surfaces above the randkluft were generally unstable and above freezing, except during cold front passages or adjacent to snowpatches.

The research reported here extended observations into the spring and the autumn periods. The spring and autumn data highlight the influence of the seasonal air temperature regime when compared with data from the summer period. Table 4.1, which summarizes data for July 23 to September 13, 1987, indicates a predominance of freeze-thaw in the 2-m zone centered on the

Table 4.1. Temperature Data for July 23 to September 13, 1987[a]

	1		2		3		4	
	A	B	A	B	A	B	A	B
Observation days	12	24	12	24	12	24	—	24
Mean daily max. temperature (°C)	8.3	6.6	8.7	7.4	2.5	7.5	—	2.2
Mean daily min. temperature (°C)	0.9	3.5	–0.4	2.6	–1.4	3.5	—	1.0
Number of daily frost cycles	3	5	5	5	8	4	—	1
Number of partial frost cycles	2	3	2	1	1	2	—	2
Frost alternation days	5	8	7	6	9	6	—	3
Frost free days	7	16	5	18	1	18	—	21
Ice days	0	0	0	0	2	0	—	0
Maximum freezing rate (°C min^{-1})	0.014	0.008	0.023	0.019	0.076	0.007	—	0.001
Maximum thawing rate (°C min^{-1})	0.016	0.008	0.076	0.016	0.078	0.015	—	0.004

[a]1A = Rock surface 1 m above randkluft, July 23–Aug. 3.
1B = Rock surface 2 m above randkluft, Aug. 20–Sept. 13.
2A = Rock surface at randkluft, July 23–Aug. 3.
2B = Rock surface 1 m above randkluft, Aug. 20–Sept. 13.
3A = Rock surface 1 m below randkluft, July 23–Aug. 3.
3B = Rock surface at randkluft, Aug. 20–Sept. 13.
4B = Below randkluft 3 m, Aug. 20–Sept. 13.

randkluft lip (thermistor leads 1–3). At 3 m depth within the randkluft (thermistor lead 4B) temperatures were relatively more stable just above the freezing point with fewer fluctuations across the freezing point than in the other locations.

Spring data, shown in Table 4.2, depict a different temperature regime in an environment dominated by winter snow (Figure 4.2). The randkluft itself is filled with snow. Rock surfaces beneath the randkluft lip or the boundary between continuous snow cover and the rockwall are constantly below freezing (thermistor lead 4A, Table 4.2). At and above the randkluft lip, the rock surface undergoes a freeze-thaw temperature regime. Table 4.2 (thermistor leads 1 A/B, 2 A/B, and 3 A/B) generally shows a positive mean daily maximum and a negative mean daily minimum temperature. A significant number of frost alternation days were recorded on the rock surface. During the same periods recorded daytime air temperatures did not always exceed the freezing point as indicated by standard meteorological observations in the region. Thus, superimposed on the seasonal and diurnal air temperature regime is a local effect on the rockface. From early April to the middle of May the recording site at the Dome Glacier headwall is rarely or briefly exposed to direct solar radiation. Rather, observations at the site indicated that reflected solar radiation and longwave radiation from nearby rock surfaces heated by direct solar radiation were sufficient to raise temperatures above the freezing point and thus create a freeze-thaw regime on the rock surface.

During fieldwork, melting around snowpatch margins on the rockface was observed. The meltwater was noted to be absorbed into the rock surface and/or to flow down the rock surface and to refreeze on contact with cold surfaces. This is a phenomenon that has been observed previously at high elevations in the Himalayas and Andes (e.g., Kuhle, 1988; Francou, 1988). Given these observations of temperature and moisture, and recognizing that the precise relationships between temperature, moisture,and rock weathering are poorly understood, the most likely location of frost shattering on the headwall in late winter and spring is on the exposed rock surfaces above the randkluft lip.

The temperature data for summer and early autumn indicate different conditions (Table 4.3). The ablation season progressed to the point that most of the winter snow cover had melted, and the randkluft had opened to create a significant moat between the glacier and the headwall (Figure 4.3). The moat ranged in width at the lip between 2 and 4 m, exposing the ice marginal rockwall to the atmospheric temperature regime. The study site is not exposed to direct solar radiation during the late summer and autumn period. The data (Table 4.3) show a relatively close correspondence between air temperatures and rock surface temperatures at all locations with the exception of lead 5 A/B. The presence of a freeze-thaw temperature regime at this location on the rockface at a time when other locations experienced above freezing temperatures, may be explained by the presence of a small snow/ice patch in the vicinity.

Table 4.2. Temperature Data for April 13 to June 23, 1988[a]

	1		2		3		4
	A	B	A	B	A	B	A
Observation days	11	39	5	39	11	39	11
Mean daily max. temperature (°C)	5.9	6.6	7.7	6.5	4.9	0.2	–2.9
Mean daily min. temperature (°C)	–0.9	–0.3	0.1	–0.2	–1.3	–1.8	–4.2
Number of daily frost cycles	3	20	1	19	4	11	0
Number of partial frost cycles	4	3	1	2	2	2	0
Frost alternation days	7	23	2	21	6	13	0
Frost free days	4	16	3	16	5	16	0
Ice days	0	0	0	2	0	10	11
Maximum freezing rate (°C min^{-1})	0.008	0.015	0.003	0.003	0.013	0.083	—
Maximum thawing rate (°C min^{-1})	0.031	0.024	0.024	0.030	0.029	0.100	—

[a] 1A = Rock surface 4 m above randkluft, April 13–April 23.
1B = Rock surface 4 m above randkluft, April 23–June 3.
2A = Rock surface 2 m above randkluft, April 13–April 18.
2B = Rock surface 2 m above randkluft, April 23–June 3.
3A = Rock surface at randkluft, April 13–April 23.
3B = Rock surface at randkluft, April 23–June 3.
4A = Rock surface 1.5 m below randkluft, April 13–April 23.

Table 4.3. Temperature Data for July 20 to October 9, 1988, and from July 20 to August 27, 1990[a]

	1		2		3		4		5		6		7	
	A	B	A	B	A	B	A	B	A	B	A	B	A	B
Observation days	31	51	31	8	31	51	31	51	31	40	31	32	31	9
Mean daily max. temperature (°C)	10.3	6.1	11.1	8.9	10.6	6.2	10.6	7.2	4.2	5.9	11.4	8.1	11.8	3.2
Mean daily min. temperature (°C)	5.6	2.1	4.7	4.6	5.8	2.4	5.9	2.6	0.4	0.9	4.6	3.1	3.0	0.6
Number of daily frost cycles	0	11	0	0	0	8	0	12	6	12	0	2	2	1
Number of partial frost cycles	0	2	0	0	0	2	0	2	1	3	0	2	1	0
Frost alternation days	0	13	0	0	0	10	0	14	7	15	0	4	3	1
Frost free days	31	37	31	8	31	38	31	37	24	22	31	28	28	8
Ice days	0	1	0	0	0	3	0	0	0	3	0	0	0	0
Maximum freezing rate (°C min^{-1})	0	0.020	0	0	0	0.020	0	0.020	0.007	0.049	0	0.020	0.054	0.006
Maximum thawing rate (°C min^{-1})	0	0.020	0	0	0	0.020	0	0.025	0.004	0.030	0	0.020	0.016	0.004

[a]1A = "Air" temperature, July 20–Aug. 19, 1988.
1B = "Air" temperature, Aug. 19–Oct. 9, 1988.
2A = Below randkluft 2 m, July 20–Aug. 19, 1988.
2B = Below randkluft 2 m, Aug. 19–Aug. 26, 1988.
3A = Rockface 3 m above randkluft, July 20–Aug. 19, 1988.
3B = Rockface 3 m above randkluft, Aug. 19–Oct. 9, 1988.
4A = Rockface 10 m above randkluft, July 20–Aug. 19, 1988.
4B = Rockface 10 m above randkluft, Aug. 19–Oct. 9, 1988.
5A = Rockface 8 m above randkluft, July 20–Aug. 19, 1988.
5B = Rockface 8 m above randkluft, Aug. 19–Sept. 27, 1988.
6A = Below randkluft 5 m, July 20–Aug. 19, 1988.
6B = Below randkluft 5 m, Aug. 19–Sept. 19, 1988.
7A = Below randkluft 10 m, July 20–Aug. 19, 1990.
7B = Below randkluft 10 m, Aug. 19–Aug. 27, 1990.

Shown in Table 4.3 are the temperature data for the July 20–August 19, 1988 and the August 19–October 7, 1988 periods. During the latter period the impact of the atmospheric freeze-thaw temperature regime is replicated in the rock surface temperature regime on the rock surface above and, to a lesser degree, within the randkluft. The data do indicate that apart from a few freeze-thaw cycles, temperature conditions deep within the randkluft are similar to those at other locations (Table 4.3) during the July-August period. With the approach of autumn and the onset of an atmospheric freeze-thaw regime, the randkluft temperature regime appears to be slightly more stable than that at surrounding locations, with fewer frost alternation days. It is characterized, therefore, as a relatively cool and more stable thermal environment.

In a freeze-thaw environment, freezing rate has been suggested as an important factor in the "frost shattering" process. The importance of freezing rate and its significance among such other factors as extreme cold, presence of moisture, and rock conditions has not been conclusively demonstrated. A criterion of 0.1°C min^{-1} freezing rate as a threshold necessary for frost shattering has been suggested (Battle, 1960). The data collected in this research indicate that such a rate of freezing is rarely, if ever, achieved. During 168 "data days" such a freezing rate was achieved only twice. These data and those from previous field studies where the freezing rate threshold is rarely if ever achieved, suggest that it may not be as significant and as general a factor as previously suggested.

Summary and conclusions

The results of this research further define the concept of a glacier-margin weathering environment in so far as temperature and moisture conditions are concerned. The previously described randkluft lip freeze-thaw zone present in the summer is not a factor in the late winter and spring. Although seasonal snow cover persists and impinges on the glacier headwall, rock surface temperatures beneath the snow remain below freezing. Rock surface temperatures above the randkluft undergo a freeze-thaw regime, even when air temperatures remain below 0°C. Above-freezing temperatures in these conditions and locations appear to be caused by the absorption of reflected shortwave and longwave radiation from nearby surfaces. In addition to liberated meltwater produced by these energy sources, this temperature regime may encourage rock weathering.

Observations over two years at the Dome Glacier headwall indicate a significant enlargement of the randkluft during the glacier ablation season. This enlargement decoupled the snow and ice of the glacier from the headwall sufficiently to expose the latter, even within the randkluft, to the atmospheric temperature regime. This occurred by late summer and early autumn in 1988 and effectively destroyed the focused randkluft lip freeze-thaw zone described in previous research (Gardner 1987). In these conditions, atmospheric temperatures appear to provide a good representation of rock surface temperatures. This conclusion may be qualified by the observation that temperatures deep in the randkluft are relatively less variable than those in the atmosphere. These observations do indicate a relatively strong annual freeze-thaw cycle at depth within the randkluft.

The freeze-thaw zonation phenomenon therefore may be summarized as follows. While winter snowcover persists in the randkluft, freeze-thaw occurs on the exposed rock surfaces of the headwall. As the randkluft draws away from the headwall and a moat begins to form, freeze-thaw continues to occur on the exposed rock surfaces in the vicinity of the randkluft lip and around the margins of snow patches on the headwall. Temperatures within the randkluft remain stable at or below freezing. As the ablation season progresses, headwall rock surface temperatures take on a pattern similar to those in the atmosphere, and freeze-thaw is more focused at the randkluft lip. Eventually, the randkluft moat widens to the point where temperatures within the randkluft, as on the rock surfaces of the headwall, take on the regime of the atmosphere. Deep (> 5 m) within the randkluft temperatures may be cooler and more stable.

These generalizations apply only to glacier margins and headwalls with a similar morphology and exposure. Given that most active cirque and other glacier headwalls in the Canadian Rockies, under present climatic conditions, have similar north to northeast exposures, this qualification is not so restrictive. These observations further suggest that freeze-thaw temperature zones impinge on a wide area of the headwall through seasonal shifts and an even wider area with longer-term changes in glacier volumes.

This and prior research indicate that such glacier-margin environments are not devoid of moisture, another important ingredient for weathering. Certainly moisture is present throughout the period of this study (mid-April to early October), and continuing research at the study site is examining rock moisture content and its variation with temperature. Solution erosional microforms and deposits on the limestones and dolomites of the study site suggest that the role of chemical weathering in the glacier-margin environment warrants further investigation as well.

Acknowledgments

The research was funded by an on-going Operating Grant (A9152) from the Natural Sciences and Engineering Research Council of Canada. Eric Mattson, Ingrid Bajewsky, Karen McColeman, and Ken MacDonald assisted with the fieldwork.

References

Battle, W. R. B., Temperature observations in bergschrunds and their relationship to frost shattering, in *Norwegian Cirque Glaciers*, edited by W. V. Lewis, pp. 83–95, Royal Geographical Society, London, 1960.

Battle, W. R. B., and W. V. Lewis, Temperature observations in bergschrunds and their relationship to cirque erosion, *Journal of Geology*, **59**, 537–545, 1951.

Coutard, J.-P., and B. Francou, Rock temperature measurements in two alpine environments: implications for frost shattering, *Arctic and Alpine Research*, **21**, 399–416, 1989.

Fahey, B. D., and T. H. Lefebvre, The freeze-thaw weathering regime at a section of the Niagara Escarpment, Ontario Canada, *Earth Surface Processes and Landforms*, **13**, 293–304, 1988.

Francou, B., Temperatures de parois rocheuses et gelifraction dans les Andes Centrals du Perou, *Bulletin Centre de Géomorphologie C.N.R.S*, **34**, 159–180, 1988.

Fukuda, M., Freezing-thawing process of water in pore spaces in rocks (in Japanese with English summary), *Low Temperature Science, Series A.*, **29**, 225–229, 1971.

Gardner J. S., Snowpatches: their influence on mountain wall temperatures and the geomorphic implications, *Geografiska Annala*, **51A**, 114–120, 1969.

Gardner, J. S., Evidence for headwall weathering zones, Boundary Glacier, Canadian Rocky Mountains, *Journal of Glaciology*, **33**, 60–67, 1987.

Hall, K., Freeze-thaw activity at a nivation site in northern Norway, *Arctic and Alpine Research*, **12**, 193–194, 1980.

Hall, K., Rock moisture content in the field and laboratory and its relation to mechanical weathering, *Earth Surface Processes and Landforms*, **11**, 131–142, 1986.

Hall, K., Daily monitoring of a rock tablet at a marine Antarctic site: moisture and weathering results, *British Antarctic Survey Bulletin*, **79**, 17–25, 1988.

Johnson, W. D., The profile of maturity in alpine glacial erosion, *Journal of Geology*, **12**, 569–578, 1904.

Kuhle, M., Topography as a fundamental element of glacial systems, *GeoJournal*, **17**, 545–568, 1988.

Lautridou, J. P., and J. C. Ozouf, Experimental frost shattering: 15 years research at the Geomorphology Centre of CNRS, *Progress in Physical Geography*, **6**, 215–232, 1982.

Lewis, W. V., A meltwater hypothesis of cirque formation, *Geological Magazine*, **75**, 249–255, 1938.

Lewis, W. V., Snowpatch erosion in Iceland, *Geographical Journal*, **94**, 153–161, 1939.

Mante, A., Evolution du champ de temperature dans une paroi rocheuse naturelle, *Bulletin Centre de Géomorphologie C.N.R.S.*, **30**, 99–139, 1985.

Matsuoka, N., The rate of bedrock weathering by frost action: field measurements and a predictive model, *Earth Surface Processes and Landforms*, **15**, 73–90, 1990.

Matthes, F. E., Glacial sculpture of the Bighorn Mountains, *U.S. Geological Survey 21st Annual Report*, 179–185, 1900.

McCabe, L. H., Nivation and corrie erosion in West Spitzbergen, *Geographical Journal*, **94**, 447–465, 1939.

McGreevy, J. P., Some perspectives on frost shattering, *Progress in Physical Geography*, **5**, 56–75, 1981.

McGreevy, J. P., and W. B. Whalley, The geomorphic significance of rock temperature variations in cold environments: a discussion, *Arctic and Alpine Research*, **14**, 157–162, 1982.

Nussbaum, F., Beobachtungen uber gletschererosion in den Alpen und in den Pyrenaen, *Proceedings of the International Congress of Geography*, 63–73, 1938.
Thorn, C. E., Alpine bedrock temperatures: an empirical study, *Arctic and Alpine Research*, **12**, 73–86, 1980.
Whalley, W. B., High altitude rock weathering processes, in *International Karakoram Project*, vol. I, edited by K. J. Miller, pp. 365–373, Cambridge University Press, Cambridge, 1984.
Whalley, W. B., and J. P. McGreevy, Weathering, *Progress in Physical Geography*, **9**, 559–581, 1985.

5 Mechanical Weathering in the Antarctic: A Maritime Perspective

Kevin J. Hall
Department of Geography, University of Natal

Abstract

Considerations of mechanical weathering in the Antarctic are usually limited to the continent and to the dry valleys in particular. However, the oft-unrecognized maritime Antarctic differs climatically from the continent, particularly in terms of moisture availability, and consequently experiences a markedly different weathering regime. Data are presented on mechanical weathering processes, with special emphasis on freeze-thaw, from a maritime Antarctic location. The interrelationship of the various weathering mechanisms is shown and the manner in which those factors controlling freeze-thaw can exert an influence on other processes is demonstrated. For the first time an attempt is made to integrate a combination of field and simulation data to deduce the actual freeze-thaw mechanism causing rock breakdown. It is shown that, compared to the continent, despite the potentially more dynamic maritime weathering environment, weathering rates are still slow.

Introduction

As a consequence of factors such as the distribution of population, the large numbers of centers of research, and the relative ease of access to study areas, the bulk of periglacial and cold region weathering investigations have been conducted in the Northern Hemisphere. In general, periglacial texts (e.g., Embleton and King, 1975; French, 1976; Washburn, 1979) as well as specific review papers (e.g., Lautridou, 1988) have paid little attention to the mountainous and cold regions of the Southern Hemisphere and to Antarctica in

Periglacial Geomorphology. Edited by J. C. Dixon and A. D. Abrahams

particular. In fact, it has been suggested (C. Thorn, personal communication, 1986) that while those working in Antarctic regions find a need to keep abreast of northern studies, this is often not mirrored by their northern-orientated counterparts with respect to research in the Antarctic.

As a generalization, the Antarctic appears to be perceived as a great ice-covered continent. Even for those geomorphologists who recognize its existence, the notion of extremeness of climate tends to put it in a unique category—one that has no parallel elsewhere. Partly as a consequence of this alien perception of Antarctic regions and partly as a result of the poorly disseminated, and often obscure, nature of research publications, very little is generally known regarding weathering studies from Antarctica. However, in reality there is a substantial body of Antarctic information pertaining both to mechanical weathering processes and to their association with certain landforms. In fact a number of original hypotheses have originated from Antarctic studies, and in several instances, observations and data bases pertaining to weathering controls exceed their northern counterparts. Central to the whole problem is the realization that although Antarctica is very cold and arid, there are gradations, both in time and space, in the severity of these elements. Consequently, mechanical weathering is neither as simple nor as extreme as it is so often portrayed. Indeed it may often be comparable to that operating in the Arctic and Sub-Arctic today and possibly over a more extensive area during the Quaternary.

Climatic considerations

Campbell and Claridge (1987, p.4) state that ". . . the climate in Antarctica is unique, by virtue of the very low temperatures and the aridity." Although technically true, this statement hides within it two important factors with respect to weathering. First, Antarctica does not constitute a single climatic entity (Blümel and Eitel, 1989); and second, although air temperatures may be low, rock temperatures can be relatively high owing to warming by direct radiation. This rock warming may facilitate the presence of unfrozen water in or at the rock margin for short periods of time.

Four major climatic divisions are recognized for the Antarctic, namely the interior Antarctic plateau, the Antarctic slope (i.e., the steeply dipping ice margin between the plateau and the coast), the Antarctic coast, and the oceanic or maritime Antarctic (Weyant, 1966; Holdgate, 1970). Without considering the effect of direct insolation on rock, there are still some significant differences between these four zones. For instance, although no rain falls on the continent proper, this form of precipitation can occur on some of the maritime Antarctic islands as well as on parts of the Antarctic Peninsula (Loewe, 1957). Snowfall is also greater within the maritime regions. On the Peninsula and some of the islands not immediately adjoining the continental coast, air temperatures may frequently rise well above freezing during the summer, while on the plateau they may reach only −10°C (Campbell and Claridge, 1987). However, the higher temperatures in the oceanic zone are usually offset by low incoming radiation receipts owing to the high incidence of cloud cover. Toward the continental coast air temperatures may still be low. However, these temperatures can be greatly affected by the amount of exposed rock, which can cause them to reach as high as +9°C at rocky locations (Phillpot, 1985).

Although there is relatively detailed information available regarding the general meteorology and climatology of the Antarctic (van Rooy, 1957; Weyant, 1966; Phillpot, 1985; Dolgin, 1986), this is not the case with respect to the conditions actually experienced by the rock. When rock temperature and moisture regimes are considered, the perception of the weathering potential changes. Considering temperature first (Hall and Walton, 1992), data indicate substantial periods above 0°C. For instance, Sekyra (1970) referenced rock surface temperature ranges in the order of +30° to −35°C at the coast and +10° to −60°C at the plateau margin. Jonsson (1985) measured temperatures at depths of 2 to 3 cm inside rock cracks and found values of +20° to +30°C when the air was −7°C. Both Jonsson (1985) and van Autenboer (1964) noted that when the heat source (i.e., the sun) is cut off, temperature changes very rapidly, up to 49°C in only three hours in the outer shell of rock. In addition, rock albedo exerts a strong influence. Kelly and Zumberge (1961, Table 1) showed that a black rock (biotite schist) was 21.1°C, while a nearby white marble was only 12.8°C, a 65% difference. Even in the extreme-climate dry valleys of Victoria Land, Friedmann, McKay, and Nienow (1987) found that in spring, while air temperatures fluctuated between −45°C and −10°C, the rock surfaces could achieve temperature of +5°C for short periods. In summer, when mean air temperatures is −5°C, the rock surface could be as much as +10°C. In fact, McKay and Friedmann (1985) showed that rock surface temperatures could be above 0°C for 10 to 12 hours per day and that diurnal freeze-thaw could occur for part of the year.

In the milder maritime Antarctic, air temperatures are higher and freeze-thaw cycles relatively frequent. For the islands of the South Shetlands group, Simonov (1977) reported a mean annual temperature of −2.9°C but showed that rock surfaces experience frequent oscillations through 0°C. Blümel (1986) monitored 122 days with freeze and thaw (Frostwechseltagen) during 1979. The high incidence of cloud cover in the maritime region limits frequent rock heating by incoming radiation, but on clear days rock temperatures of +20°C have been recorded (Simonov, 1977). Similar results have been found for the South Orkneys (Chambers, 1967; Walton, 1982), where data also clearly indicate the influence of aspect (Hall and Walton, 1992).

Thus temperatures conducive to both freeze-thaw and thermal stress fatigue and suitable for thermal changes to salts occur within the Antarctic regions. However, for freeze-thaw to occur moisture must be present. Yet the general perception of the climate would appear to preclude this, even though it is recognized that rain can fall during summer both on the Peninsula and on some of the islands.

Although water is relatively sparse, it is far from nonexistent. Along much of the ice-free coastal margin of both the continent and the islands, loss of sea ice during summer facilitates tidal wetting of the coastal zone. Exposure of the wetted rock to subzero air temperatures means the potential for freeze-thaw weathering is high. In the same vein, wetting of rock can take place around the short-lived rivers, lakes, and ponds in the ice-free areas of both the continent and the islands. Shallow ponds, fed by snow melting in contact with heated rock, can form at the base of cliffs (Armitage, 1905). A number of authors have cited the formation of water as a result of snow coming into contact with rock warmed by the sun (Priestley, 1914; van Autenboer, 1962; Bardin, 1964). Friedmann, McKay, and Nienow (1987) also showed that water exists for short periods during the summer in the surface layers of rocks in the dry valleys of the McMurdo region. At the other end of the scale, Debenham (1921) found streams up to an altitude of ca. 100 m that grew to become rivers closer to the coast. More recently, Mosley (1988) gave details of some of these rivers in Victoria Land, such as the Onyx, which may flow for up to 90 days each year. In some locations, such as the Vestfold Hills, the area can become ". . . awash with water, soggy and inaccessible by foot because of the rapidly flowing streams in summer" (Pickard, 1986, p. 344).

Thus Antarctica, although experiencing an extreme climate, does produce conditions conducive to a whole range of mechanical weathering processes. Certainly the area in which these processes operate is limited, and many processes are constrained to a short time period. However, despite these limitations, mechanical weathering is operative, produces a number of distinctive features, and has some modes of operation that are optimum under these climatic conditions.

Recent texts that cover weathering in the Antarctic (e.g., Campbell and Claridge, 1987) provide relatively extensive detail with respect to the continent but almost completely ignore the maritime component. There is, however, a substantial body of work from this region (see Hall and Walton, 1992) which offers a different perspective on mechanical weathering. For example, although a number of authors (e.g., Souchez, 1967; Sekyra, 1970; Selby, 1971, 1972; Campbell and Claridge, 1987) argue strongly against the action of freeze-thaw weathering on most of the continent, the same is not the case for the oceanic islands. As the maritime Antarctic experiences more precipitation than the continent and has a milder climate with frequent oscillations through 0°C, freeze-thaw weathering (gelifraction) is generally considered to be very active (Holtedahl, 1929; Olsacher, 1956; Corté and Somoza, 1957; Dutkiewicz, 1982; Blümel, 1986). Thus the aim of this paper is

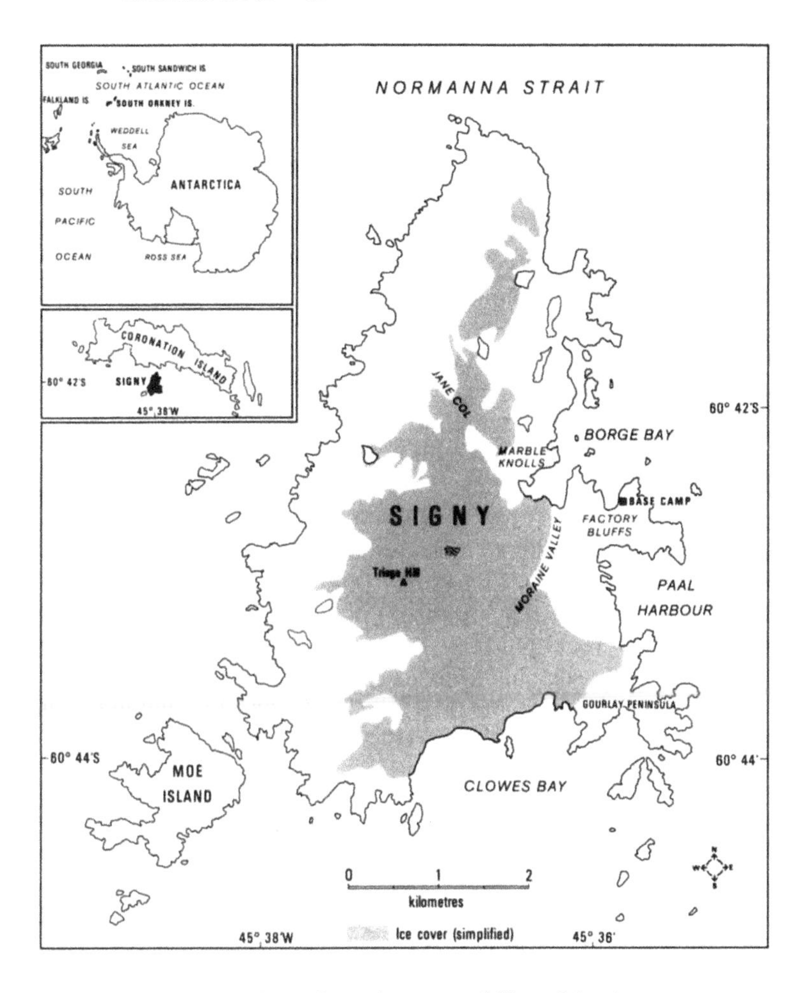

Figure 5.1. Location map of Signy Island

to offer a synthesis of recent studies from Signy Island (60°43′S, 45°38′W), one of the smaller islands in the South Orkney group (Figure 5.1). This island is a representative maritime Antarctic location and the site of a major British scientific station. It is studied here as a means of illustrating the nature of physical weathering within the maritime Antarctic. In addition, an attempt is made to identify the possible mechanisms causing rock breakdown. This is important for although many advances have been made with respect to our understanding of freeze-thaw weathering, we are still unable to make judgments regarding the actual mechanism of breakdown involved (H. French, personal communication, 1989).

Signy Island: background

Early studies on Signy Island (e.g., Mathews and Maling, 1967), as in most of the rest of the Maritime Antarctic (Araya and Hervé, 1972), clearly viewed freeze-thaw as the major weathering process. However, although freeze-thaw does constitute a process in its own right, it usually operates synergistically with several other mechanical processes (Figure 5.2), Singleton (1979) being one of the few to note this. In addition, although most researchers have clearly stated that the observed weathering is due to "freezing action," rarely, if ever, have data been supplied to corroborate this judgment. In fact, process determination requires background information on the controlling factors (Figure 5.3), and without this information it is virtually impossible to determine either the spatial and temporal operation of the various weathering processes or their interoperation with freeze-thaw.

In an effort to overcome these limitations, and so derive a meaningful estimation of process interaction plus weathering rates for Signy Island, background data pertaining to the controlling field conditions were first obtained. These data related to geology (Mathews and Maling, 1967; Storey and Meneilly, 1985), rock temperatures (Walton, 1977, 1982, personal communication, 1984), rock-moisture content (Hall, 1986a), rock moisture chemistry (Hall, Verbeeck, and Meiklejohn, 1986), the physical properties of the local quartz-micaschist bedrock (Hall, 1987), and thermal gradients and rates of change of temperature within the rock (Hall and Hall, 1991). On the basis of these field data it was possible to perform computer simulations

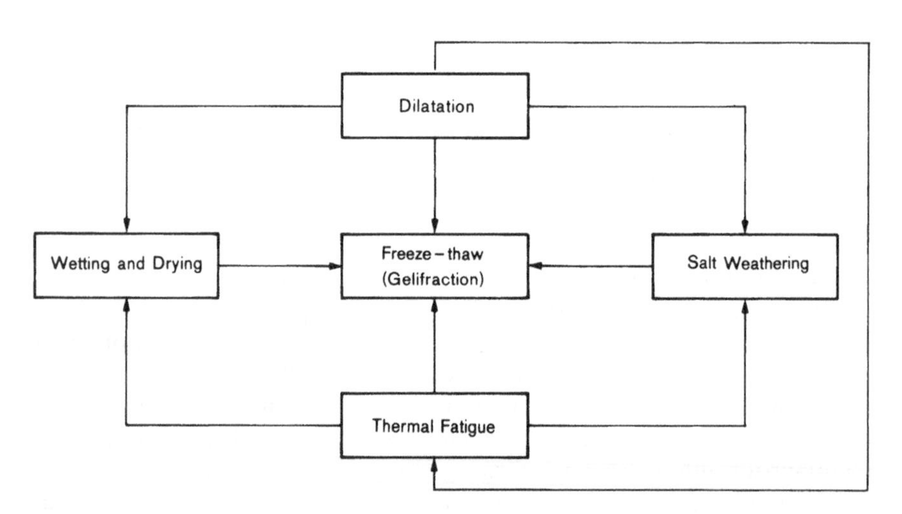

Figure 5.2. A flow chart showing the relationship of freeze-thaw weathering to the other mechanical weathering processes

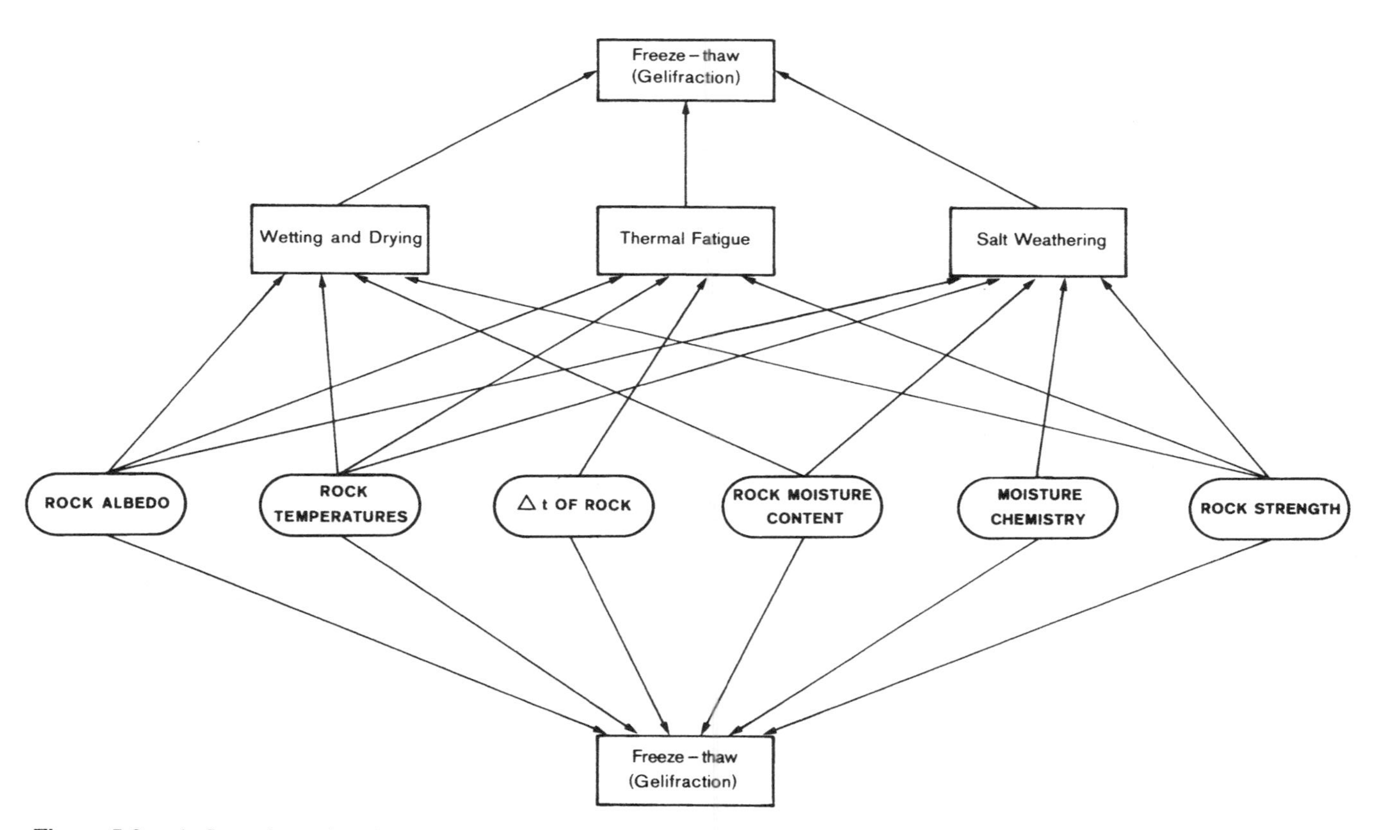

Figure 5.3. A flow chart showing the role of the controlling factors with respect to freeze-thaw and the other mechanical weathering processes

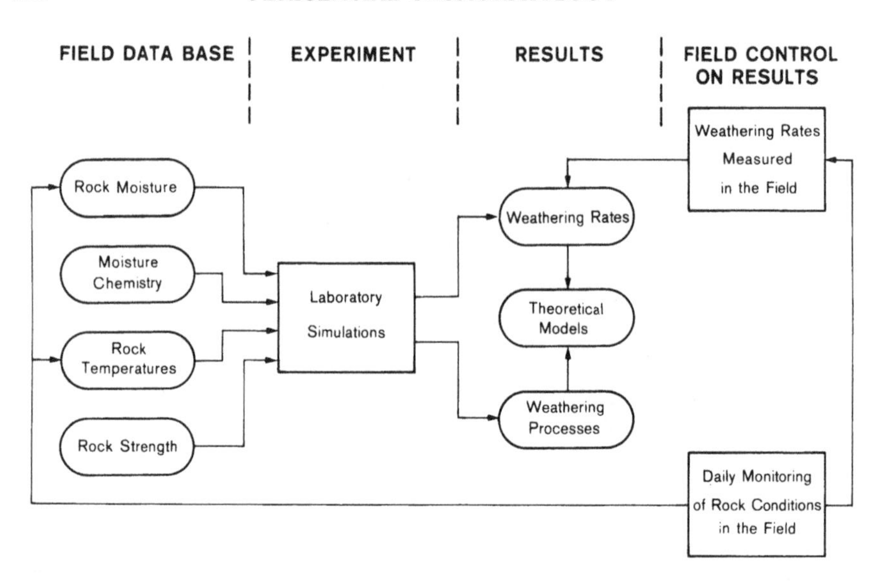

Figure 5.4. A flow chart showing the framework within which the Signy Island studies were undertaken

(Hall, Cullis, and Morewood, 1989) that replicated real world conditions and so provided information on both weathering processes and rates (Hall, 1986b, 1988a,b). As a further control on the simulations and as a verification that the derived weathering rates were of the correct order of magnitude, two long-term field experiments were initiated. The first monitored daily changes in rock moisture content, mass loss, and environmental conditions for one year (Hall, 1988c), while the second consisted of retesting rock tablets left in the field for varying lengths of time (Hall, 1990). The framework within which these various approaches were integrated is shown graphically in Figure 5.4.

In summary, the field data describe an environment with a small amplitude of mean monthly temperatures; the mean of the warmest month is greater than 0°C and that of the coldest month is −9°C (Collins, Baker, and Tilbrook, 1975). Radiation receipts are diminished by the extensive cloud cover but can be high on cloud-free days. Permafrost is present over most of the island, but the active layer is usually more than 1 m thick (Chambers, 1967). Temperature oscillations across 0°C are fairly frequent (>40 yr^{-1}), and on days with radiative heating rock temperatures can exceed 30°C. Although precipitation occurs fairly frequently, the annual water equivalent is small. However, precipitation can occur in the form of rain and, as humidity is high and summer temperatures not frequently below 0°C, much of this water becomes available to the rock. This moisture availability was reflected in data on interstitial rock moisture content. These data showed that from a total of 155 rocks sampled, 40.4% had moisture contents in excess of 50% and 17% in excess of 90% saturated. The degree of saturation was largely constrained by physical location, with the highest moisture contents occurring in positions close to melting snow or associated with snow-melt rivulets. Situations away from such favorable conditions had commensurately lower moisture contents. Data on daily changes in moisture content indicate that it can be high during time of freezing temperature; thus the potential for freeze-thaw weathering does exist. The maritime location of the site is reflected in the chemistry of the interstitial rock moisture, which showed a mean NaCl molarity of 0.47 ($s = 0.1$).

The rock itself, quartz-micaschist, is strongly anisotropic, and this is reflected in rock strength data. The mean compressive strength normal to schistosity is 2.3 MN m^{-2}, while that parallel to schistosity is only 0.50 MN m^{-2} (i.e., 78% weaker). As tensile strength (i.e., that applicable to most weathering forces) is approximately 20% of the compressive strength (Szlavin, 1974), the rock is classified as "low" to "medium" in strength on the engineering rock strength scale of Broch and Franklin (1972) when measured normal to schistosity but "low" to "very low" in strength when stressed parallel to schistosity. In a similar vein, calculations (Hall, 1986c) of the stress intensity factor K_{IC} (an index of fracture toughness) showed low values, varying from 1.4891 (normal to schistosity) to 1.2000 (parallel to schistosity).

Processes

With respect to freeze-thaw weathering, the derived values of K_{IC} allow indirect calculation of the pressures required for crack propagation due to ice formation. Knowing that ice pressure in saturated rock increases with negative temperatures at a rate of 1.14 MPa deg^{-1} (Hallet, 1983), it is possible to calculate the temperature required to generate any given pressure. These values (Hall, 1986c, Table II) indicate that for saturated rocks, cracks in the size range 0.1 m to 0.005 m with an overburden of between 1 m and 150 m of bedrock are able to be affected by frost action on Signy Island. Propagation of smaller cracks requires temperatures lower than −29°C, which are rare in this region. It should be noted that these calculations are valid only for saturated rock; the determinant equations do not strictly apply to nonsaturated rock. Nonetheless, a maximum possible threshold for potential frost action has been established.

With respect to the action of freeze-thaw, it is the combination of temperature, direction of freeze penetration, moisture content, and moisture chemistry together with the anisotropic nature of the rock that define the manner and rate of operation of this process. There is a clear distinction, often not explicitly recognized by many workers, between omni- and undirectionally frozen rock. A saturated omnidirectionally frozen block lost more than 57 times more mass than a saturated unidirectionally frozen block subject to the same temperature changes. The greater mass loss for the omnidirectionally frozen block reflected not only a "closed" versus an "open" system but also the influence of anisotropy.

The orientation of the plane of schistosity with respect to freeze penetration affects the manner in which freezing takes place and hence the rate of breakdown. When schistosity is normal to the freezing front, two possibilities arise. First, if the rock has a high moisture content, water can be forced away from the freezing centers into the rock to generate high hydrostatic pressures (the "hydrofracture" of Powers, 1945). Second, if moisture contents are low and the freezing rate is favorable, water may be drawn to the freezing center to produce strong tensile forces (Hallet, 1983). Both these possibilities may produce extensive breakdown of the rock. On the other hand, if schistosity is parallel to freeze penetration, then breakdown is limited to situations in which the moisture content along any laminae is greater than or equal to 91%. This high moisture content is necessary, as water migration between laminae is not possible. Consequently, as the laminae are frozen consecutively, only those that have ice growth in excess of the volume available experience damage. Thus an unconstrained block will be affected by omnidirectional freezing that can generate large internal stresses (Hallet, 1983). In addition, the freeze penetration will be asymmetrical owing to the effects of schistosity.

Irrespective of schistosity, rock damage can occur only if temperatures are conducive and water is present. In the absence of suitable temperatures for an adequate length of time and the availability of water to actually freeze, breakdown cannot occur by means of the freeze-thaw mechanism. Temperature data from the field indicate that in the presence of moisture, freezing will occur. However, it has been found that the nature of the temperature conditions will influence the manner in which freezing takes place. To date three main forms of freezing have been identified. First, with a relatively slow decline in environmental temperature (ca. 1°C h^{-1}) there occurs a rapid, large-scale (ca. 80% of the water freezes) transformation of water to ice subsequent to extensive supercooling of the interstitial water. Second, with a more rapid fall in temperature (≥3°C h^{-1}) there is a slow, progressive freeze from the outer margin of the rock inward. Third, when the salinity of the interstitial rock moisture is high (≥0.5 M NaCl), a progressive but rapid freeze (intermediate between the other two forms), with no sign of supercooling, takes place. Constraining these three forms of freezing are the requirements (1) that temperatures be less than or equal to −3°C, and (2) that these temperatures be maintained for at least 10 hours. These requirements are similar to those suggested by Lautridou (1971).

Thus with schistosity normal to freeze penetration and a slow rate of temperature decline, it is possible for hydrofracture or cavitation-induced nucleation of ice to take place (Hodder, 1976), dependent upon the actual rate of phase change. With a more rapid decline in environmental temperature, ice lens may grow, as proposed by Hallet (1983). With schistosity parallel to freeze penetration, either the simple 9% volumetric growth concomitant with phase change or frost bursting (Michaud, Dionne, and Dyke, 1989) can occur. The rate of temperature decline and the degree of saturation control whether either of these two processes can take place.

Moisture exerts a powerful influence on the nature of the freeze-thaw mechanism. It is not just the degree of saturation but also the distribution and chemistry of that water that are influential. Although the available data from Signy Island show that rocks in opportune locations may be greater than 50% saturated, so far it has not been possible to determine the distribution of the water within the rock. Ultimately this is of major importance, for the concentration of moisture in the outer margin may produce saturated conditions in that zone while the rock as a whole is well below 91% saturated. As a consequence of this moisture gradient, the actual freeze mechanism may differ from that presumed upon the basis of total rock moisture content.

With respect to moisture chemistry, the presence of salts in solution has a direct effect upon freeze-thaw in addition to their role in salt weathering. As already discussed, the nature of the freeze is directly affected by high (≥0.5 M NaCl) saline levels. In addition, the presence of salts depresses the freezing point (by 0.9°C for 0.25 M NaCl and 1.9°C for 0.5 M NaCl). Thus temperatures must be much lower for freezing to occur, and thawing takes place at solution temperatures below 0°C. This then means that freeze-thaw can take place without the temperature of the interstitial rock moisture ever going above 0°C.

All of the above is synthesized in Figure 5.5, which shows the various combinations of temperature, moisture, and rock conditions together with the resulting possible freeze-thaw mechanisms. Thus for the Signy Island situation it is now possible to make, for the first time, a reasonable judgment as to the possible form that freeze-thaw weathering may take based upon the extant environmental and rock conditions. However, as stated at the beginning, freeze-thaw does not work on its own, and thus it is now pertinent to consider how the elements shown in Figure 5.5 fit within, and integrate with, the other mechanical weathering processes.

The maritime location of Signy Island results in relatively high levels of NaCl in the interstitial rock water. This has a number of repercussions with respect to both freeze-thaw and salt weathering (Figure 5.6) in addition to the effects noted above. In freeze-thaw itself the role of salt is complex and little understood, although it has been shown that "... the extent to which salts enhance or inhibit frost weathering ... varies with both the type and concentration of the salt, the intensity of the freeze regime, and the rate of freezing and thawing" (Jerwood, Robinson, and Williams, 1990, p. 619). With low moisture levels and high concentrations of salts it is possible that the freezing point depression may facilitate a longer period of water mobility and

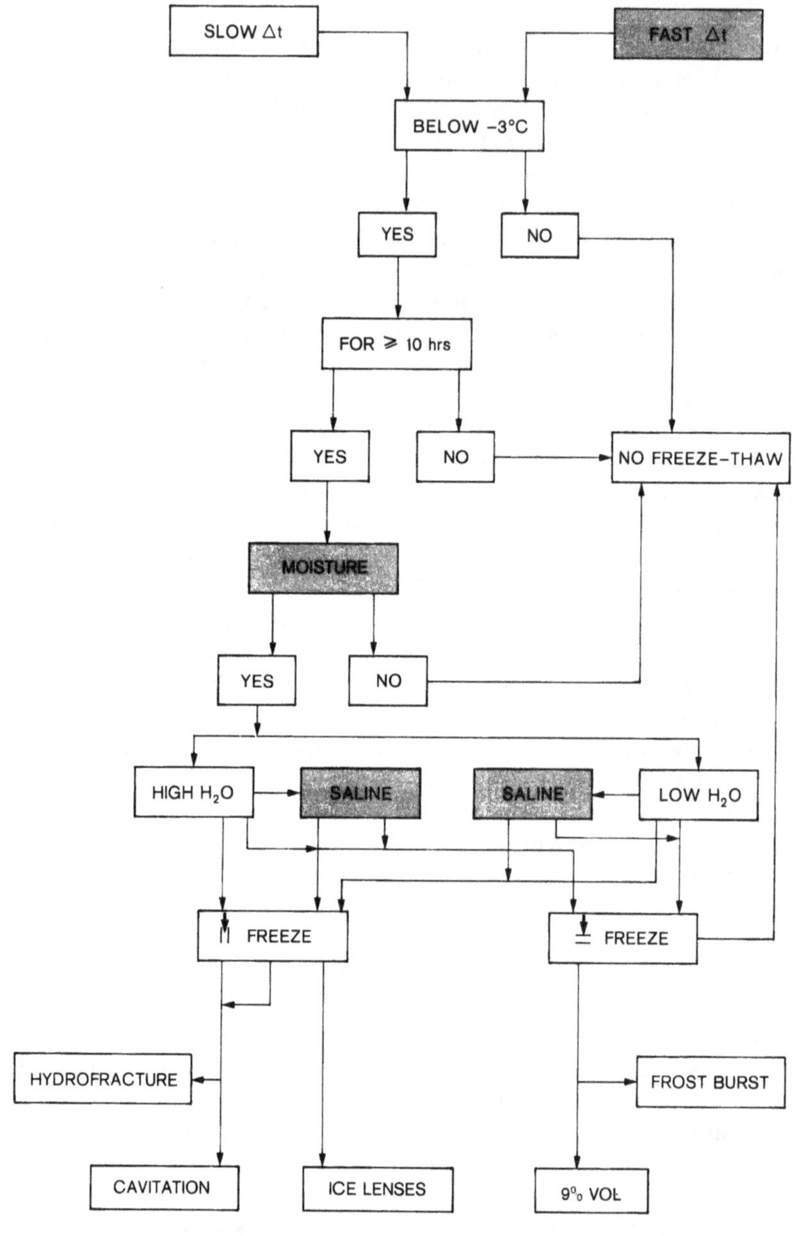

Figure 5.5. A flow chart of freeze-thaw weathering that allows deduction of the possible freeze-thaw mechanism. The shaded boxes (Saline, Fast Δt, and Moisture) are the starting points for the more detailed flow charts referring to salt weathering, thermal fatigue, and wetting and drying presented in Figures 5.6, 5.7, and 5.8. The symbol = indicates where freeze penetration is transverse to schistosity, and the symbol II where freeze penetration is parallel to schistosity

thus enhance the growth of ice lenses in the quartz-micaschist (Figure 5.5). With high moisture levels, according to Jerwood, Robinson, and Williams (1990), intense freezing regimes (≤−30°C) and low salt concentrations result in the eutectic temperature being reached (−21.1°C for NaCl), at which point a solid cryohydrate is formed and high internal rock stresses are created. With high salt concentrations and low intensity freezes (ca. −10°C), destruction is limited, as not all the water freezes and there is little or no crystallization of salt. Although no detailed simulations have been undertaken on the quartz-micaschist, the findings of Jerwood, Robinson, and Williams (1990) appear to agree with some preliminary tests which show that for non-saturated samples subject to omnidirectional freezing down to −19°C the greatest amount of breakdown occurred to the sample in a nonsaline solution. The next greatest amount of damage was found for a 1.0 M NaCl solution (the others being 0.25 M, 0.5 M and 0.75 M NaCl). However, the damage in this case may have been caused by salt crystallization resulting from drying of parts of formerly wetted rock during the thaw phase.

Salts brought into the rock in solution can precipitate out during warm periods, particularly when the rock is heated by incoming radiation, such that crystallization pressures may cause rock breakdown (Figure 5.6). Once the salts are present in the rock, usually close to the outer margin of the rock from whence the water is evaporated, they can cause further damage either by thermal expansion during times of high radiation receipts or by hydration resulting from high humidity (Figure 5.6). Thus many of the elements that affect the freeze-thaw process (such as rate of change of temperature, freeze amplitude, duration of freeze, heating during thaw phase) can also directly or indirectly affect saline rock moisture, and thereby produce salt weathering during the thaw phase or create an element of salt weathering within the freeze-thaw process itself.

The changes of temperature, both positive and negative, which are necessary for freeze-thaw to take place and which also affect salt weathering, can themselves cause rock breakdown (Figure 5.7). If the rate of change of temperature Δt is rapid (≥2°C min^{-1}), then the rock is subject to thermal stresses. Although this rate is very high, far in excess of that usually considered with respect to freeze-thaw, two situations can favor its occurrence. First, during subzero air temperatures the outer shell of a rock may be subject to intense warming by incoming radiation. This can cause a steep temperature gradient (and thus tensile stresses) and high values of Δt (≤9°C min^{-1}) for short periods of time (ca. 3 to 4 min) at the rock margin. The heating, and hence the expansion, will not be uniform about the rock, and so heated faces trying to expand will be buttressed by unheated surfaces, thereby accentuating tensile stresses. Second, when the heat source is rapidly removed, the rock is subject to extremely rapid cooling and, thus, compressive stress. During these conditions rates of temperature change ranging from 2°C min^{-1}

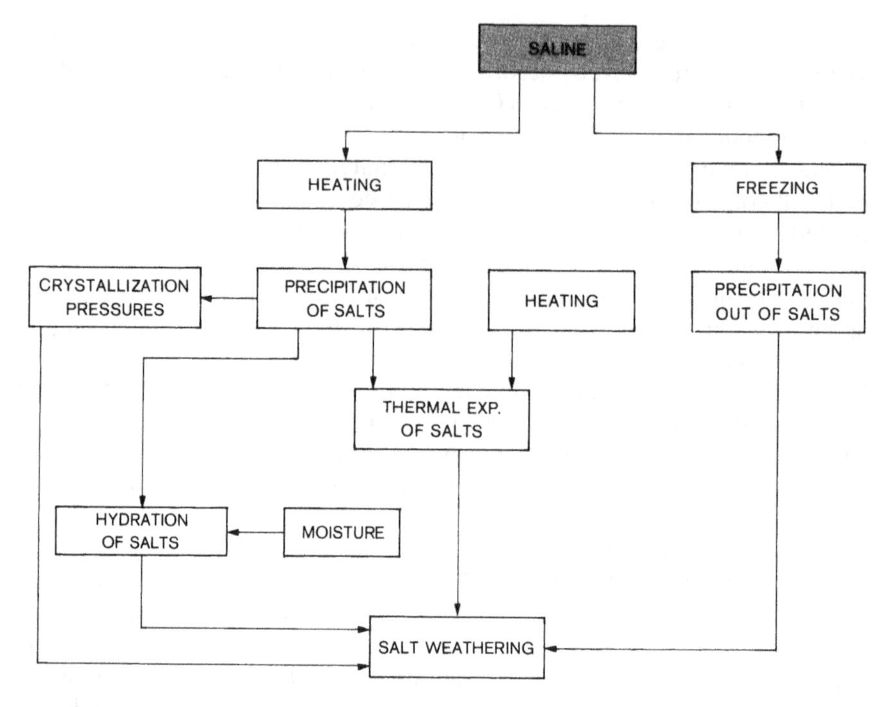

Figure 5.6. A flow chart showing the possible mechanisms associated with salt weathering

to 7°C min^{-1} have been recorded for four minutes. In these cases, the rate of thermal change is equal to, or in excess of, that suggested to be the threshold for the initiation of thermal stress fatigue (Richter and Simmons, 1974; Yatsu, 1988).

Although radiative heating of the rock during times of subzero air temperatures does not occur very often on Signy Island owing to the high incidence of cloud cover, available data suggest that when it does take place it affects the outer 2 cm of rock. Data from the laboratory simulations indicate that during these times of high positive Δt, above-zero temperatures penetrate to a depth of ca. 1 cm. This then means that, should water be present in the outer margin, freeze-thaw weathering will take place within the zone affected by thermal stress fatigue. Thus freeze-thaw weathering and thermal stress fatigue will operate both independently and synergistically such that it may be difficult to discern the role played by either in causing surface cracking and flaking of the rock.

The presence of water and, more particularly, fluctuations in water through time can cause mechanical weathering. The variations in moisture content that rock undergoes during the wetting → freezing → thawing → drying

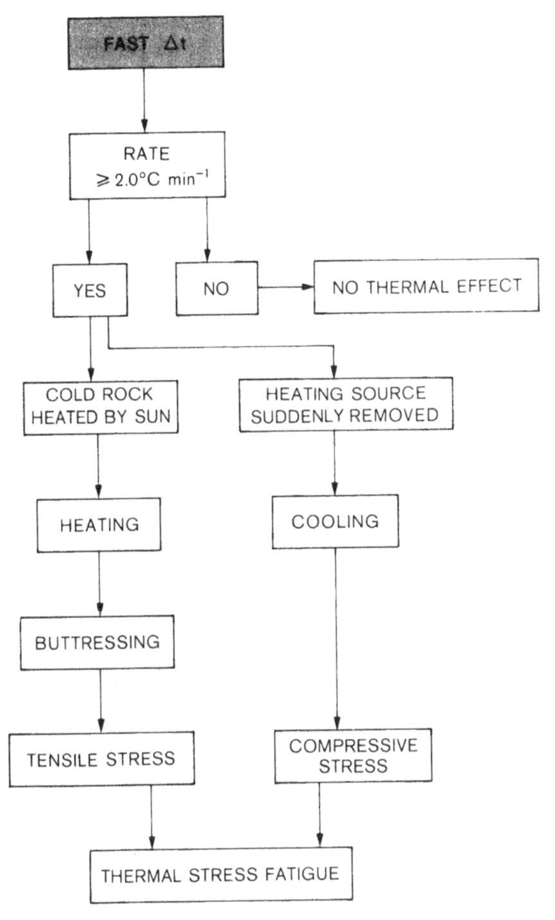

Figure 5.7. A simplified flow chart of the thermal fatigue weathering mechanism

sequence or during snowmelt and evaporation within periods of above-zero temperatures can promote the mechanical weathering process "wetting and drying" (Figure 5.8). Laboratory simulations on quartz-micaschist utilizing ultrasonic monitoring of rock conditions show that the presence of water weakens the bonding strength of the rock and that during drying elastic strain recovery is not always total. Thus, the wetted part of the rock "... will experience a decrease in strength resulting from diminished elasticity due to loss of bonding strength" (Hall, 1988b) which can lead to the formation of microfractures. It also means that the strength available to resist failure during freezing is diminished. Again, like thermal stress fatigue, the zone within which wetting and drying occur is the same as that affected by freeze-thaw, and the two work synergistically.

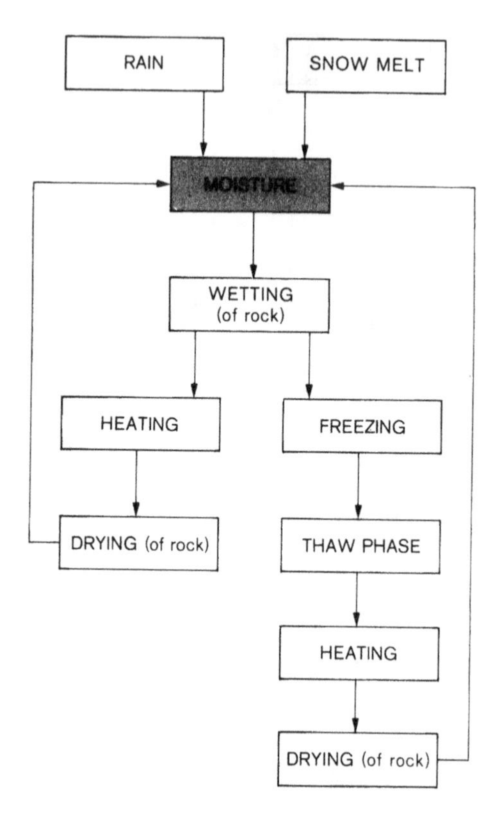

Figure 5.8. A simplified flow chart of the wetting and drying mechanism

Synthesis

Figures 5.5. to 5.8 show that process interaction takes place such that an element controlling one process (e.g., temperature change within freeze-thaw) can exert an influence on another process (e.g., thermal stress fatigue). Thus, as shown simplistically in Figure 5.2, given the presence of the required factors (e.g., moisture, salt, rapid Δt) all the mechanical weathering processes can operate both independently and synergistically. This means that the conditions facilitating freeze-thaw weathering can also cause wetting and drying, salt weathering and/or thermal fatigue. The maritime Antarctic situation, as exemplified by Signy Island, may be ideal for such combinations of mechanisms. Compared to the continent, precipitation is relatively large, can even occur in the form of rain, and is saline. Thus, unlike on the continent, exposed rock is frequently wetted. Although temperatures are not severe, freeze-thaw cycles are frequent, particularly during the spring to autumn period. Winter temperatures, however, approach −30°C and both give rise to a prolonged, severe freeze and exceed the eutectic point of NaCl, thereby allowing the formation of a potentially rock-damaging cryohydrate. Although cloud cover is usually fairly extensive, clear days occur on which radiation receipts are high and rock heating (and subsequent cooling) may be dramatic.

Perhaps the most significant result is that, for the quartz-micaschist, it is now possible, based on knowledge of the controlling factors, to determine the actual form of the freeze-thaw mechanism (Figure 5.5). In addition, based on the same data it is possible to gain some insight into the role of the other mechanical weathering processes (Figures 5.6, 5.7, and 5.8). Thus, for the first time the background data have been obtained and utilized to determine, with some degree of certainty, the actual weathering process(es). Now it is possible to get some idea not only of the temporal and spatial variability in the freeze-thaw mechanism but also of its effect upon, and interaction with, the other mechanical weathering processes.

As discussed above and illustrated in Figure 5.5, the situation is far from simple. Cold temperatures alone need not imply the operation of freeze-thaw; no moisture may have been present or the freeze may not have been of sufficient amplitude or duration. Thus the argument is not that Signy Island experiences a particularly destructive regime, but rather that in any consideration of weathering simplistic judgments should not be made with respect to what is the operative process. In fact, available data on weathering rates on Signy Island (Hall, 1988c, 1990) show them to be relatively slow with cut rock tablets exhibiting an extrapolated mass loss of only 2% per 100 years. This rate for omnidirectionally frozen, relatively wet samples may be as much as 50 times greater than for unidirectionally (relatively dry) frozen bedrock!

Having just argued for the operation of a number of mechanical weathering processes and recognizing that chemical and biological weathering components also aid breakdown, the finding of slow weathering rates may seem contradictory. However, this is not so for several reasons. First, the data base with respect to weathering rates only pertains (so far) to five years and the extrapolation to 100 years was linear, whereas the reality is more likely exponential (but the data base is not yet sufficient to model this). Second, cut rock blocks were used for the weathering experiments, and these may show initially slower rates than naturally occurring blocks. Third, and perhaps most important, the discussions with respect to process interaction were not put within a temporal or spatial framework. The reality is that while the maritime Antarctic may be more dynamic an environment than the continent with respect to weathering, this is still relative, and rates are substantially slower than in most arctic or alpine situations. Also, weathering often operates through fatigue, and it is the repetition of stressful conditions that ultimately leads to failure. In the maritime Antarctic the required combination of conditions (e.g., Figure 5.5) necessary to exert stress does not occur all that frequently, and so weathering rates remain slow.

Conclusions

Weathering conditions in the maritime Antarctic are somewhat different from those of the Antarctic continent with, perhaps, the most significant factor being the greater availability of moisture in the maritime region. Studies undertaken on Signy Island, a representative maritime Antarctic location, have shown the potential lines of interaction between the major weathering processes. More importantly, with respect to the process of freeze-thaw it has been possible to determine which mechanism is actually operative and how the various factors that influence this mechanism also affect other weathering processes. Because the various processes are both temporally and spatially interactive, simplified judgments as to the cause of weathered debris are no longer viable. Despite the potential range of weathering processes, it appears (from the available data) that weathering rates in the maritime Antarctic, although faster than on the continent, are still relatively slow.

Acknowledgments

My sincere thanks to R. Laws and D. W. H. Walton for allowing my participation in the studies on Signy Island. I have benefited from discussions on weathering with a number of people and would particularly like to thank Jim McGreevy, Jean-Pierre Lautridou, Matti Seppälä, David Walton, Norikazu Matsuoka, and Barry Fahey. Funding for this work has been generously provided by the University of Natal and the Foundation for Research and Development. The figures were redrawn by the University Cartographic Unit and the typing was undertaken by Alida Hall, both of whom are thanked for their patience.

References

Araya, R., and F. Hervé, Periglacial phenomena in the South Shetland Islands, in *Antarctic Geology and Geophysics*, edited by R. Adie, pp. 105–109, Universitetsforlaget, Oslo, 1972.

Armitage, A. B., *Two Years in the Antarctic*, 315 pp., Edward Arnold, London, 1905.

Bardin, V. I., O rel'efoobrazuiuschel deiatel'nosti talykh vod v gorakh Zemli Korolevy Mod (in Russian), *Moskovsky Universitet, Vestnik, Seriya Geografii*, **3**, 90–91, 1964.

Blümel, W. D., Beobachtungen zur Verwitterung an vulkanischen Festgesteinen von King-George-Island (S-Shetlands/W-Antarktis), *Zeitschrift für Geomorphologie, Supplement Band*, **61**, 39–54, 1986.

Blümel, W. D., and B. Eitel, Geoecological aspects of maritime-climatic and continental periglacial regions in Antarctica (S-Shetlands, Antarctic Peninsula and Victoria-Land), *Geoökodynamik*, **10**, 201–214, 1989.

Broch, E., and J. A. Franklin, The point-load strength test, *International Journal of Rock Mechanics and Mining Science*, **9**, 669–697, 1972.

Campbell, I. B., and G. Claridge, *Antarctica: Soils, Weathering Processes and Environments*, 368 pp., Elsevier, Amsterdam, 1987.

Chambers, M. J. G., Investigations of patterned ground at Signy Island, South Orkney Islands: II. Temperature regimes in the active layer, *British Antarctic Survey Bulletin*, **10**, 71–83, 1967.

Collins, N. J., J. H. Baker, and P. J. Tilbrook, Signy Island, Maritime Antarctic, *Ecological Bulletin*, **20**, 345–374, 1975.

Corté, A. E., and A. L. Somoza, Observaciones criopedologicas y glaciologicas en las Islas Decepcion, Media Luna y Melchior, *Instituto Antartico Argentino, Publicacion*, **4**, 65–131, 1957.

Debenham, F., Recent and local deposits of McMurdo Sound Region, *British Antarctic ("Terra Nova") Expedition, 1910. Natural History Report, Geology*, **1**(3), 63–100, 1921.

Dolgin, I. M., *Climate of Antarctica*, 213 pp., A. A. Balkema, Rotterdam, 1986.

Dutkiewicz, L., Preliminary results of investigations on some periglacial phenomena on King George Island, South Shetlands, *Biuletyn Peryglacjalny*, **29**, 13–23, 1982.

Embleton, C., and C. A. M. King, *Periglacial Geomorphology*, 203 pp., Edward Arnold, London, 1975.

French, H. M., *The Periglacial Environment*, 309 pp., Longmans, London, 1976.

Friedmann, E. I., C. P. McKay, and J. A. Nienow, The cryptoendolithic microbial environment in the Ross Desert of Antarctica: satellite-transmitted continuous nanoclimate data, 1984 to 1986, *Polar Biology*, **7**, 273–287, 1987.

Hall, K. J., Rock moisture content in the field and the laboratory and its relationship to mechanical weathering studies, *Earth Surface Processes and Landforms*, **11**, 131–142, 1986a.

Hall, K. J., Freeze-thaw simulations on quartz-micaschist and their implications for weathering studies on Signy Island, Antarctica, *British Antarctic Survey Bulletin*, **73**, 19–30, 1986b.

Hall, K. J., The utilization of the stress intensity factor (K_{IC}) in a model for rock fracture during freezing: an example from Signy Island, the Maritime Antarctic, *British Antarctic Survey Bulletin*, **72**, 53–60, 1986c.

Hall, K. J., The physical properties of quartz-micaschist and their application to freeze-thaw weathering studies in the Maritime Antarctic, *Earth Surface Processes and Landforms*, **12**, 137–149, 1987.

Hall, K. J., A laboratory simulation of rock breakdown due to freeze-thaw in a Maritime Antarctic environment, *Earth Surface Processes and Landforms*, **13**, 369–382, 1988a.

Hall, K. J., The interconnection of wetting and drying with freeze-thaw: some new data, *Zeitschrift für Geomorphologie, Supplement Band*, **71**, 1–11, 1988b.

Hall, K. J., Daily monitoring of a rock tablet at a Maritime Antarctic site: moisture and weathering results, *British Antarctic Survey Bulletin*, **79**, 17–25, 1988c.

Hall, K. J., Mechanical weathering rates on Signy Island, Maritime Antarctic, *Permafrost and Periglacial Processes*, **1**, 61–67, 1990.

Hall, K. J., A. Cullis, and C. Morewood, Antarctic rock weathering simulations: Simulator design, application and use, *Antarctic Science*, **1**, 45–50, 1989.

Hall, K., and A. Hall, Thermal gradients and rates of change of temperature in rock at low temperatures: New data and significance for weathering, *Permafrost and Periglacial Processes*, **2**, 103–112. 1991.

Hall, K. J., A. Verbeek, and I. A. Meiklejohn, The extraction and analysis of solutes from rock samples with some comments on the implications for weathering studies: an example from Signy Island, Antarctica, *British Antarctic Survey Bulletin*, **70**, 79–84, 1986.

Hall K. J., and D. W. H. Walton, *Rock Weathering in Cold Regions: Background Synthesis and Applied Aspects*, Cambridge University Press, Cambridge, in press, 1992.

Hallet, B., The breakdown of rock due to freezing: a theoretical model, *Proceedings of the 4th International Conference on Permafrost*, 433–438, 1983.

Hodder, A. P. W., Cavitation-induced nucleation of ice: A possible mechanism for frost-cracking in rocks, *New Zealand Journal of Geology and Geophysics*, **19**, 821–826, 1976.

Holdgate, M. W., Vegetation, in *Antarctic Ecology*, **2**, edited by M. W. Holdgate, pp. 729–732, Academic Press, London, 1970.

Holtedahl, O., On the geology and physiography of some Antarctic and sub-Antarctic islands, *Scientific Results of the Norwegian Antarctic Expeditions 1927–1928 and 1928–1929, Det Norsk Videnskaps-Akademi I Oslo*, **3**, 172 pp., 1929.

Jerwood, L. C., D. A. Robinson, and R. B. G. Williams, Experimental frost and salt weathering of chalk—I, *Earth Surface Processes and Landforms*, **15**, 611–624, 1990.

Jonsson, S., Geomorphological and glaciological observations in Vestfjella, Dronning Maud Land, in *Report of the Norwegian Antarctic Research Expedition (NARE) 1984/85. Norsk Polarinstitutt Rapportserie*, edited by O. Orheim, **22**, 83–88, 1985

Kelly, W. C., and J. H. Zumberge, Weathering of a quartz diorite at Marble Point, McMurdo Sound, Antarctica, *Journal of Geology*, **69**, 433–446, 1961.

Lautridou, J.-P., Conclusions générales des experiences de Gélifraction experimentale, *Bulletin du Centre de Géomorphologie du CNRS*, **10**, 63–84, 1971.

Lautridou, J.-P., Recent advances in cryogenic weathering, in *Recent Advances in Periglacial Geomorphology*, edited by M. J. Clark, pp. 34–47, John Wiley, Chichester, England, 1988.

Loewe, F., Precipitation and evaporation in the Antarctic, in *Meteorology of the Antarctic*, edited by M. P. van Rooy, pp. 71–90, Weather Bureau, Pretoria, 1957.

Mathews, D. H., and D. H. Maling, The geology of the South Orkney Islands: I. Signy Island, *Scientific Reports of the Falkland Islands Dependencies Survey*, **25**, 1–32, 1967.

McKay, C. P., and E. I. Friedmann, Cryptoendolithic microbial environment in the Antarctic cold desert; temperature variations in nature, *Polar Biology*, **4**, 19–25, 1985.

Michaud, Y., J.-C., Dionne, and L. D. Dyke, Frost bursting: a violent expression of frost action in rock, *Canadian Journal of Earth Sciences*, **26**, 2075–2080, 1989.

Mosley, M. P., Bedload transport and sediment yield in the Onyx River, Antarctica, *Earth Surface Processes and Landforms*, **13**, 51–67, 1988.

Olsacher, J. 1956. Contribucion al conocimiento geologico de la Isla Decepcion, *Instituto Antartico Argentino, Publicacion*, **2**, 1–78, 1988.

Phillpot, H. R., Physical Geography—Climate, in *Key Environments—Antarctica*, edited by W. N. Bonner and D. W. H. Walton, pp. 23–61, Pergamon, Oxford, 1985.

Pickard, J., The Vestfold Hills: A window on Antarctica, in *Antarctic Oasis*, edited by J. Pickard, pp. 333–351, Academic Press, Sydney, 1986.

Powers, T. C., A working hypothesis for further studies of frost resistance of concrete, *Journal of the American Concrete Institute*, **16**, 245–22, 1945.

Priestly, R. E., *Antarctic Adventure*, 382pp., T. Fisher Unwin, London, 1914.

Richter, D., and G. Simmons, Thermal expansion behavior of igneous rocks, *International Journal of Rock Mechanics and Mining Science and Geomechanics Abstracts*, **11**, 403–411, 1974.

Sekyra, J., Forms of mechanical weathering and their significance in the stratigraphy of the Quaternary in Antarctica, *Symposium on Antarctic Geology and Solid Earth Geophysics*, 669–674, 1970.

Selby, M. J., Slopes and their development in an ice-free, arid area of Antarctica, *Geografiska Annaler*, **53A**, 235–245, 1971.

Selby, M. J., Antarctic tors, *Zeitschrift für Geomorphologie*, **13**, 73–86, 1972.

Simonov, I. M., Physical-geographic description of the Fildes Peninsula (South Shetland Islands), *Polar Geography*, **1**, 223–242, 1977.

Singleton, D. G., Physiography and glacial geomorphology of the central Black Coast, Palmer Land, *British Antarctic Survey Bulletin*, **49**, 21–32, 1979.

Souchez, R., Gélivation et évolution des versants en bordure de l'Inlandsis d'Antarctide orientale, *Congrés et Collogues de l'Université de Liege*, **40**, 291–298, 1967.

Storey, B. C., and A. W. Meneilly, Petrogenesis of metamorphic rocks within a subduction-accretion terrane, Signy Island, South Orkney Islands, *Journal of Metamorphic Geology*, **3**, 21–42, 1985.

Szlavin, R., Relationships between some physical properties of rock determined by laboratory tests, *International Journal of Rock Mechanics and Mining Science*, **11**, 57–66, 1974.

van Autenboer, T., Ice mounds and melt phenomena in the Sør-Rondane, Antarctica, *Journal of Glaciology*, **4**, 349–354, 1962.

van Autenboer, T., The geomorphology and glacial geology of the Sør-Rondane, Dronning Maud Land, Antarctica, *Mededelingen van de Koninklijke Vlaamse Academie voor Wetenschappen, Letteren en Schone Kunsten van Belgie*, **8**, 1964.

van Rooy, M. P., *Meteorology of the Antarctic*, 240 pp., Weather Bureau, Pretoria, 1957.

Walton, D. W. H., Radiation and soil temperatures 1972–74: Signy Island Terrestrial Reference Site, *British Antarctic Survey Data*, **1**, 49 pp., 1977.

Walton, D. W. H., The Signy Island Terrestrial Reference Sites: XV, microclimatic monitoring, 1972–1974, *British Antarctic Survey Bulletin*, **55**, 111–126, 1982.

Washburn, A. L., *Geocryology*, 406 pp., Arnold, London, 1979.

Weyant, W. S., The Antarctic climate, in *Antarctic Soils and Soil Forming Processes*, edited by J. C. F. Tedrow, pp. 47–59, American Geophysical Union, Antarctic Research Series, 8, 1966.

Yatsu, E., *The Nature of Weathering*, 624 pp., Sozosha, Tokyo, 1988.

6 Miniature Sorted Stripes in the Páramo de Piedras Blancas (Venezuelan Andes)

Francisco L. Pérez
Department of Geography, University of Texas

Abstract

Miniature pebble-sorted (PS) and cobble-sorted (CS) stripes were studied in an Andean paramo at elevations from 4300 to 4560 m. Stripes are oriented downslope but curve around outcrops. Trough cobbles have their longest axis parallel to the slope. PS stripes are found on gently sloping granite and gneiss sites; CS stripes are restricted to steep taluses of amphibolitic schist. Both stripe types have short wavelengths, but CS stripes are nearly twice as wide (10–12 cm) as PS stripes. Sorting indices indicate pronounced lateral particle segregation only within a shallow superficial layer about 5 cm thick. Gravel and coarse sand are sorted into troughs; finer soil grains are concentrated in ridges. Temperature records during the dry season indicate a frost penetration into the soil of 7 to 13.5 cm. Measurements of painted gravel and marbles showed rapid superficial movement rates ranging from 2.9 to 9.5 cm yr^{-1}; isolated outliers moved faster, about 21.4 cm yr^{-1}. Tracers were laterally sorted into troughs during their descent. Only a few markers at a depth of 5 cm and none at a depth of 10 cm were disturbed during a 4-year period, indicating that frost effectiveness is restricted to a shallow, superficial layer. Needle ice appears to be the main agent of stripe formation. Pebbles migrate initially to the soil surface, then are laterally sorted into troughs and ridges. Stripe spacing is presumably set by elongated convection cells. Fine-soil ridges attain higher water content and remain moist longer than coarse troughs. Thus they are more prone to needle-ice formation. As fine strips are repeatedly heaved, remaining pebbles are ejected from them; these roll laterally and downwards into adjacent troughs, where they collect.

Periglacial Geomorphology. Edited by J. C. Dixon and A. D. Abrahams

Introduction

Sorted stripes are "patterned ground with a striped pattern and a sorted appearance due to parallel lines of stones and intervening strips of finer material oriented down the steepest available slope" (Washburn, 1956, p. 836). Other common characteristics of sorted stripes (Troll, 1958; Price, 1972, 1981; Washburn, 1980) are as follows: (1) stripes commonly develop by the downslope extension of sorted polygons or the linear alignment of soil buds occupying level areas; (2) stripes are found on gradients of more than 3° but rarely on slopes steeper than ca. 30°; (3) stripes curve around blocks and outcrops; (4) stone strips are depressed with respect to fine-soil bands; (5) stripes have a shallow depth of sorting, and the largest stones are near the surface; (6) bands of fines are usually wider than stony ones; (7) stones in troughs are oriented parallel to the stripe and on edge; and (8) size of stripe increases with stone size. However, it seems that actual measurements of many of these characteristics have seldom been obtained. Sorted stripes are common in periglacial areas. Both large and small forms occur, with the latter being more frequently found in high tropical and subtropical mountains, where needle ice is a major agent of particle sorting and stripe formation (Troll, 1958; Lliboutry, 1961; Hastenrath, 1973; Price, 1981; Francou, 1984).

Miniature sorted stripes have been reported from several mountains of East Africa (Zeuner, 1949; Troll, 1958; Furrer and Freund, 1973; Hastenrath, 1973, 1974, 1978), New Guinea (Löffler, 1975), the Kun Lun mountains of China (Francou, Federmeyer, Kuang-Yi, and Yaning, 1990), Hawaii (Noguchi, Tabuchi, and Hasegawa, 1987), the Mexican volcanoes (Heine, 1977), the Andes of Peru and Bolivia (Troll, 1958; Graf, 1971, 1973, 1976; Hastenrath, 1977; Francou, 1984, 1986, 1988), and Chile (Lliboutry, 1955, 1961). Small sorted stripes are also common in sub-Antarctic islands (Hall, 1979, 1983; Heilbronn and Walton, 1984a,b), where temperature regimes are similar to those of tropical mountains. In the Venezuelan Andes, Schubert (1975, 1979a, 1980) and Pérez (1985, 1987a) have found different kinds of miniature sorted stripes in the Páramo de Piedras Blancas, a high-elevation area strongly affected by needle-ice formation.

Despite numerous publications on sorted stripes, only a few studies have dealt with their morphological or pedological characteristics (Ballard, 1973; Furrer and Freund, 1973; Hall, 1979; Heilbronn and Walton, 1984a; Noguchi, Tabuchi and Hasegawa, 1987); and data on these characteristics from the equatorial Andes are particularly scarce (Graf, 1973). This study will focus on two types of miniature sorted stripes found in the Páramo de Piedras Blancas, Venezuela. The goals are (1) to describe the main morphological characteristics of these stripes and the environmental conditions of their sites; (2) to investigate the pedological properties and the degree of lateral particle sorting of the stripes; (3) to report on field measurements of rates of movement and patterns of sorting on these stripes; (4) to discuss the role of needle-ice activity on stripe formation.

The study area

The Páramo de Piedras Blancas is in the Venezuelan Andes at 8°52′ N and 70°55′ W (Figure 6.1). The study sites are between 4300 and 4560 m elevation. Yearly precipitation is about 800 mm and is distributed in two distinct seasons: more than 100 mm per month fall during the rainy season (May to August), but less than 20 mm per month during the dry season (Schubert, 1975, 1979a). Precipitation occurs primarily as rain, although snow and hail are also common. Snow cover is thin (less than 4 cm) and discontinuous and usually melts within a few hours. Owing to its equatorial location and altitude, Piedras Blancas shows pronounced diurnal temperature fluctuations. Air temperatures commonly reach minima of −5 to −11° C and maxima of 15 to 23° C in the dry season; the daily amplitude is somewhat smaller during the rainy season (Schubert, 1975). The area experiences about 325 to 350 daily freeze-thaw cycles per year (Pérez, 1987a). This periglacial regime causes

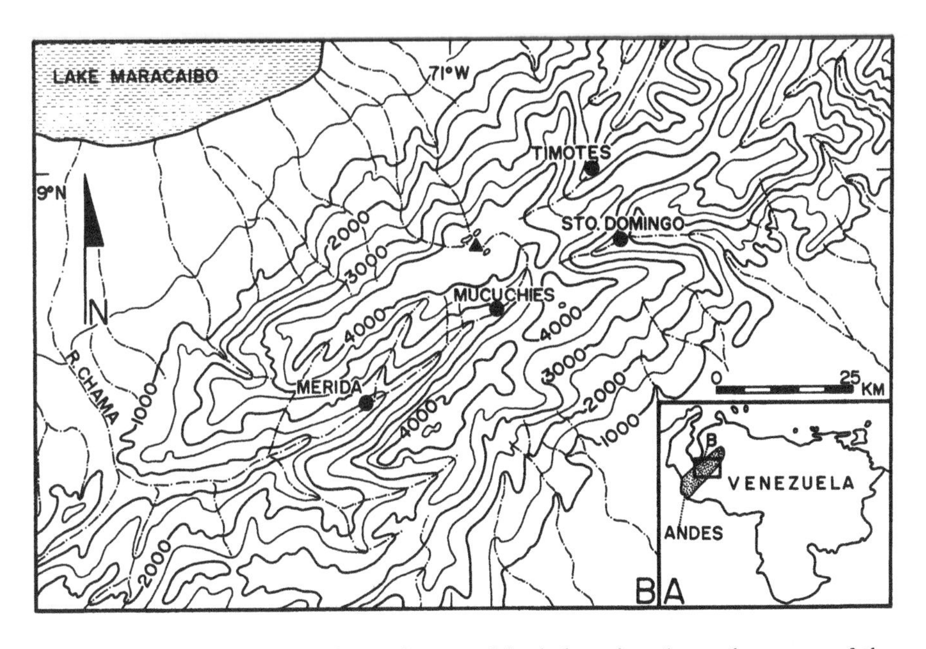

Figure 6.1. (A) Location of the study area. Stippled section shows the extent of the Venezuelan Andes. (B) Topographic map of the Venezuelan Andes near Mérida. Dots indicate major cities; triangle indicates the Páramo de Piedras Blancas. Contour interval is 500 m

recurrent needle-ice growth, which is widespread during the rainy season and more localized but intense during the dry season (Pérez, 1984).

Soils in Piedras Blancas have developed on igneous and metamorphic rocks. They are commonly shallow and have a low organic matter content and a coarse texture with a large proportion of sand. The fine fraction is mostly silt, with little clay (Pérez, 1984, 1986a). This texture is highly conducive to ice segregation and needle-ice formation (Meentemeyer and Zippin, 1981). Thus, the soil surface of the high paramo is repeatedly disturbed by needle-ice activity. Soil grains and pebbles lying on or near the ground surface are rapidly uplifted and transported by frequent formation of needle ice. Several types of sorted and non-sorted ground patterns, which seem to be caused primarily by needle ice, are common in the highest paramo areas (Schubert, 1975, 1979a, 1980; Pérez, 1984, 1985, 1987a, 1991). The vegetation of the high paramo is characterized by giant Andean rosettes (*Coespeletia timotensis*), occurring in dense woodlands between 4000 and 4500 m and as isolated individuals up to 4600 m (Pérez, 1987b). The upper elevations are occupied by the Periglacial Desert (Monasterio, 1980), a zone of mostly bare ground with only a few scattered low plants, primarily *Draba chionophila* (Pfitsch, 1988).

Methods

FIELD METHODS

The area was visited in 1985, 1987, and 1989/1990. Stripes were investigated at six locations between 4300 and 4560 m. Slope aspect, gradient, and altitude were measured at each site with compass, clinometer, and altimeter. Soil depth was measured with a 5-cm diameter soil auger. Rock type was ascertained from nearby outcrops. Prevailing stripe orientation was determined at sites 4 and 6 by placing a 30-cm ruler in 50 troughs and measuring their orientation. The size and shape of 50 clasts gathered from troughs at sites 4 and 6 were quantified from measurements of the three main particle axes (*a*, *b*, *c*) (Krumbein, 1941) with a vernier caliper. The orientation of the longest (*a*) axis was measured for the clasts of site 6. Stripe morphology was assessed from two 60-cm long cross-sections at sites 2 and 5; these were drawn on plexiglass plates after excavation down to the C horizon. Three soil profiles were also dug and examined at sites 1, 4, and 6.

Paired soil samples from ridge and trough positions were taken from stripes in all sites. A total of 90 "surface" samples was collected from the upper 5 cm of the soil; 40 samples were also taken between 5 and 10 cm depth at sites 4 and 5. Samples were carefully gathered with a 33-cm^3 plastic cylinder and a narrow trowel. Soil moisture content of ridges and troughs was measured in the field in December 1985 (site 2) and December 1987 (site 4), by collecting six and ten pairs of samples, respectively; soils were kept in hermetically sealed containers. Air and soil temperatures were recorded at site 1 during seven consecutive days in December 1985. Three mechanical thermographs, cross-calibrated prior to going to the field, were used; their probable measurement error is plus or minus 0.5° C. Air temperatures (10 cm above the ground surface) were recorded under a white-painted shelter which allowed free air circulation. Soil temperatures were taken at 7 and 14 cm depth. Measurements were obtained on bare plots away from rosette plants, as these can significantly affect both air and soil temperatures (Pérez, 1989).

Stripe movement and sorting were studied at sites 1 and 3 starting December 22, 1985. Two types of markers were utilized: painted gravel particles and 15-mm-diameter glass marbles (Pérez, 1987c,d; Mackay and Mathews, 1974). The average size (± standard deviation) of seeded particles (n = 25) was 3.8 ± 0.29 mm × 2.4 ± 0.18 mm ($a \times b$ axes). The smallest grain measured was 1.4 × 1.1 mm, and the largest 5.7 × 3.8 mm. Therefore the painted particles were classified as coarse sand to very fine gravel. Three colors of gravel were placed at different depths (blue at the surface, red at 5 cm, and green at 8–10 cm). A total of 1062 marbles of different colors was also buried at 5 and 8–10 cm depth. Gravel and glass markers were placed along 70 to 100-cm-long lines parallel to the contours. The line end-points were permanently marked with steel bars firmly inserted in the ground. Painted gravel was seeded on a 4-cm-wide, 5-mm-thick strip along each line (Pissart, 1973, 1982), and marbles were placed contiguously along a straight furrow. Eleven plots at site 1 and ten at site 4 contained a total of 21 surface gravel lines and 57 buried lines (42 marble, 15 gravel). Plot numbers are preceded by the letter M or G, which designates the marker type used in subsurface lines. All experimental plots were inspected on December 30, 1985, and on January 3, 1990. Clear plastic sheets were laid over the plots, and the areas containing gravel markers outlined. The stripe patterns of all plots at site 3 were also photographed. Marbles that had shifted downslope were located, and their distance of transport and depth of burial measured.

LABORATORY METHODS

Field moisture content of soil samples was determined gravimetrically. Soil samples were oven-dried for 48 hours at 105° C and then weighed. Dry soil colors were determined with the Munsell color charts; *p*H was measured by electrode in a 2 : 1 (water : soil) paste. Dry and wet soil consistence were assessed following the procedures in Soil Survey Staff (1975). Organic matter content was measured with the loss of weight on ignition. Particle size distribution was then obtained by sieving the mineral soils through a 21-mesh series, from 64 mm ($-6\ \phi$) to 0.090 mm ($3.5\ \phi$). Smaller grain sizes were determined by the hydrometer method. Water retention of surface soils samples was measured by thoroughly saturating about 25 mkg of soil in covered, ribbed-glass funnels lined with filter paper. Samples were left to equilibrate in a room with stable temperature (19–21° C) and relative humidity (55–60%) until free water drainage from the funnels had ceased (Smith and Atkinson, 1975). This was empirically determined to occur about 15 hours after saturation. At this point, soils were considered to have attained their field capacity (Pitty, 1979), and their water content was determined gravimetrically.

Results and discussion

TYPES OF STRIPES

There are several types of miniature ground stripes in the high Páramo de Piedras Blancas. The most common pattern is striated soil (Figure 6.2A), occasionally found as low as 3500 m but primarily above 4000 m. Striated soil consists of narrow bands of soil particles aligned not downslope, but in the direction of the early morning sun's rays. Striated soil is caused by repeated lifting of soil grains by needle ice and is the only nonsorted striped pattern in this paramo. Its characteristics have been studied in detail elsewhere (Schubert, 1973; Pérez, 1984).

Sorted stripes occur only above 4200 to 4300 m. These are essentially parallel to the slope and are composed of intervening bands of fine soil and pebbles or cobbles. Schubert (1975, 1979a, 1980) designated these stripes as nonsorted (*bandas no-escogidas*), despite the fact that his description clearly indicates sorting. In this paper, stripes are divided into two types according to the largest particles in the coarse bands. Following Wentworth's (1922) size classes, stripes consisting of gravel particles whose long axes are shorter than 64 mm are called pebble-sorted (Figures 6.2B, C), whereas those comprising clasts longer than 64 mm are designated as cobble-sorted (Figure 6.2D). In addition to miniature stripes, larger linear patterns containing sizable clasts are found on steep, high-paramo slopes. Some of these stripes are generated by the convergence of rocks rolling from upslope into elongated depressions (Pérez, 1988a, 1991). Others form because blocks obstruct the descent of superficial debris. As clasts are dammed upslope from boulders, they are deflected laterally toward the block edges, where stones extend downslope as elongated bands (Francou, 1986). These larger stripe patterns will not be considered here.

GENERAL SITE CHARACTERISTICS

Pebble-sorted (PS) stripes were found on several locations from 4300 to 4510 m elevation; cobble-sorted (CS) stripes are less common and were observed only on talus slopes above ca. 4500 m. Sites studied span this

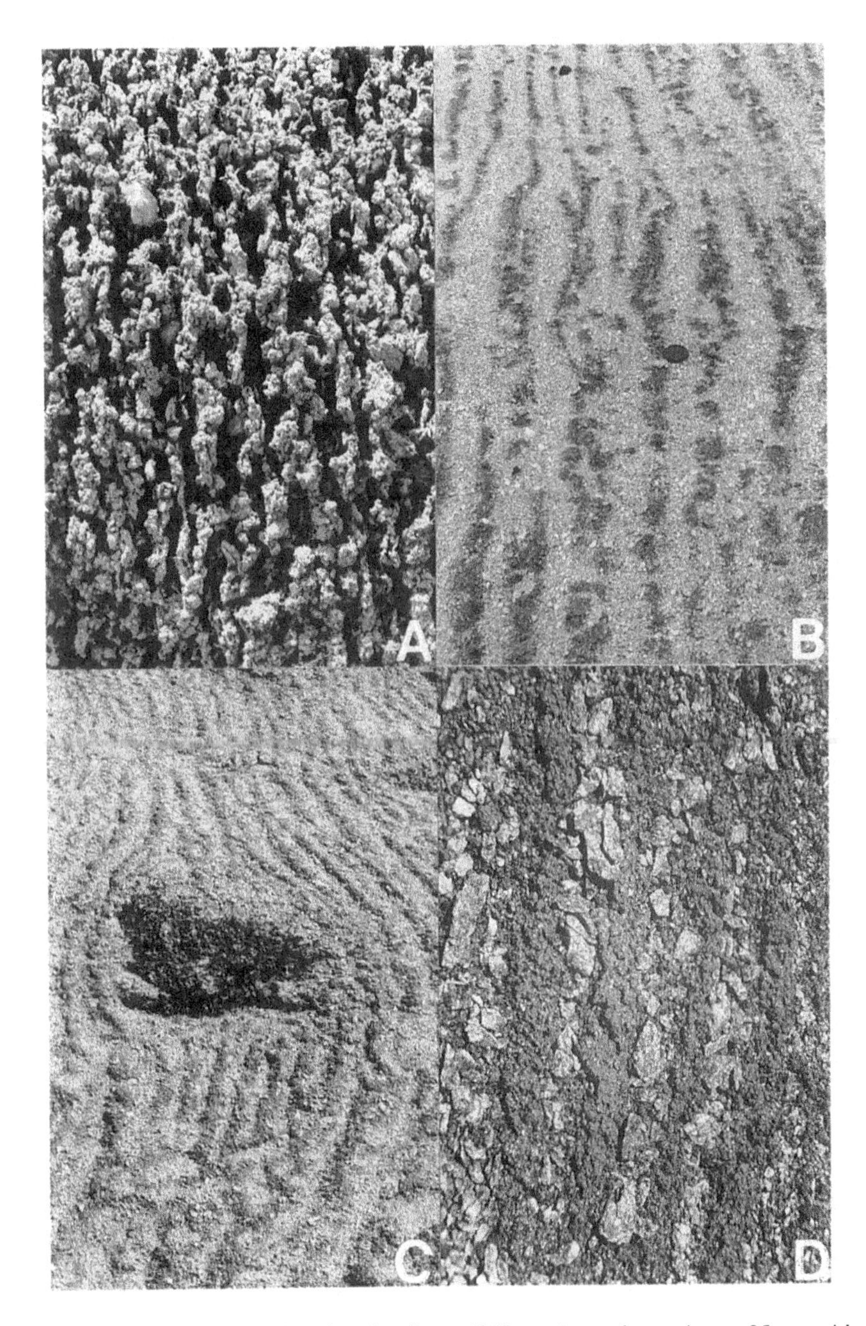

Figure 6.2. (A) Nonsorted, striated soils at 4260 m. Area shown is ca. 25 cm wide; downslope is to the left. (B) Pebble-sorted stripes near site 3 at 4480 m. Downslope is towards the bottom. The disk is 55 mm across; area shown is ca. 80 cm wide. (C) Pebble-sorted stripes at site 1 at 4300 m; downslope is toward the bottom. Note how the stripes bend around the outcrop; area shown is ca. 70 cm wide. (D) Cobble-sorted stripes near site 6 at 4570 m. Note how the longest axes of the stones are oriented downslope (toward the bottom). Area shown is ca. 55 m wide

Table 6.1. Selected Characteristics of Stripe Sites

Site Characteristic	Site Number					
	1	2	3	4	5	6
Altitude (m)	4300	4310	4470	4510	4535	4560
Aspect	N 315°W	N 350°W	S 128°E	S 116°E	S 171°E	S 165°E
Average slope angle	9°	7°	23°	12°	31°	34°
Soil depth (cm)	13–22	6–9	20–28	8–10	10–12	⩽ 49
Stripe type[a]	PS	PS	PS	PS	CS	CS
Dominant rock types	Pegmatitic granite		Gneiss & quartzite		Amphibolitic schist	

[a]PS = pebble-sorted stripes, CS = cobble-sorted stripes.

altitudinal range (Table 6.1). PS stripes are associated with pegmatitic granite at sites 1 and 2 and with granitic gneiss and quartzite at sites 3 and 4. CS stripes are restricted to areas with fine-grained, amphibolitic schist clasts, although some gneiss fragments were also found in stripe troughs. Slope angle is gentle to moderate in PS stripe sites, which never exceed 25°; CS stripes occupy only steep slopes. No preferential aspect was noticed for stripes, their occurrence being apparently a function of suitable rock material and slope angle. Sites 1 and 2 have a northwest orientation, whereas all other sites face southeast. At sites 3 and 4, PS stripes have formed by the gradual coalescence of soil buds (Heine, 1977), which are small isolated domes of fine soil surrounded by coarser particles. Buds occur on gentle (about 6° or less) upper-slope sections, and as gradient increases they become aligned with the slope to form stripe ridges (Furrer and Freund, 1973; Pérez, 1987a). It seems that pebbles descend from bud to stripe areas on this slope, because soils are slightly coarser along stripe troughs than around soil buds (Pérez, unpublished data). The main source of coarse particles for PS stripes at sites 1 and 2, both located near a ridge, was in situ rock weathering. CS stripes occupy mid- and low-slope positions, and are considerably affected by clast transport from a sizable talus upslope.

STRIPE AND CLAST MORPHOLOGY

Both kinds of stripes are aligned parallel to the slope (Figures 6.3A, B). Rayleigh tests (Doornkamp and King, 1971; cf. Hall, 1979) indicate a strongly preferential orientation in both stripe types. This orientation departs only 0.9° (115.1°) from the slope direction of PS stripes and 1.7° (163.3°) from the slope direction of CS stripes. The high L% values (Curray, 1956, Figure 4) of these

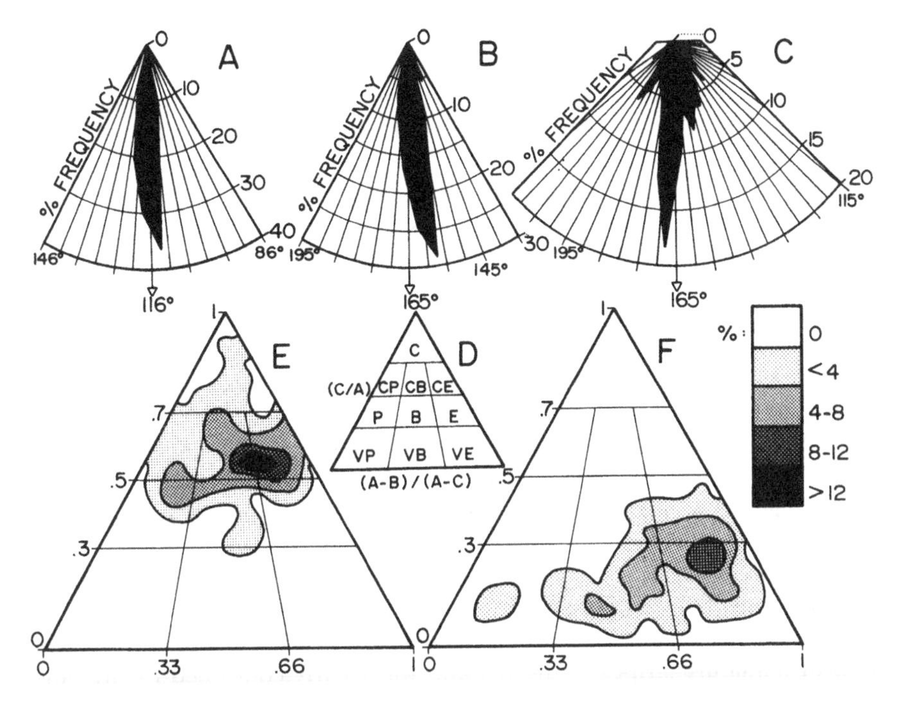

Figure 6.3. Morphological characteristics of stripes. Diagrams based on 50 stripes or particles. Frequency interval for A, B, and C is 5°. Arrows indicate downslope direction. (A) Percent frequency of orientations of pebble-sorted stripes at site 4. (B) Percent frequency of orientations of cobble-sorted stripes at site 6. (C) Percent frequency of orientations (*a* axes) of particles from troughs of cobble-sorted stripes at site 6. (D) Key for triangular particle-shape diagrams: axial indices and shape classes: C = compact, CP = compact platy, CB = compact bladed, CE = compact elongate, P = platy, B = bladed, E = elongate, VP = very platy, VB = very bladed, VE = very elongate (Sneed and Folk, 1958). (E) Contoured diagram (Pérez, 1987e) for particles from troughs of pebble-sorted stripes at site 2. Shading indicates particle density as a percentage of the total. (F) Contoured diagram for particles from troughs of cobble-sorted stripes at site 6

statistics (95.4 for PS stripes, 93.7 for CS stripes) show that both orientations are highly significant ($P < 0.0001$). The slope alignment of PS stripes is disrupted in the vicinity of blocks and outcrops, where stripes gently bend around the obstacles, giving a raked appearance to the ground surface (Figure 6.2C); this is not observed in CS stripes.

Clasts occupying the troughs of CS stripes are also strongly aligned with the slope (Figure 6.2D). A Rayleigh test for the *a* axis of 50 particles shows that the preferred orientation (164.3°) departs only 0.7° from the slope direction. The scattering about the main axis, however, is greater than for stripe orientation, as a few clasts are aligned oblique or perpendicular to the slope line (Figure 6.3C). The considerably lower value of L% (38.0) reflects this axial dispersion, but the average orientation is still significant ($P < 0.001$). Although no data were gathered on particle fabric, it was observed that some clasts rest on their edge. Most, however, do not and simply lie on the trough surface with their *b* axis parallel or slightly oblique to the ground.

Clast orientation was not measured in PS stripes, where pebbles are usually quite small. A comparison of particle shape between trough pebbles and cobbles (Figures 6.3D, E, F) shows some significant differences: particles in PS stripes are compact and equidimensional, while those in CS stripes are mostly thin, elongated slabs. Axial indices of sphericity $\{[c^2 / (a \times b)]^{0.333}\}$ (Sneed and Folk, 1958) and elongation (a/b) also indicate sharp differences. Pebbles have an average sphericity of 0.749 ± 0.08 and an elongation of 1.344 ± 0.2; the corresponding values for cobbles were 0.467 ± 0.14 and 2.281 ± 0.62. These differences in shape result from the two main lithologies found at stripe sites. The well-developed foliation of the amphibolitic schist facilitates the production of relatively large, thin slabs, which accumulate in CS troughs. The coarse-grained granitic rocks, on the other hand, tend to break into small, equant fragments. Thus most pebbles in the PS troughs are compact quartz grains. Similar associations of particle shape and lithology have been observed in larger talus clasts at Piedras Blancas (Pérez, 1986b). The distribution of miniature stripes in the paramo was clearly controlled by lithology, which played a decisive role in defining clast morphology.

SOIL PROFILE CHARACTERISTICS

In both stripe types, fine-soil bands are elevated a few centimeters above the coarse-grained troughs. The average wavelength of PS stripes (site 2) is about 6 to 7 cm; CS stripes (site 5) are larger, with a mean size of 10 to 12 cm (Figures 6.2 and 6.4). No systematic difference in width between ridges and troughs is evident; these are nearly equal in size in most cases. Both kinds of stripes are shallow, with the sorted layer extending into the soil only 4 to 6 cm. Most stripe sites have extremely shallow soils (Table 6.1). Cross-sections (Figure 6.4) of paramo stripes are very similar to those from other tropical mountains (Lliboutry, 1961; Furrer and Freund, 1973; Graf, 1973; Hastenrath, 1973). PS stripes occupy shallow, weakly developed Lithic Cryochrepts with a thin A1 horizon overlying an AC layer and a weathered, hard C horizon formed by in situ rock weathering (Pérez, 1987c, 1991) (Table 6.2). CS stripes are on sandy-skeletal Typic Cryorthents with a simple AC/C profile and lithological discontinuities associated with sharp changes in particle size distribution (Lee and Hewitt, 1982). These indicate that the profiles have evolved on the steep talus primarily by processes of truncation and gravitational cumulation.

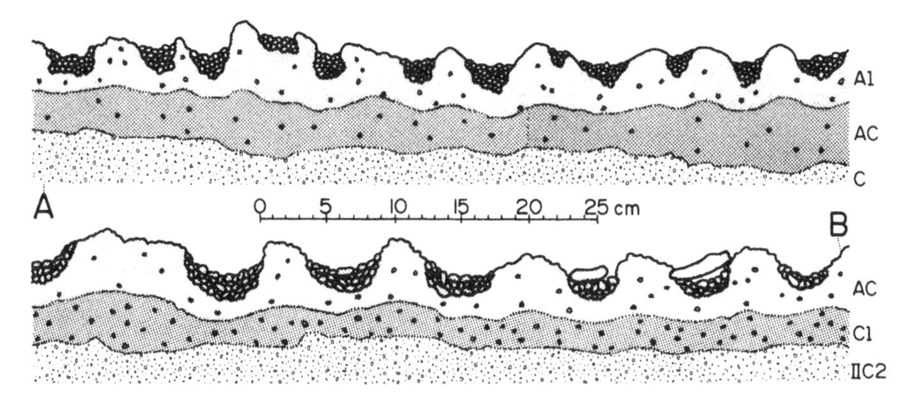

Figure 6.4. Transversal cross-sections of (A) pebble-sorted and (B) cobble-sorted stripes in sites 2 and 5, respectively. Letters designate soil horizons (see Table 6.2 for further explanation)

In both profile types, visible sorting of materials is restricted to the loamy-sand uppermost horizon, which is thickest beneath stripe ridges. In addition to the lateral sorting and color differences, this horizon can be readily identified in the field by its fine crumb structure and low bulk density. These characteristics are most noticeable along ridges, which are covered at the surface by minute, puffy earth lumps (nubbins) (Washburn, 1980). The high porosity and superficial nubbin texture are both produced by recurrent needle-ice activity (Pérez, 1984).

PROPERTIES OF STRIPE SOILS

Surface samples

The most evident pedological characteristic of stripes is the pronounced lateral particle sorting of a superficial layer into fine-soil ridges and coarse-grained troughs. In all paired surface samples, the ridge was always considerably finer than the adjacent troughs. Ridges usually lack particles larger than 8 to 10 mm, but troughs often have a high proportion of large pebbles or cobbles. Ridge and trough soil populations are strikingly different at all sites, and their graphic envelopes show practically no overlap. However, the particle size distributions are strongly site-specific; thus some plots have coarser ridge and trough soils than others (Figure 6.5). Within-site variability of ridge samples is generally lower than that of trough soils (Figure 6.5A, B, C, E). Textural variability of CS stripes increases with greater sample size (Figure 6.5E, F). Differences between samples are statistically significant ($P < 0.001$ or 0.005) for both gravel (> 2 mm) and fines (< 0.063 mm) fractions

(Table 6.3). Ridge and trough samples from PS stripes contain a lower percentage of both gravel and fines than do those from CS stripes. The first difference apparently results from the continuous addition of coarse particles to CS stripes by talus shift; the second may be ascribed to the fine-grained nature of the schist associated with CS stripes.

Organic matter content of the soils is low in all positions, around 2%, but is higher in CS stripes and in ridge soils (Table 6.3). The difference between troughs and ridges is significant ($P < 0.001$) only in PS stripes. These trends probably result from the association of fine particles with organic matter, as coarse, dry soils allow for its rapid oxidation and loss (Pitty, 1979, p. 75). Soil color varies consistently between troughs and ridges. CS stripes have lighter ridges (by about 1 value unit) than troughs, where dark schist gravel-sized fragments collect (Table 6.3). PS stripes show an opposite trend because light, off-white quartz fragments gather along troughs. CS-stripe soils are dominated by gray hues, whereas soils of PS stripes are light brown.

Subsurface samples

Subsurface soils do not show the sorting observed in surface samples. Even though in most cases the proportions of gravel and fines follow the same trends as at the surface (i.e., greater gravel content and lower percentage of fines in troughs), the differences between troughs and ridges are never broad or significant (Figure 6.6). CS stripes still have a higher gravel percentage at depth than PS stripes, but their fines content is slightly lower (Table 6.3). All

Table 6.2. Soil Profile Descriptions[a]

A Horizon	Depth (cm)	Description
A1	0–5	Grayish brown, 10YR 5/2; loamy sand, 22.5% fines, 20.6% gravel; fine crumb structure, low bulk density, and high porosity, generated by needle-ice activity (nubbin layer); loose to soft dry consistence; slightly sticky, nonplastic wet consistence; 3.3% organic matter; *p*H = 5.3. Abrupt wavy boundary to:
AC	5–13	White, 10YR 8/2; loamy sand, 23.3% fines, 18.0% gravel; single grain structureless; soft dry consistence; nonsticky, nonplastic wet consistence; 0.7% organic matter; *p*H = 5.2. Clear irregular boundary to:
C	13–16	Light gray, 10YR 7/2; weathered, coarse stony layer of gravel-sized particles; extremely hard dry consistence and high bulk density. Abrupt smooth boundary to:
R	16+	Leucocratic, pegmatitic granite with large crystals (~ 5 mm diameter) of muscovite (± 20%), quartz (± 25%), and potassic feldspars (orthoclase and microcline, ± 55%).

Table 6.2. (*continued*)

B Horizon	Depth (cm)	Description
A1	0–5	Light brownish gray, 10YR 6/2.5; loamy sand, 14.9% fines, 9.2% gravel; fine crumb structure, low bulk density, and high porosity, generated by needle-ice activity (nubbin layer); loose to soft dry consistence; nonsticky, nonplastic wet consistence; 2.2% organic matter; *p*H = 4.9. Abrupt smooth boundary to:
AC	5–10	Light gray, 10YR 7/2; loamy sand, 27.3% fines, 17.0% gravel; single grain structureless; soft dry consistence; slightly sticky, slightly plastic wet consistence; 2.3% organic matter; *p*H = 4.6. Abrupt wavy boundary to:
C	10–13	Very pale brown, 10YR 7/3; weathered, coarse stony layer of gravel-sized particles; extremely hard dry consistence and high bulk density. Abrupt smooth boundary to:
R	13+	Granitic gneiss, with weakly defined foliation, and with a tendency for the biotite to become oriented in bands. Biotite (± 15%), quartz (± 25%), and feldspars (plagioclase and orthoclase, ± 60%).

C Horizon	Depth (cm)	Description
AC	0–5	Gray to light gray, 10YR 6/1; loamy sand, 18.8% fines, 29.4% gravel, fine crumb structure, low bulk density, and high porosity, generated by needle-ice activity (nubbin layer); loose to soft dry consistence; slightly sticky, slightly plastic wet consistence; 2.1% organic matter; *p*H = 5.5. Abrupt wavy boundary to:
C1	5–38	Gray to light gray, 10YR 6/1; sand, 4.6% fines, 63.9% gravel; single grain structureless; soft dry consistence; nonsticky, nonplastic wet consistence; 1.5% organic matter; *p*H = 5.7. Gradual irregular boundary to:
IIC2	38–49+	Light gray, 10YR 7/1; sandy loam, 35.8% fines, 31.0% gravel; fine granular structure; slightly hard dry consistence; sticky, plastic wet consistence; 2.3% organic matter; *p*H = 5.8.

[a]Terminology follows Soil Survey Staff (1975). Fines include silt and clay (< 0.063 mm); colors are for dry soils. All profiles were examined below stripe ridges. Samples for R horizons are from nearby outcrops. (A) Site 1: Lithic Cryochrept. Upper-slope section near Mifafi Creek with less than 5% plant cover (*Coespletia timotensis, Hinterbubera imbricata*, and *Senecio sclerosus*). (B) Site 4: Lithic Cryochrept. Upper-slope section near Pico Los Nevados with less than 1% plant cover of *Arenaria musciformis*, *Coespeletia timotensis*, and *Draba chionophila*. *(C) Site 6: Typic* Cryorthent, sandy-skeletal, with ca. 5% small fragments of melanocratic, amphibolitic schist. Steep, nearly bare talus on Pico Los Nevados. Only a few *Coespeletia timotensis* are present. Outcrops are of fine-grained schist with well-developed foliation. The rock is composed of hornblende and actinolite (± 60%) in acicular crystals forming alternate bands with a lighter material (plagioclase, ± 40%). Nearby outcrops of granitic gneiss are similar in composition to that described for site 4.

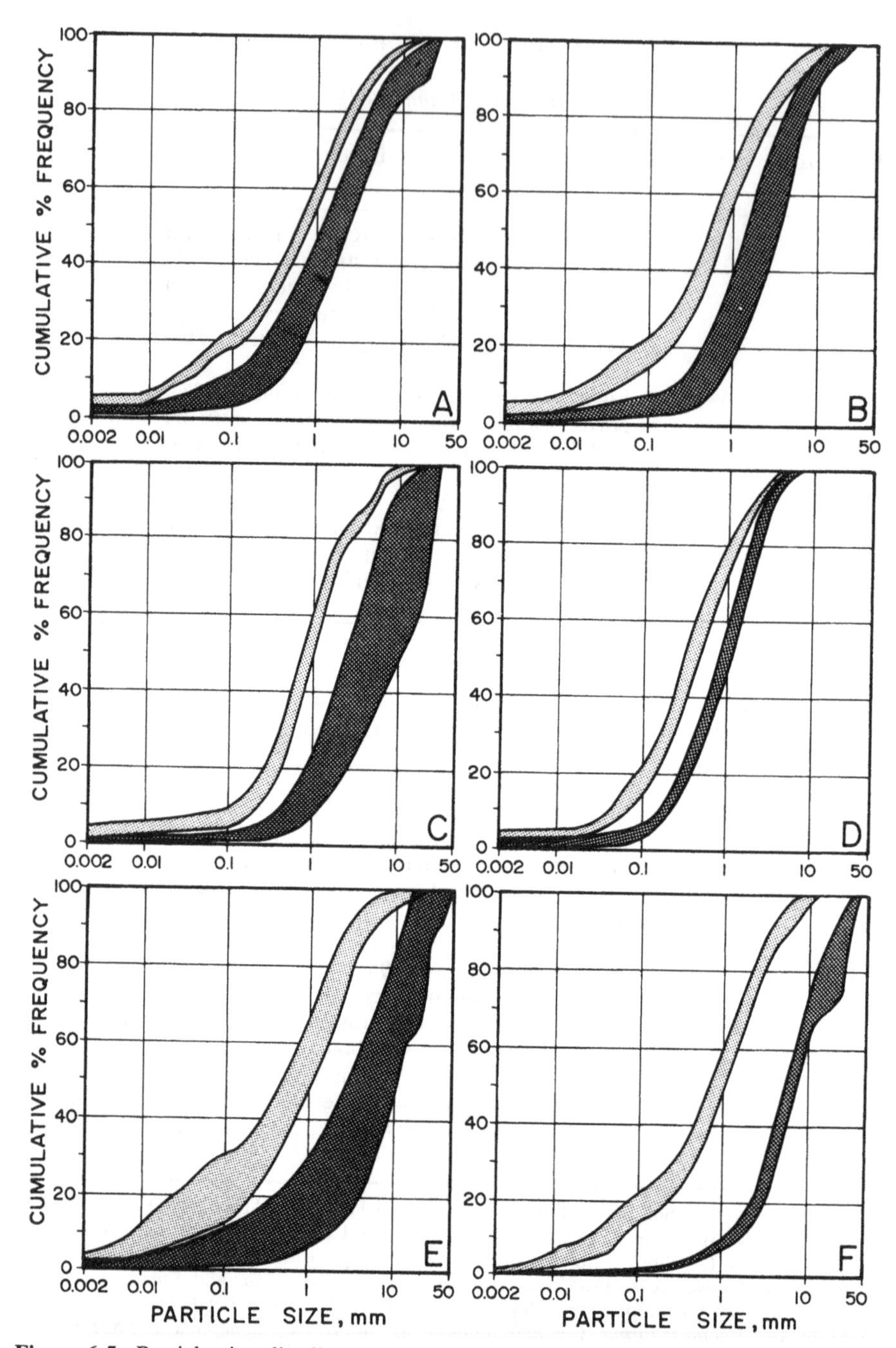

Figure 6.5. Particle size distributions for surface soils. Graphic envelopes represent several samples. Number of sample pairs is given in parenthesis. Dark shading signifies trough soils and light shading ridge soils. (A) Pebble-sorted stripes at site 1 (5). (B) Pebble-sorted stripes at site 2 (10). (C) Pebble-sorted stripes at site 3 (5). (D) Pebble-sorted stripes at site 4 (5). (E) Cobble-sorted stripes at site 5 (15). (F) Cobble-sorted stripes at site 6 (5)

Table 6.3. Average Values for Selected Soil Properties of Paramo Stripes[a]

Soil Property	Sampling Level	Stripe Type and Sampling Position			
		Pebble-sorted		Cobble-sorted	
		Ridges	Troughs	Ridges	Troughs
% Gravel (> 2 mm)	Surface	18.9 ± 5.32[b]	46.1 ± 17.33	24.8 ± 5.82[b]	73.1 ± 10.38
	Subsurface	15.0 ± 1.12	16.5 ± 2.31	50.9 ± 8.25	56.5 ± 14.40
% Fines (< 0.063 mm)	Surface	15.4 ± 4.97[b]	6.5 ± 3.49	19.6 ± 6.23[c]	12.7 ± 6.50
	Subsurface	20.2 ± 2.34	17.6 ± 3.95	11.9 ± 5.46	13.3 ± 7.94
% Organic matter	Surface	2.2 ± 0.57[b]	1.4 ± 0.57	2.3 ± 0.43	2.1 ± 0.4
	Subsurface	2.2 ± 0.17	2.1 ± 0.12	2.2 ± 0.42	2.2 ± 0.43
Color	Surface	10YR 5/2 Grayish brown	10YR 6.5/2 Light brownish gray	10YR 7/1 Light gray	10YR 6/1 Gray

[a]Average values are given ± the standard deviation. Significance values are for Mann–Whitney U test comparisons between ridge and trough sampling positions. Sample size: pebble-sorted, surface = 25 pairs; pebble-sorted, subsurface = 10 pairs; cobble-sorted, surface = 20 pairs; cobble-sorted, subsurface = 10 pairs.
[b]$P < 0.001$.
[c]$P < 0.005$.

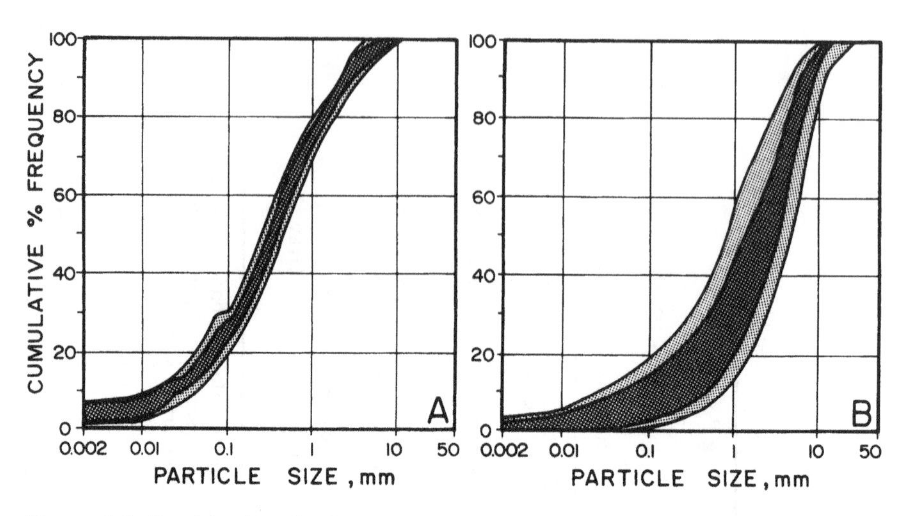

Figure 6.6. Particle size distributions for subsurface soils. Each diagram shows 10 sample pairs. Dark-shaded envelopes indicate samples below both ridges and troughs; light-shaded envelopes indicate only samples below troughs. (A) Pebble-sorted stripes at site 4. (B) Cobble-sorted stripes at site 5

subsurface samples of PS stripes closely resemble those of surface ridges. Depth samples from CS stripes have a texture intermediate between that of surface ridges and surface troughs (compare Figures 6.5E and 6.6B). Within-site variability of ridge samples is lower than that of soils below troughs. Data for PS stripes suggest that gravel has been sorted out of the subsurface layer to the surface, and from there into troughs (Table 6.3). In comparison, the "average" content of surficial gravel in CS stripes is nearly equal to that of the subsurface horizon; thus no vertical gravel sorting is discernible. Organic matter content is similar in ridges and troughs and for both stripe types; values are about the same as in surface samples.

DEGREE OF PARTICLE SORTING IN STRIPES

Relative particle sorting between paired samples was assessed in detail for different soil fractions with an index previously utilized for patterned ground analysis (Ballantyne and Matthews, 1983; Pérez, 1991). The numerical sorting index SI is defined by $SI = RS/TS$, where RS is the percentage (by weight) of a given fraction in the ridge sample, and TS is the percentage (by weight) of a given fraction in the trough sample. $SI = 1$ indicates no lateral sorting of a given fraction; $SI < 1$ indicates that a given fraction is deficient in the ridge compared with the paired sample from the adjacent trough; $SI > 1$ indicates that a given fraction is deficient in the trough; and $SI = 0$ indicates "perfect" sorting with all material of a given size laterally sorted into the trough. Sorting indices for the two stripe populations were calculated for 12 particle size classes with a class interval of 1ϕ.

The results for the surface samples indicate several interesting trends. In both stripe types, coarse gravel particles are concentrated in troughs. Sorting becomes more pronounced in the largest size classes, which are lacking in ridge soils (Figure 6.7B, D). Conversely, the soil fraction (< 2 mm) is strongly sorted into the ridges, particularly in the case of CS stripes. This result, however, is mostly a function of the considerably higher mass of the gravel fraction compared to the soil. To eliminate the confounding influence caused by the weight of the gravel component, sorting indices for the soil size classes were calculated based on the amount of material passing the 2 mm sieve instead of total sample weight (Ballantyne and Matthews, 1983). These data indicate that soil fractions greater than 0.5 mm (coarse and very coarse sand (Soil Survey Staff, 1975)) are also sharply concentrated in the troughs of both stripe types. Smaller size classes predominate in the ridges. This trend is more pronounced for the smaller fractions, particularly in PS stripes (Figure 6.7A, C). Vertical bars on the diagrams show confidence intervals ($P < 0.005$) for a *t* distribution. These can be used as statistical tests of comparison with $SI = 1$. As all indices are different from 1, all size classes of surface samples display significant lateral sorting.

Subsurface samples show virtually no evidence of lateral segregation in either stripe type. Only the gravel fraction from 16 to 32 mm is concentrated below the troughs of CS stripes (Figure 6.7D). All the other size classes have indices around 1 and confidence intervals which overlap this value, thus indicating no sorting. In summary, this analysis clearly shows that lateral particle sorting is restricted to a shallow superficial soil layer about 5 cm thick, and that, in addition to the visible concentration of coarse particles into troughs, there is a strong segregation of fine grains into ridges. This sorting, unexpectedly, is more pronounced in PS stripes (Figure 6.7A).

TEMPERATURE FLUCTUATIONS

The progression of air temperatures shows maxima of 11 to 18° C and minima of −2° to −9.8° C; thus the diurnal amplitude varied from 13 to 28° C (Figure 6.8). Narrower daily fluctuations were recorded during cloudy conditions (December 18/19). Clouds restrict nightly heat radiation losses to the atmosphere and reduce solar radiation input during the day. Broader amplitudes were observed under clear skies (December 22/23). Even higher maxima, up to 25° C, are attained on south-facing slopes in the dry season (Pérez, 1986a, 1987a) due to greater incoming solar radiation. The diurnal cycle of soil temperature is closely related to that of air temperature, but both soil maxima and minima showed a lag time of 1.5 to 2 hours at −7 cm and 3 to 3.5 hours at

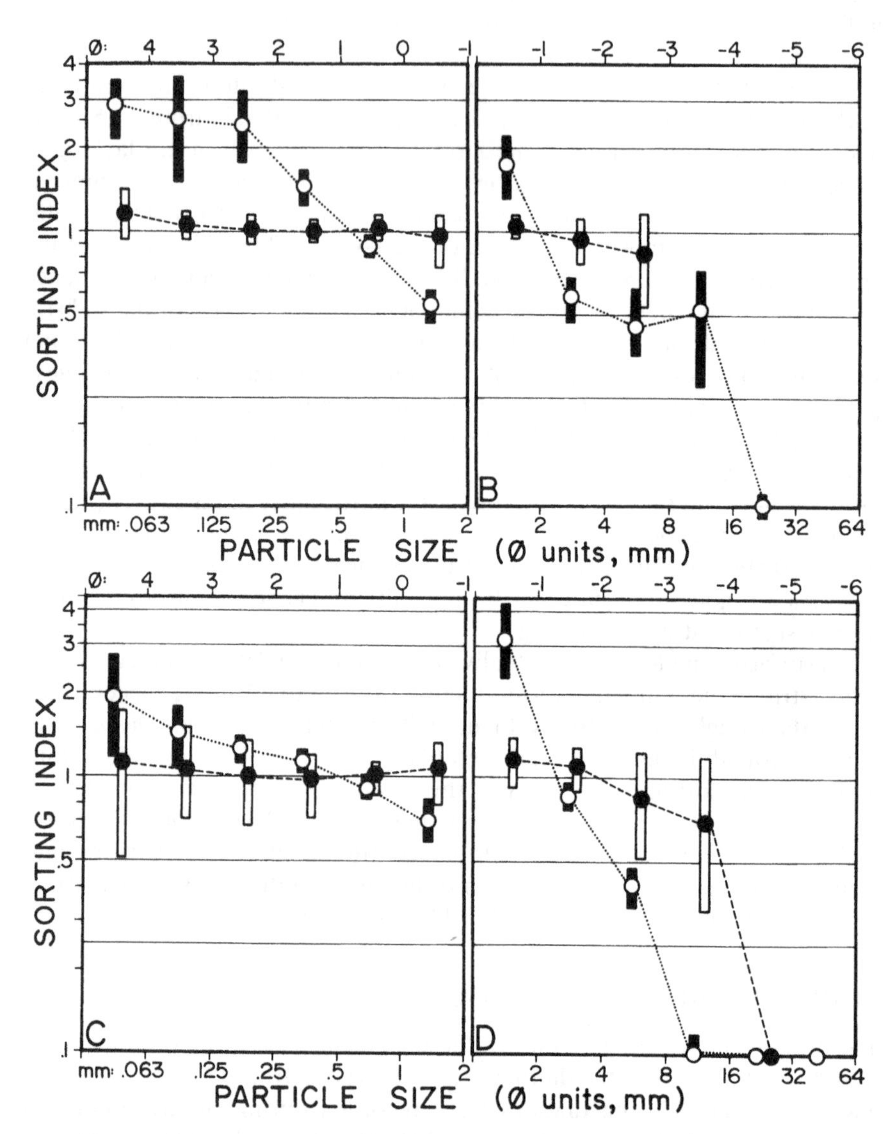

Figure 6.7. Sorting indices for different soil fractions. Size-class interval is 1 phi (ϕ) unit. Open circles represent surface soils and solid circles subsurface soils. Bars indicate the statistical confidence intervals ($P < 0.005$). (A) Soil fraction (< 2 mm) from pebble-sorted stripes. (B) Whole sample from pebble-sorted stripes. (C) Soil fraction from cobble-sorted stripes. (D) Whole sample from cobble-sorted stripes. See text for further explanation

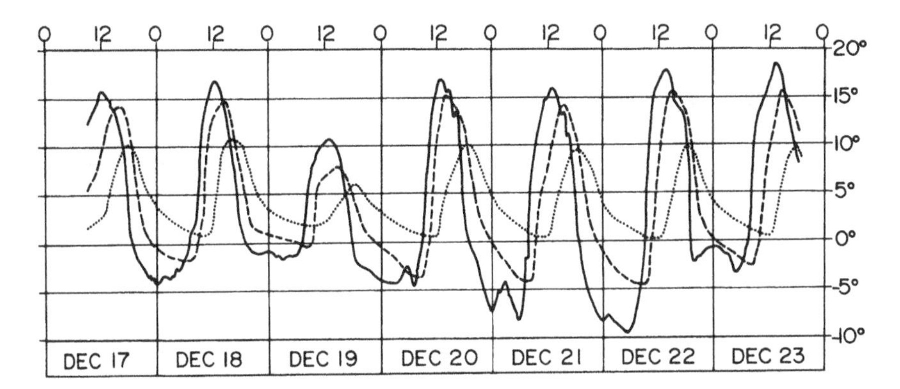

Figure 6.8. Temperature fluctuations at site 1, December 1985. Solid line shows air temperatures at +10 cm; dashed line shows soil temperatures at −7 cm; and dotted line shows soil temperatures at −14 cm

−14 cm. The daily range at −7 cm was 8 to 20° C, whereas that at −14 cm was 6 to 10° C.

Freezing occurred every night at −7 cm but was not recorded at −14 cm. Deeper freezing took place under clear skies; the diagrams indicate a frost penetration of 7.5 to 13 cm. This is slightly greater than previous data show (Pérez, 1984, 1989; Pfitsch, 1988) and can be ascribed to extremely dry soils at the time of measurement. Shallower frost penetration is to be expected in moist soils (Lliboutry, 1955; Pérez, 1984) because the latent heat of fusion of soil water is slowly released as the water freezes. The minimum frost depth was attained during conditions resembling those normally observed during the wet season. This depth (7.5 cm) compares well with data from nearby site 2 (only 35 m away from site 1), where Schubert (personal communication, 1984) buried small vials filled with water and found that frost penetrated only 6 cm during the rainy season (April 1975) (Schubert, 1979a, 1980).

It appears that stripes at sites 1 and 2 are repeatedly affected by freezing both during the dry and the rainy seasons. Miniature stripe sites at higher altitudes should experience even lower temperatures, as the environmental lapse rate of the area is −0.66° C/100 m. However, sites 1 and 2 are near the Mifafi valley bottom, and marked cold air drainage in this valley (Pfitsch, 1988; Pérez, 1989) tends to negate, and even reverse, the expected drop of temperature with elevation.

MARKER MOVEMENT

After two years surface gravel lines at site 1 had a slightly diffuse appearance but remained nearly unchanged; they had shifted downslope only 5 to 15 cm. Subsurface gravel and marble lines had not moved. The site was left undis-

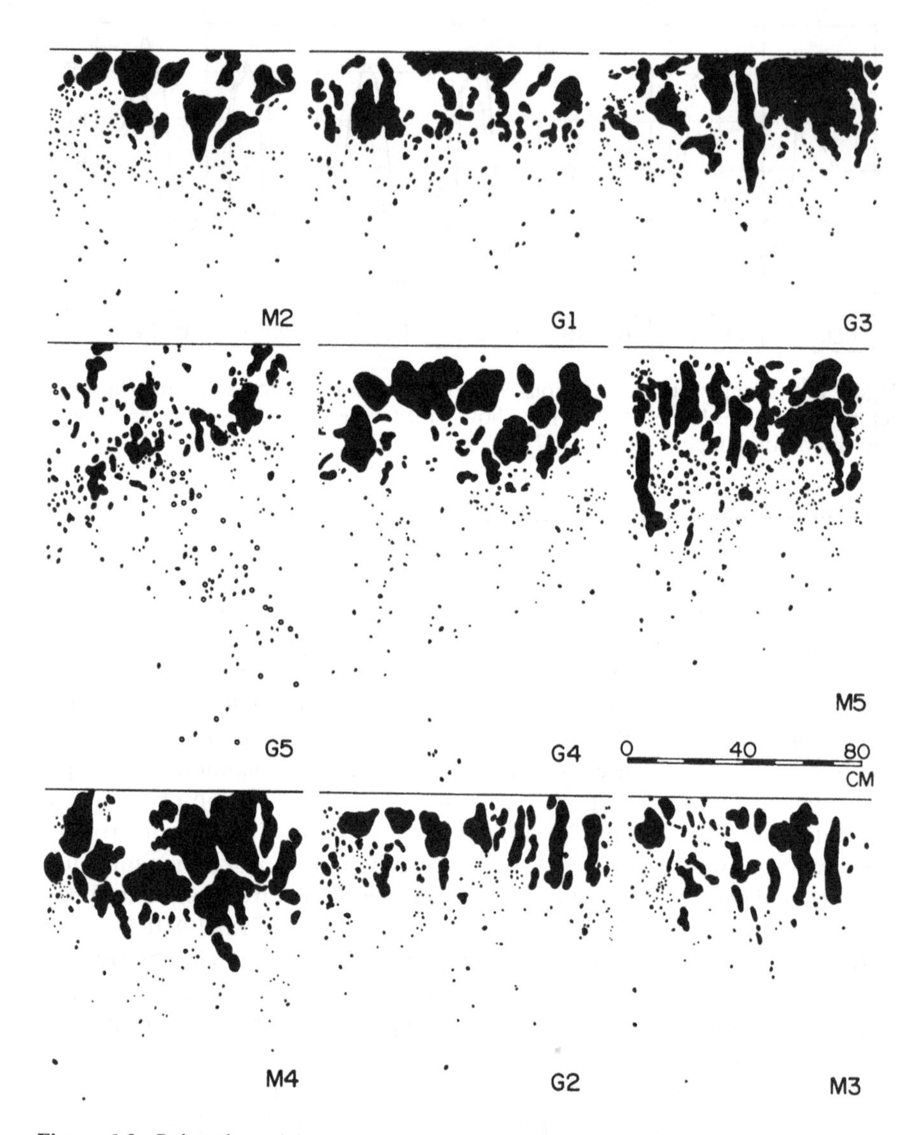

Figure 6.9. Painted-particle movement on nine plots with pebble-sorted stripes at site 3 from December 1985 to December 1987. Drawings were made by direct photoreduction from plastic sheet overlays drawn in the field. Horizontal line shows the initial position of painted particles. Dark shading shows areas where blue particles were concentrated; isolated dots indicate outliers. Open circles indicate red particles (only on plot G5)

turbed, and no overlays were taken. In contrast, all surface gravel lines at site 3 displayed considerable particle movement. The difference between sites is attributed to the much steeper gradient at site 3. Subsurface lines remained mostly stationary; a few red gravel particles (seeded at −5 cm) had moved only in plot G5. During the descent, painted grains had spread and segregated into several irregular patches; in addition, numerous isolated outliers were visible (Pérez, 1987c). Outliers were usually large pebbles that had shifted downslope faster than the main marker masses (Figure 6.9). Plot M1, not included in Figure 6.9, exhibited no defined gravel masses, but only several isolated pebbles.

Grain movement was calculated by dividing lines into five equal-length sections and randomly taking 10 measurements within each segment. Five measurements were obtained from the distal (fastest-moving) edges of the main grain masses; the other five were taken from their proximal (slowest-moving) edges. Thus data sets consist of 25 maximum-shift and 25 minimum-shift measurements per line. Maximum, minimum, and "average" shift rates (obtained by combining the first two data sets) were calculated (Table 6.4).

Table 6.4. Mean Movement Rates for Blue Gravel Particles (placed on the soil surface) and Red Gravel Particles (placed at 5 cm depth) on Plots with Pebble-sorted Stripes[a]

		Main Particle Mass			Outliers
Plot Number	Slope Angle (degrees)	Maximum Rates ($cm\ yr^{-1}$)	Minimum Rates ($cm\ yr^{-1}$)	Average Rates ($cm\ yr^{-1}$)	Average Rates ($cm\ yr^{-1}$)
Blue Particles					
M1	25.0	—	—	—	27.5 ± 1.33
M2	20.0	9.7	1.5	3.76 (2.6–5.3)	24.4 ± 0.88
M3	22.0	12.4	2.5	5.53 (4.0–7.5)	23.0 ± 1.14
M4	22.0	20.8	2.9	7.76 (5.3–11.3)	33.6 ± 1.37
M5	22.0	19.7	4.6	9.51 (7.5–12.1)	34.3 ± 0.86
G1	21.0	11.7	2.3	5.14 (3.6–7.4)	29.2 ± 0.96
G2	21.0	11.5	2.9	5.81 (4.5–7.5)	21.7 ± 0.89
G3	18.5	13.8	0.6	2.92 (1.8–4.8)	24.8 ± 0.86
G4	22.0	17.3	5.1	9.42 (7.6–11.6)	34.6 ± 1.26
G5	21.0	13.9	5.0	8.32 (6.2–11.2)	36.1 ± 1.23
Mean rates ($cm\ yr^{-1}$)		14.09 (13.3–14.9)	2.58 (2.2–3.0)	6.03 (5.38–6.74)	21.39 ± 0.27
Red Particles					
G5	21.0	—	—	—	35.8 ± 3.25

[a]Period of measurement: December 1985 to December 1987. Confidence intervals for geometric movement rates of the main particle mass are at $P < 0.05$. The standard errors for the arithmetic movement rates of outliers are shown.

Distributions of particle displacements were tested for normality using their skewness and kurtosis (Jones, 1969). Most were not normally distributed, so the data were log-transformed (Caine, 1968). The distributions then tested as normal ($P < 0.05$). Distance of transport was measured for each outlier. Number of identified blue outliers per plot varied from 79 to 281. A total of 1497 blue and 26 red outliers was measured. As the distribution of outlier travel distances tested as normal ($P < 0.05$), no transformation was needed.

Average movement of grain masses varied widely between plots from 2.92 cm yr^{-1} to 9.15 cm yr^{-1}. The mean rate for all measurements was 6.03 cm yr^{-1} (Table 6.4). Outliers shifted considerably faster on all plots: they had a mean rate of 21.39 cm yr^{-1}. Rates of movement were compared using Mann–Whitney *U* tests. These tests showed that outliers descended significantly faster ($P < 0.001$) than the distal edges of the grain masses. Displacement rates are similar, though somewhat lower, to rates previously measured in Piedras Blancas. C. Schubert (personal communication, 1984) measured stripe movement at site 2 of this study for a 4.5-year period (1975 to 1980), and found maximum rates of particle masses of about 14 cm yr^{-1}. He also observed that isolated pebbles moved about twice as fast and became sorted into stripe troughs. Some lines remained largely stationary and were only slightly diffused. Pérez (1987c) also measured maximum rates of 7.3 to 25.4 cm yr^{-1} at a nearby site during 1981 and 1982. Some variation in rates could be ascribed to different methods and measurement periods; shorter measurement time is commonly associated with greater descent rates (Caine, 1981; Pérez, 1987d). Slope angle is also an important factor affecting movement rates. Pissart (1973, Figure 9) demonstrated that pebble transport by needle ice on sorted stripes is strongly related to gradient. Rates of paramo tracers fit his linear regression well.

In January 1990, subsurface tracers were inspected and the areas below the lines raked. Buried gravel and marble tracers had remained stationary. Only nine marbles from line M2 (site 3), seeded at −5 cm, had shifted downslope from 12 to 28 cm. The data set proved to be normally distributed ($P < 0.05$) with an average movement of 17.94 ± 5.31 cm, which corresponds to a rate of movement of 4.49 cm yr^{-1}. All marbles were in stripe troughs, either at the surface or covered by 10 to 18 mm of soil. Only a few scattered blue gravel particles could be found in troughs on either site. Some red pebbles were also visible on three plots, but their distance of transport was not measured. Two reasons can account for the disappearance of gravel tracers: burial (as shown by the marbles) and paint fading. Fading is a common problem (C. Schubert, personal communication, 1984; Pérez, 1987c) and apparently affects different paint colors or types selectively. This study indicates that fading intensifies with longer periods of time. Marker burial and incorporation into the soil also increases markedly with time, especially if tracers are small and move rapidly (Caine, 1981; Pérez, 1987a).

The above results indicate several significant trends of soil disturbance: (1) surface particles moved rapidly on steep slopes but slowly on gentle gradients; (2) gravel tracers buried at 5 cm were slightly disturbed during the first two years of the study and only a few of the larger, heavier marbles (about 5 mkg each) buried at the same depth moved after four years; and (3) no markers buried at 8 to 10 cm were ever disturbed. Clearly, the soil layer moving downslope is extremely shallow, as disturbance decreases rapidly with depth (Pérez, 1985, 1987a).

SPATIAL SORTING OF MARKERS

During their descent, painted grains were laterally sorted into troughs. As noted above, this was evident in outlying pebbles and marbles. Grain masses showed the same pattern and often became segregated into elongated patches that corresponded essentially with stripe troughs. Overlay comparisons with photographs indicated that spatial sorting was more pronounced in plots with clearly defined stripes (Figure 6.9, G3, M5, G2, M3); ill-defined stripes were associated with more irregular and diffuse patches (Figure 6.9, M2, G4, M4). Figure 6.10 shows two plots with clear lateral sorting. About 80 to 90% of the tracer masses and most of the outliers coincide with troughs. This was to be expected because seeded particles were within the coarse sand to very fine gravel size classes, and these fractions have been shown to concentrate in troughs.

SOIL MOISTURE CONTENT AND NEEDLE-ICE GROWTH

Soil moisture content is a critical variable of needle-ice growth. Previous field studies (Pérez, 1984, 1986a) suggest that needle-ice formation in Piedras Blancas occurs only in moist soils at or above field capacity, and that ice growth rapidly ceases as soils become desiccated. Meentemeyer and Zippin (1981, Figure 3) empirically determined the minimum moisture level required for needle-ice growth and found that this increases as soils become coarser. A similar diagram contrasting the water content of surficial stripe soils at field capacity with their percentage of fines (Figure 6.11) shows a tendency for greater moisture to occur in finer soils. Consequently, most ridge samples contain more water than paired trough samples. Moisture variability may also have resulted from differences in organic matter content (Table 6.3) or proportion of grains finer than 0.02 mm, as these factors affect soil water retention considerably (Pérez, 1987b; Pitty, 1979, p. 133). Adding to the scattergram the line of "minimum moisture" needed for needle-ice formation (Meentemeyer and Zippin, 1981, Figure 3) reveals an interesting pattern. All ridge samples but one lie at or above this line within the zone of needle-ice growth. Conversely, the majority of trough soils fall below the line in the area

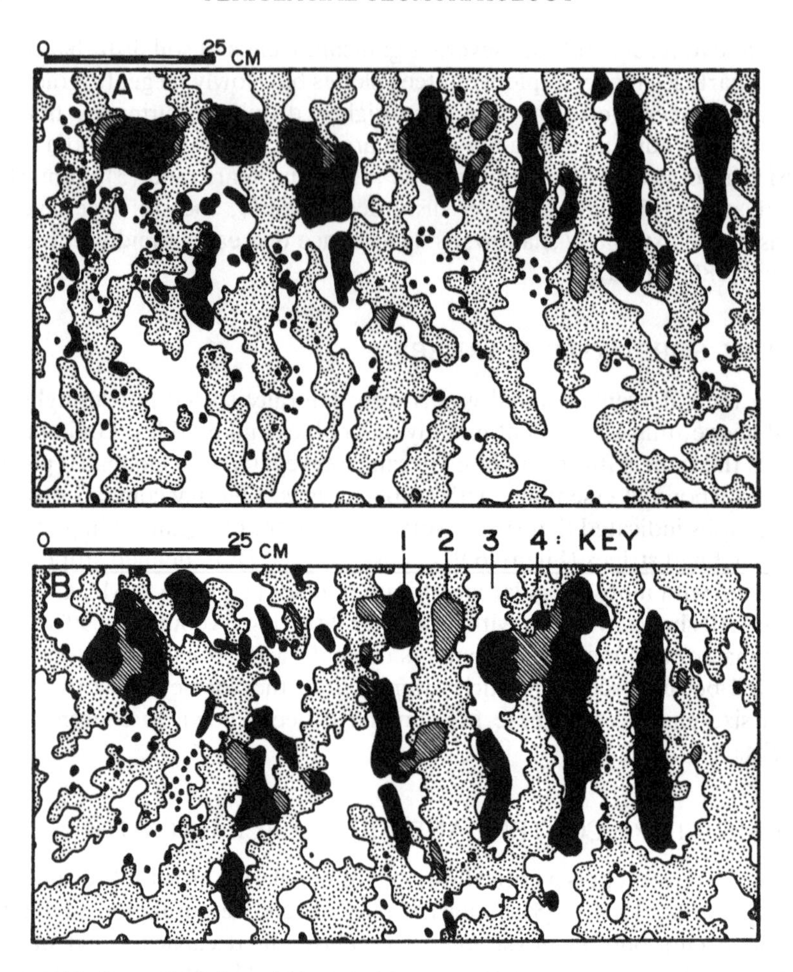

Figure 6.10. Lateral sorting of blue particles on areas with pebble-sorted stripes (site 3) after two years of movement (December 1985 to December 1987). Diagrams drawn from photographs and plastic sheet overlays. Key: 1 = painted particles on troughs, 2 = painted particles on ridges, 3 = trough areas, and 4 = ridge areas. (A) Plot G2. (B) Plot M3

of no needle-ice growth. Three of the four trough soils well above this boundary are from CS stripes, which have greater amounts of fines and organic matter than troughs of PS stripes. This diagram does not imply that needle ice will not form in troughs. Instead, it suggests (1) that ridge soils would allow needle-ice growth more readily than trough soils with similar moisture content; and (2) that needle-ice formation will cease in most trough soils only a few hours after their being moistened, while growth can probably continue in ridge soils for longer periods of time. Meentemeyer and Zippin

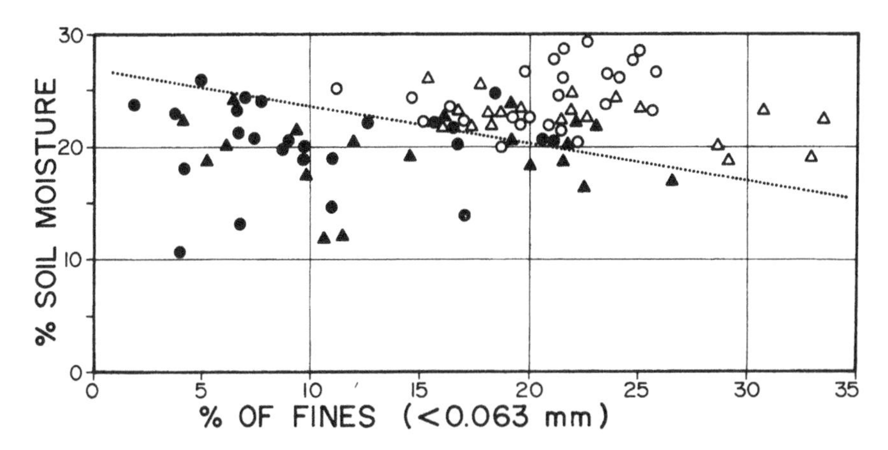

Figure 6.11. Relationship between proportion of fines (particles < 0.063 mm) and soil water content at or near field capacity for 90 surface soil samples. Key: solid symbols = trough soils; open symbols = ridge soils. Circles = pebble-sorted stripes; triangles = cobble-sorted stripes. Dotted line shows the minimum soil moisture required to produce needle ice, thus indicating the boundary between conditions leading to needle-ice growth (above the line) and to no needle-ice growth (below the line) (after Meentemeyer and Zippin, 1981, Figure 3)

(1981, p. 118) found that soils with less than 7.9% fines do not produce needle ice even if saturated. All paramo samples below this value (Figure 6.11) are from troughs. These soils will not support needle-ice growth even during the rainy season, when slopes near sites 3 and 4 reach moisture levels up to 33.5% at the surface (Pérez, 1987b).

Soil water content of stripes is normally too low during the dry season to allow needle-ice formation, but field data show that ridge soils are even then more moist than trough soils. On December 18, 1985, the ridges of PS stripes at site 2 had an average moisture of 6.9 ± 1.46%, while troughs had a value of 0.92 ± 0.46%. Similar water contents were found at site 4 on December 30, 1987, when ridges had 6.34 ± 1.68% moisture, but troughs had only 1.25 ± 0.81%. In both cases, a Mann–Whitney *U* test indicated that the differences were significant ($P < 0.005$ in 1985, $P < 0.001$ in 1987). This result supports the idea that ridges remain moist, and thus allow needle-ice growth for longer periods of time than do troughs. Field observations also indicate that stripe troughs remain largely free of needle ice. Needle ice causes significant seedling disturbance and mortality, and paramo plants display different germination strategies to avoid it (Pérez, 1987c). *Draba chionophila* (rosette), the vascular plant found at highest altitudes in the Venezuelan Andes (Azócar, Rada, and Goldstein, 1988), is common on stripe sites, where it is always rooted in trough pebbles (Figure 6.12). Excavation of rosettes shows extensive root systems (40 to 50 cm long) extending upslope

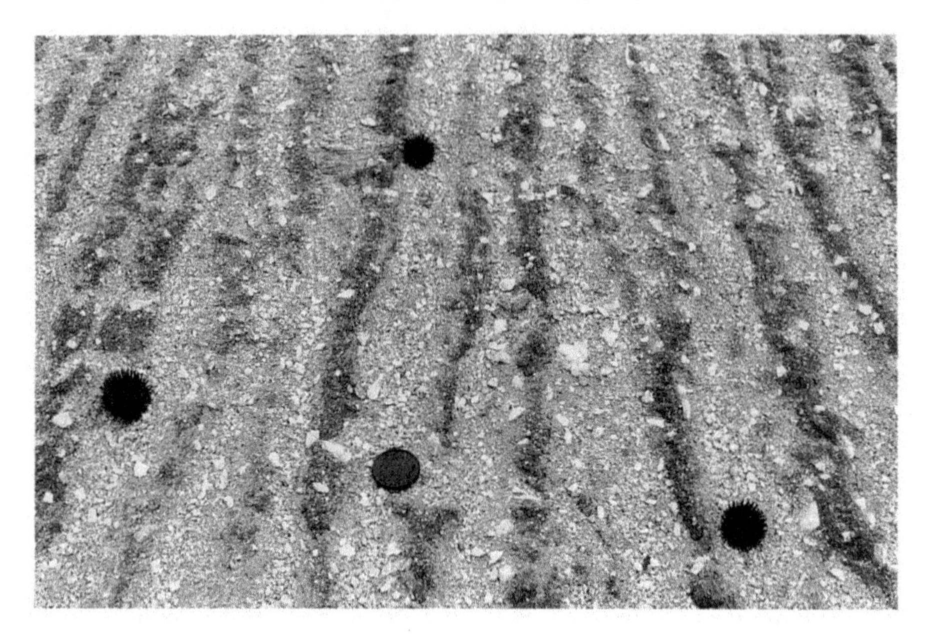

Figure 6.12. Pebble-sorted stripes at 4485 m. The three small rosettes of *Draba chionophila* are rooted in the coarse-pebble troughs. The disk is 55 mm in diameter. Photo taken on January 3, 1990

(an indicator of pronounced soil creep, Kershaw and Gardner, 1986) but commonly concentrated beneath finer ridges. This finding suggests that *Draba* avoids needle-ice heaving at the surface by germinating in troughs but exploits the greater and longer-lasting moisture supply of fine-grained ridges. Numerous observations elsewhere (Troll, 1958, p. 49; Lliboutry, 1961, Figure 3; Hastenrath, 1973, p. 175; Heilbronn and Walton, 1984a, p. 24; Noguchi, Tabuchi and Hasegawa, 1987, p. 339; Francou, 1988, Photo 9) also indicate that needle-ice growth is restricted or lacking along troughs of miniature sorted stripes.

Final remarks

The spatial distribution of miniature stripes in the Venezuelan Andes is closely dependent on slope gradient and parent material. These parameters define whether pebble-sorted or cobble-sorted stripes develop at a particular site. The correspondence of slope direction with both stripe and clast orientation indicates that the patterns are significantly affected by the downslope shift of particles. The basic mechanism proposed for the development of miniature stripes in the Andean paramo is similar to that outlined by other authors (Lliboutry, 1961; Pissart, 1973, 1982; Price, 1981). Lateral spacing of miniature stripes could presumably be set initially by the formation of downslope-elongated convection cells within the soil (Ray, Krantz, Caine, and Gunn, 1983). These would provide a regular pattern of parallel bands with contrasting thermal characteristics and eventually result in the generation of alternating fine- and coarse-grained bands. Coarse particles within a shallow superficial soil layer migrate first to the ground surface, then are laterally sorted into troughs and ridges. This dual sorting is primarily effected by needle-ice growth. Fine-soil bands attain higher water content and remain moist longer than adjacent coarse strips. Thus they are more prone to needle-ice formation. As fine strips are frequently heaved, the remaining pebbles within them are laterally ejected (Pissart, 1972). Pebbles creep and roll downwards into adjacent troughs, where they collect. Trough soils are too coarse to allow frequent needle-ice formation. Therefore pebbles and cobbles remain in them. Because stripe ridges result at several sites from the gradual linear coalescence of fine soil buds, this general sequence of development is probably applicable also to the formation of soil buds.

Paramo stripes are repeatedly affected by freezing. Frost penetration is not a factor limiting pattern depth. Instead, depth of stripe sorting corresponds closely with a shallow soil layer about 5 cm thick, where needle-ice disturbance is most effective (Jahn and Cielinska, 1974; Price, 1981). This is clearly supported by the experimental data of tracer movement and sorting: small particles at the surface are rapidly transported by needle ice, but larger and deeper grains are disturbed only infrequently. Particles on steep slopes become sorted into stripes faster than on gentle slopes because grain transport by needle ice increases rapidly with gradient. However, the role of soil moisture on stripe formation should be more thoroughly investigated in the future, as studies elsewhere (Francou, 1988, p. 512) indicate that sorted stripes on dry sites may not grow back for several years after being experimentally destroyed. Lack of moisture may have restricted marker movement at site 2, which was extremely dry on all occasions it was visited.

In addition to needle ice, other geomorphic agents may be involved in stripe genesis. Similar sorted stripes in the Peruvian Andes are affected by creep induced by the growth and thawing of granular ice (*cryoreptation*) (Francou, 1988; Van Vliet Lanoë and Francou, 1988). Although this type of ice was not investigated in the paramo stripes, granular ice crystals have been observed at Piedras Blancas (Pérez, 1986a, 1987c) and probably play a role in stripe movement. However, this role is almost certainly secondary to that of needle ice, as soil displacement by granular ice in the puna is an order of magnitude lower than that by needle ice (Van Vliet Lanoë and Francou, 1988, p. 59). Several authors (Price, 1981; Francou, 1984, 1988; Kelletat, 1985; Selby, 1985) have commented on the probable role of surface wash in the formation of "periglacial" stripes, but have also mentioned the difficulties in evaluating this process without directly witnessing it (Kelletat, 1985). No evidence for runoff was found at the stripe sites, but coarse-grained troughs of larger patterns in this paramo seem to be subjected to a moderate wash of fines (Pérez, 1991). Therefore the role of runoff on miniature-stripe formation should not be discounted.

Miniature sorted stripes are one of the most common features of the periglacial paramo, which extends at present between ca. 3600 and 4700 m elevation and occupies an area in excess of 1200 km^2 of the Venezuelan Andes (Schubert, 1979a,b). In the past, miniature stripes must have been much more abundant there. During the maximum of the glacial Late Pleistocene advance (Mérida Glaciation), the periglacial zone descended 1200 m to about 2400 m elevation in the southeastern, and to 2900 m in the northwestern, parts of the Venezuelan Andes (Schubert, 1975, 1980). At that time the periglacial zone covered an estimated area of 2200 km^2 (Schubert, 1979b). More recently, much of Piedras Blancas has experienced a periglacial-desert climate. During the Little Ice Age, from about 1290AD to the late nineteenth century, the region experienced a temperature drop of more than 2° C, which would have lowered the elevation of climatic belts by at least 200 m (Rull, Salgado-Labouriau, Schubert, and Valastro, 1987). Greater frost intensity during this period resulted in the formation of small talus-derived rock glaciers, now inactive, at elevations above ca. 4400 m (Pérez, 1988b; C. Schubert, personal communication, 1989). Thus, miniature sorted stripes should also have developed during the Little Ice Age over an area significantly larger than at present.

The formation of miniature sorted stripes has some important implications for soil development. The gradual heave of clasts towards the ground surface and their concentration there, followed by segregation into alternating bands of coarse and fine particles, would produce a distinct soil profile, even if the original parent material were perfectly homogeneous. Thus needle ice and frost heave and creep should be considered important agents of proanisotropic pedoturbation in the high paramo (Johnson, Watson-Stegner, Johnson and, Schaetzl, 1987), as they allow the formation of a shallow "stone maximum" at the soil surface (Fitzpatrick, 1987).

Sorted-stripe development may also influence plant distribution in the periglacial paramo. Few studies have examined vegetation patterns in relation to sorted stripes. Heilbronn and Walton (1984b) found that different plant species colonized coarse and fine bands. They also noted that some life-forms, such as small rosettes with a thick root ball or taproot, were more successful colonizers of frost-heaved areas. Their success is due to efficient ground insulation beneath the compact rosettes, which reduces frost heaving (Heilbronn and Walton, 1984b; Gradwell, 1955). *Draba chionophila* fits well the description above (Pittier, 1926); its morphological characteristics may help to explain the ability of this plant to occupy sorted stripes. Intense needle-ice activity provides a basic habitat for the small frost-resistant *Draba* rosettes (Azócar, Rada, and Goldstein, 1988), which are able to invade actively forming microstripes. In contrast, giant Andean rosettes are unable to survive frost heave on this substrate unless anchored by large rocks (Pérez, 1987c). In colonizing frost-disturbed stripes, *Draba* presumably escapes competition from other paramo plants; this germination strategy restricts it essentially to sorted microstripes (Pfitsch, 1988).

More work remains to be done on miniature paramo stripes. Further research should focus on the initial stages of pattern genesis and on the length of time needed for complete stripe regeneration after thorough raking and mixing of all soil material above the C horizon. Specific points to be investigated include the extent of vertical particle sorting over longer time periods. To this end, many subsurface markers remain in the paramo, and will be inspected during future visits.

Acknowledgments

I sincerely thank C. Schubert (IVIC, Caracas) for kindly sharing ideas, photographs, and unpublished data on particle movement and for critically reviewing the initial version of this chapter. My field assistants, H. Hoenicka B., A. Lucchetti, J. M. Pérez, and O. Vera, provided invaluable help, often under difficult conditions. J. L. Pérez identified and described the rocks. B. Francou (Caen), S. Hastenrath (Wisconsin), and A. Pissart (Liège) kindly provided several relevant publications. Funds were contributed by the Research Institute, University of Texas, Austin. My father, F. Pérez Conca, and E. F. Pérez helped with travel logistics. I. L. Bergquist (Texas), D. L. Johnson (Illinois), V. G. Meentemeyer (Georgia), and two anonymous reviewers contributed many valuable suggestions and greatly improved on an earlier manuscript draft. My appreciation goes to all.

References

Azócar, A., F. Rada, and G. Goldstein, Freezing tolerance, in *Draba chionophila*, a "miniature" caulescent rosette species. *Oecologia (Berlin)*, **75**, 156–160, 1988.

Ballantyne, C. K. and J. A. Matthews, Desiccation cracking and sorted polygon development, Jotunheimen, Norway, *Arctic and Alpine Research*, **15**, 339–349, 1983.

Ballard, T. N., Soil physical properties in a sorted stripe field, *Arctic and Alpine Research*, **5**, 127–131, 1973.

Caine, T. N., The log-normal distribution and rates of soil movement: An example, *Revue de Géomorphologie Dynamique*, **18**, 1–7, 1968.

Caine, T. N., A source of bias in rates of surface soil movement as estimated from marked particles, *Earth Surface Processes and Landforms*, **6**, 69–75, 1981.

Curray, J. R., The analysis of two-dimensional orientation data, *Journal of Geology*, **64**, 117–131, 1956.

Doornkamp, J. C., and C. A. M. King, *Numerical Analysis in Geomorphology*, 372 pp., St. Martin's Press, New York, 1971.

Fitzpatrick, E. A., Periglacial features in the soils of north east Scotland, in *Periglacial Processes and Landforms in Britain and Ireland*, edited by J. Boardman, pp. 153–162, Cambridge University Press, Cambridge, 1987.

Francou, B., Données preliminaires pour l'étude des processus périglaciaires dans les hautes Andes du Perou, *Revue de Géomorphologie Dynamique*, **33**, 113–126, 1984.

Francou, B., Dynamiques périglaciaires et Quaternaire dans les Andes centrales, *Rapports Scientifiques et Techniques, Centre de Géomorphologie, C.N.R.S., Caen*, *2*, 63 pp., 1986.

Francou, B., *L'Éboulisation en Haute Montagne*, 696 pp, Centre de Géomorphologie, C.N.R.S., Caen, 1988.

Francou, B., L. Federmeyer, L. Kuang-Yi, and C. Yaning, Explorations Géomorphologiques dans l'Ouest des Kun Lun (Xinjiang, Chine), *Rapports Scientifiques et Techniques, Centre de Géomorphologie, C.N.R.S., Caen*, *5*, 91 pp., 1990.

Furrer, G., and R. Freund, Beobachtungen zum subnivalen Formenschatz am Kilimanjaro. *Zeitschrift für Geomorphologie, Supplement Band*, **16**, 180–203, 1973.

Gradwell, M. W., Soil frost studies at a high country station—II, *New Zealand Journal of Science and Technology*, **37**, 267–275, 1955.

Graf, K., Die Gesteinsabhängigkeit von Solifluktionsformen in der Ostschweiz und in den Anden Perus und Boliviens, *Geographica Helvetica*, **3**, 160–162, 1971.

Graf, K., Vergleichende Betrachtungen zur Solifluktion in verschiedenen Breitenlagen, *Zeitschrift für Geomorphologie, Supplement Band*, **16**, 104–154, 1973.

Graf, K., Zur Mechanik von Frostmusterungsprozessen in Bolivien und Ecuador, *Zeitschrift für Geomorphologie*, **20**, 417–447, 1976.

Hall, K., Sorted stripes oriented by wind action: some observations from Sub-Antarctic Marion Islands, *Earth Surface Processes*, **4**, 281–289, 1979.

Hall, K., Sorted stripes on Sub-Antarctic Kerguelen Island, *Earth Surface Processes and Landforms*, **8**, 115–124, 1983.

Hastenrath, S., Observations on the periglacial morphology of Mts Kenya and Kilimanjaro, East Africa, *Zeitschrift für Geomorphologie, Supplement Band*, **16**, 161–179, 1973.

Hastenrath, S., Glaziale und periglaziale Formbildung in Hoch-Semyen, Nord-Äthiopien, *Erdkunde*, **28**, 176–186, 1974.

Hastenrath, S., Observations on soil frost phenomena in the Peruvian Andes, *Zeitschrift für Geomorphologie*, **21**, 357–362, 1977.

Hastenrath, S., On the three-dimensional distribution of subnival soil patterns in the high mountains of East Africa, *Erdwissenschaftliche Forschung*, **11**, 458–481, 1978.

Heilbronn, T. D., and D. W. H. Walton, The morphology of some periglacial features on South Georgia and their relationship to the environment, *British Antarctic Survey Bulletin*, **64**, 21–36, 1984a.

Heilbronn, T. D., and D. W. H. Walton, Plant colonization of actively sorted stone stripes in the Subantarctic, *Arctic and Alpine Research*, **16**, 161–172, 1984b.

Heine, K., Zur morphologischen Bedeutung des Kammeises in der subnivalen Zone randtropischer semihumider Hochgebirge, *Zeitschrift für Geomorphologie*, **21**, 57–78, 1977.

Jahn, A., and M. Cielinska, The rate of soil movement in the Sudety Mountains, *Abhandlungen Akademie Wissenschaften Göttingen, Mathematisch-Physikalische Klasse*, **29**, 86–101, 1974.

Johnson, D. L., D. Watson-Stegner, D. N. Johnson, and R. J. Schaetzl, Proisotropic and proanisotropic processes of pedoturbation, *Soil Science*, **143**, 278–292, 1987.

Jones, T. A., Skewness and kurtosis as criteria of normality in observed frequency distributions, *Journal of Sedimentary Petrology*, **39**, 1622–1627, 1969.
Kelletat, D., Patterned ground by rainstorm erosion on the Colorado Plateau, Utah, *Catena*, **12**, 255–259, 1985.
Kershaw, L. J., and J. S. Gardner, Vascular plants of mountain talus slopes, Mt. Rae area, Alberta, Canada, *Physical Geography*, **7**, 218–230, 1986.
Krumbein, W. C., Measurement and geological significance of shape and roundness of sedimentary particles, *Journal of Sedimentary Petrology*, **11**, 64–72, 1941.
Lee, W. G., and A. E. Hewitt, Soil changes associated with development of vegetation on an ultramafic scree, northwest Otago, New Zealand, *Journal of the Royal Society of New Zealand*, **12**, 229–242, 1982.
Lliboutry, L., L'origine des sols striés et polygonaux des Andes de Santiago (Chili), *Comptes Rendus des Séances, Académie des Sciences, Paris*, **240**, 1793–1794, 1955.
Lliboutry, L., Phénomènes cryonivaux dans les Andes de Santiago (Chili), *Biuletyn Peryglacjalny*, **10**, 209–224, 1961.
Löffler, E., Beobachtungen zur periglazialen Höhenstufe in den Hochgebirgen von Papua New Guinea, *Erdkunde*, **29**, 285–292, 1975.
Mackay, J. R., and W. H. Mathews, Movement of sorted stripes, the Cinder Cone, Garibaldi Park, B. C., Canada, *Arctic and Alpine Research*, **6**, 347–359, 1974.
Meentemeyer, V., and J. Zippin, Soil moisture and texture controls of selected parameters of needle ice growth, *Earth Surface Processes and Landforms*, **6**, 113–125, 1981.
Monasterio, M., Las formaciones vegetales de los páramos de Venezuela, in *Estudios Ecológicos en los Páramos Andinos*, edited by M. Monasterio, pp. 93–158, Universidad de Los Andes, Mérida, 1980.
Noguchi, Y., H. Tabuchi, and H. Hasegawa, Physical factors controlling the formation of patterned ground on Haleakala, Maui, *Geografiska Annaler*, **69A**, 329–342, 1987.
Pérez, F. L., Striated soil in an Andean paramo of Venezuela: its origin and orientation, *Arctic and Alpine Research*, **16**, 277–289, 1984.
Pérez, F. L., Surficial talus movement in an Andean paramo of Venezuela, *Geografiska Annaler*, **67A**, 221–237, 1985.
Pérez, F. L., The effect of compaction on soil disturbance by needle ice growth, *Acta Geocriogénica*, **4**, 111–119, 1986a.
Pérez, F. L., Talus texture and particle morphology in a North Andean paramo, *Zeitschrift für Geomorphologie*, **30**, 15–34, 1986b.
Pérez, F. L., Downslope stone transport by needle ice in a high Andean area (Venezuela), *Revue de Géomorphologie Dynamique*, **36**, 33–51, 1987a.
Pérez, F. L., Soil moisture and the upper altitudinal limit of giant paramo rosettes, *Journal of Biogeography*, **14**, 173–186, 1987b.
Pérez, F. L., Needle-ice activity and the distribution of stem-rosette species in a Venezuelan paramo, *Arctic and Alpine Research*, **19**, 135–153, 1987c.
Pérez, F. L., Soil surface roughness and needle ice-induced particle movement in a Venezuelan paramo, *Caribbean Journal of Science*, **23**, 454–460, 1987d.
Pérez, F. L., A method for contouring triangular particle shape diagrams, *Journal of Sedimentary Petrology*, **57**, 763–765, 1987e.
Pérez, F. L., The movement of debris on a high Andean talus, *Zeitschrift für Geomorphologie*, **32**, 77–99, 1988a.
Pérez, F. L., A talus rock glacier in the Venezuelan Andes, *Zeitschrift für Gletscherkunde und Glazialgeologie*, **24**, 149–159, 1988b.
Pérez, F. L., Some effects of giant Andean stem-rosettes on ground microclimate, and

their ecological significance, *International Journal of Biometeorology*, **33**, 131–135, 1989.

Pérez, F. L., Particle sorting due to off-road vehicle traffic in a high Andean paramo, *Catena*, **18**, 239–254, 1991.

Pfitsch, W. A., Microenvironment and the distribution of two species of *Draba* (Brassicaceae) in a Venezuelan paramo, *Arctic and Alpine Research*, **20**, 333–341, 1988.

Pissart, A., Vitesse des mouvements de pierres dans des sols et sur des versants périglaciaires au Chambeyron (Basses Alpes), *Congrés et Colloques de L'Université de Liège*, **67**, 251–268, 1972.

Pissart, A., L'origine des sols polygonaux et striés du Chambeyron (Basses-Alpes), *Bulletin de la Societé Géographique de Liège*, **9**, 33–53, 1973.

Pissart, A., Experiences de terrain et de laboratoire pour expliquer la genèse de sols polygonaux decimetriques triés, *Acta Geomorphologica Carpatho-Balcanica*, **15**, 39–47, 1982.

Pittier, H., *Manual de las Plantas Usuales de Venezuela*, 458 pp., L. Comercio, Caracas, 1926.

Pitty, A. F., *Geography and Soil Properties*, 287 pp., Methuen, London, 1979.

Price, L. W., The periglacial environment, permafrost, and man, *Resource Papers, Association of American Geographers*, **14**, 88 pp., 1972.

Price, L. W., *Mountains and Man: A Study of Process and Environment*, 506 pp. University of California Press, Berkeley, 1981.

Ray, R. J., W. B. Krantz, T. N. Caine, and R. D. Gunn, A model for patterned ground regularity, *Journal of Glaciology*, **29**, 317–337, 1983.

Rull, V., M. L. Salgado-Labouriau, C. Schubert, and S. Valastro, Late Holocene temperature depression in the Venezuelan Andes: Palynological evidence, *Palaeogeography, Palaeoclimate, Palaeoecology*, **60**, 109–121, 1987.

Schubert, C., Striated ground in the Venezuelan Andes, *Journal of Glaciology*, **12**, 461–468, 1973.

Schubert, C., Glaciation and periglacial morphology in the northwestern Venezuelan Andes, *Eiszeitalter und Gegenwart*, **26**, 196–211, 1975.

Schubert, C., La zona del páramo: morfología glacial y periglacial de los Andes de Venezuela, in *El Medio Ambiente Páramo*, edited by M. L. Salgado-Labouriau, pp. 11–27, Centro de Estudios Avanzados, Caracas, 1979a.

Schubert, C., Glacial sediments in the Venezuelan Andes, in *Moraines and Varves. Origin, Genesis, Classification*, edited by C. Schlüchter, pp. 43–49, Balkema, Rotterdam, 1979b.

Schubert, C., Aspectos geológicos de los Andes Venezolanos: historia, breve síntesis, el Cuaternario y bibliografia, in *Estudios Ecológicos en los Páramos Andinos*, edited by M. Monasterio, pp. 29–46. Universidad de Los Andes, Mérida, 1980.

Selby, M. J., *Earth's Changing Surface*, 607 pp., Clarendon Press, Oxford, 1985.

Smith, R. T., and K. Atkinson, *Techniques in Pedology. A Handbook for Environmental and Resource Studies*, 213 pp., Elek Science, London, 1975.

Sneed, E. D., and R. L. Folk, Pebbles in the lower Colorado River, Texas: A study in particle morphogenesis, *Journal of Geology*, **66**, 114–150, 1958.

Soil Survey Staff, *Soil Taxonomy*, 754 pp., USDA, Soil Conservation Service, Agriculture Handbook 436, 1975.

Troll, K., Structure soils, solifluction, and frost climates of the Earth, *U.S. Army Snow, Ice, and Permafrost Research Establishment Translation*, **43**, 121 pp., 1958.

Van Vliet Lanoë, B., and B. Francou, Etude micromorphologique et dynamique comparative de sols stries et autres petites formes fluantes superficielles en milieu

Arctique, Alpin et Andin, *Bulletin Centre de Géomorphologie du C.N.R.S., Caen*, **34**, 47–63, 1988.
Washburn, A. L., Classification of patterned ground and review of suggested origins, *Geological Society of America Bulletin*, **67**, 823–865, 1956.
Washburn, A. L., *Geocryology: A Survey of Periglacial Processes and Environments*, 406 pp., John Wiley, New York, 1980.
Wentworth, C. K., A scale of grade and class terms for clastic sediments, *Journal of Geology*, **30**, 377–392, 1922.
Zeuner, F. E., Frost soils on Mount Kenya, and the relation of frost soils to aeolian deposits, *Journal of Soil Science*, **1**, 20–30, 1949.

[illegible] *J.* [illegible]
[illegible] *Sediment. Petrol.* [illegible]
[illegible] *Sediments* [illegible]
[illegible] John Wiley, New York [illegible]
Wentworth, C. K.: A scale of grade and class terms for cl[illegible] *J. Geology*, 30, [illegible]
[illegible] *Journal of Sedim*[illegible]

7 A Model of Water Movement in Rock Glaciers and Associated Water Characteristics

John R. Giardino
Departments of Geography & Geology, Texas A & M University

John D. Vitek
Graduate College & School of Geology, Oklahoma State University

Joseph L. DeMorett
Ocean Drilling Program, Texas A & M University

Abstract

A rock glacier physically and chemically influences the water which passes through it and acts as a concentrating rather than a filtering mechanism. A model for the hydrological system of rock glaciers includes direct precipitation, runoff from adjacent slopes, ice and snow from avalanching, groundwater, and initial glacial and/or periglacial ice. The primary outputs of the hydrological system are surface runoff, subsurface discharge, subsurface seepage, sublimation, and evaporation. Four subsystems, cliff and talus, surface, subsurface, and groundwater, constitute the cascading system. The model reveals many intricate pathways through a rock glacier and the necessity for careful assessment of observed differences between the quality of water input and output.

Because the nature of the internal characteristics of rock glaciers must be acquired primarily from external data, water exiting rock glacier systems was analyzed. It was noted that exiting water had higher ion concentrations than entering forms; water input had total dissolved loads ranging from 1.5 to $6.0\,mg\,l^{-1}$, whereas outputs had total dissolved loads ranging from 8.0 to $54.0\,mg\,l^{-1}$; silica concentrations ranged from $0–0.1\,mg\,l^{-1}$ in inputs to $1.9–6.0\,mg\,l^{-1}$ in outputs; alkalinity values for inputs ranged from 1.3 to $2.7\,mg\,l^{-1}$ $CaCO_3$, whereas outputs ranged from 18.1 to $44.4\,mg\,l^{-1}$ $CaCO_3$;

Periglacial Geomorphology. Edited by J. C. Dixon and A. D. Abrahams

and total hardness values for inputs were 0.0 mg l^{-1} $CaCO_3$, whereas outputs ranged between 21.0 and 50.1 mg l^{-1} $CaCO_3$.

Water flowing through three rock glaciers decreased in *p*H. Inputs into the rock glaciers had *p*H values ranging between 6.4 and 7.0, whereas outputs had *p*H values ranging between 7.2 and 8.5. Peak discharge of the three rock glaciers occurred between 0900 and 1200 hours daily, before daily maximum temperature occurred.

Introduction

Streams flowing from rock glaciers indicate that these features operate as aquifers. Because rock glaciers consist of a matrix of angular rocks, fine sediments, and ice, a more or less porous medium exists. This internal makeup of a rock glacier is conducive to the formation and growth of ice and to the entrapment of water in a liquid state. Ice in rock glaciers, either interstitially or as discrete lenses, provides one source of water for stream flow. A large number of alpine rock glaciers have streams flowing from frontal slopes.

Within some alpine watersheds, rock glaciers have a major impact on the hydrology of the streams. Although rock glaciers have been studied for over a hundred years (Steenstrup, 1883), only recently has attention been directed toward their hydrological significance (Corte, 1976a,b; Johnson, 1981; Evin, 1983, 1984; Haeberli, 1985; Gardner and Bajewsky, 1987; Bajewsky and Gardner, 1989; Muhll and Haeberli, 1990). Attributes of the drainage basin, including the presence of rock glaciers, influence water quality to varying degrees. Johnson (1981) suggested that understanding the hydrology of rock glaciers could lead to a better explanation of their internal characteristics.

Viewing the hydrological characteristics of a rock glacier as a cascading system provides a model from which to evaluate how water moves through the system and changes in quality. Accurate depiction of a hydrologic model may help answer the following questions: How do rock glaciers influence the flow to and the chemistry of alpine streams? What relationships exist between water and ice within rock glaciers? What is the residence time for water within a rock glacier? How does the chemistry of the water change as it passes through a rock glacier? As with many areas of science, numerous questions exist and few definitive answers.

Study objectives

In order to provide some answers to the previous questions, our study had three principal objectives: (1) to identify input, output, storage, and flow-pathways of the hydrological system for rock glaciers; (2) to examine the chemistry of precipitation entering, water and ice within, and water exiting rock glaciers; and (3) to determine the extent to which rock glaciers influence the water chemistry of alpine streams.

Background studies

Numerous observations of water interacting with rock glaciers have been reported. Although Capps (1910) and Tyrrell (1910) observed water in and near rock glaciers, such water did not become the focus of research until 15 years ago. Others noted the impact of water on rock glaciers, including Potter (1969) who described a series of melt-water channels (ca. 0.5 m wide and 34 m deep) that were eroded into ice beneath the debris cover. White (1971) observed water flowing over an ice core approximately 5 m beneath the surface of the Arapaho Rock Glacier.

Brown (1925) observed seasonal ponds in surface depressions of a rock glacier in the Hurricane Basin in the San Juan Mountains. Wahrhaftig and Cox (1959) described rock glaciers having surface streams flowing and often disappearing into conical pits. Several of the rock glaciers in our study area display seasonal ponds (Figures 7.1 and 7.2). Johnson (1978) monitored several ponds on rock glaciers in the Grizzly Creek area of the Yukon and

Figure 7.1. An intermittent pond 8 m in length at approximately 3840 m elevation on the Beasley Peak Rock Glacier, Colorado. The photograph was taken on August 6, 1979

Figure 7.2. An intermittent pond 30 m in length at approximately 3780 m elevation on the Dutch Creek Rock Glacier, Colorado. The photograph was taken on July 1, 1987

concluded that the surface drainage of a rock glacier was regulated to some degree by internal drainage characteristics.

Discharge from rock glaciers typically varies from a few liters per minute to several hundred liters per minute (Domaradzki, 1951; Schweizer, 1968; Jackson and MacDonald, 1980; Haeberli, 1985; Bajewsky and Gardner, 1989). Giardino (1983) and Giardino and Vick (1987) hypothesized that internal water pressures occasionally become sufficient to expel water and rock with great force from the toe of a rock glacier. Although the input, storage, and output of water associated with rock glaciers have been reported, the pathway has not been modeled. Giardino (1979) first suggested that the hydrology of a rock glacier be viewed as a system.

Hydrological characteristics of rock glaciers were first suggested by Corte (1976a, 1976b). Johnson (1978) used dye in three different types of rock glaciers (i.e., moraine, talus-derived, and avalanche) to show that flow times were slow and variable. Johnson (1981) used water chemistry, dye flow times, and discharge volumes to infer internal characteristics of a rock glacier.

Evin and Assier (1983a,b) conducted chemical analyses of the waters flowing from rock glaciers in the southern French Alps. Discharge measurements and dye introductions indicated that maximum runoff occurred from 3 to 20 hours after the rock glacier received a large input of water. Evin (1984) conducted chemical analyses on water flowing from the frontal slopes of different types of rock glaciers (i.e., active ice-cemented, ablation complex, and fossil) and found distinct chemical signatures in the discharge of each type. Haeberli (1985) monitored runoff from the Gruben Rock Glacier in Switzerland and concluded that precipitation was the major influence on discharge.

Gardner and Bajewsky (1987) and Bajewsky and Gardner (1989) compared sediment transport characteristics of a stream issuing from the Hilda Rock Glacier with a stream flowing from the nearby Boundary Glacier in the Canadian Rockies. Although both streams had similar discharges, a comparison of the hydrological and chemical characteristics of the two streams showed marked differences: the rock glacier-fed stream had a more constant, but lower maximum discharge, a higher dissolved load, and a lower sediment load. They concluded that rock glaciers are hydrologically and geomorphologically conservative systems.

Hydrological model

Based upon the suggestions put forth by Haeberli (1985) and other observations of water movement relative to rock glaciers, a model of the flow of water through a rock glacier was constructed (Figure 7.3). The movement of water through a rock glacier can occur along two pathways: near subsurface flow (just below the surface) and a deep subsurface flow. The initiation and maintenance of two flow paths require an impermeable layer within the rock

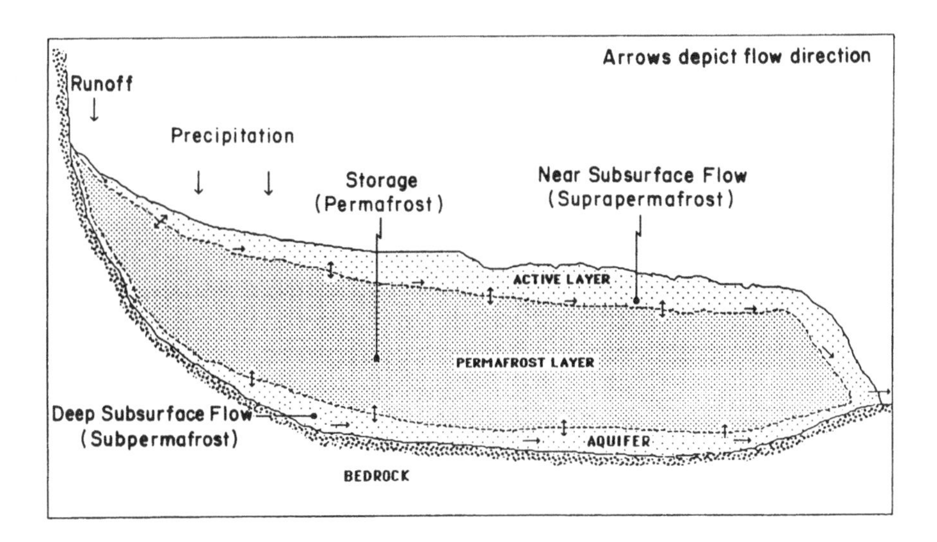

Figure 7.3. Diagram of water in a rock glacier (after Haeberli, 1985)

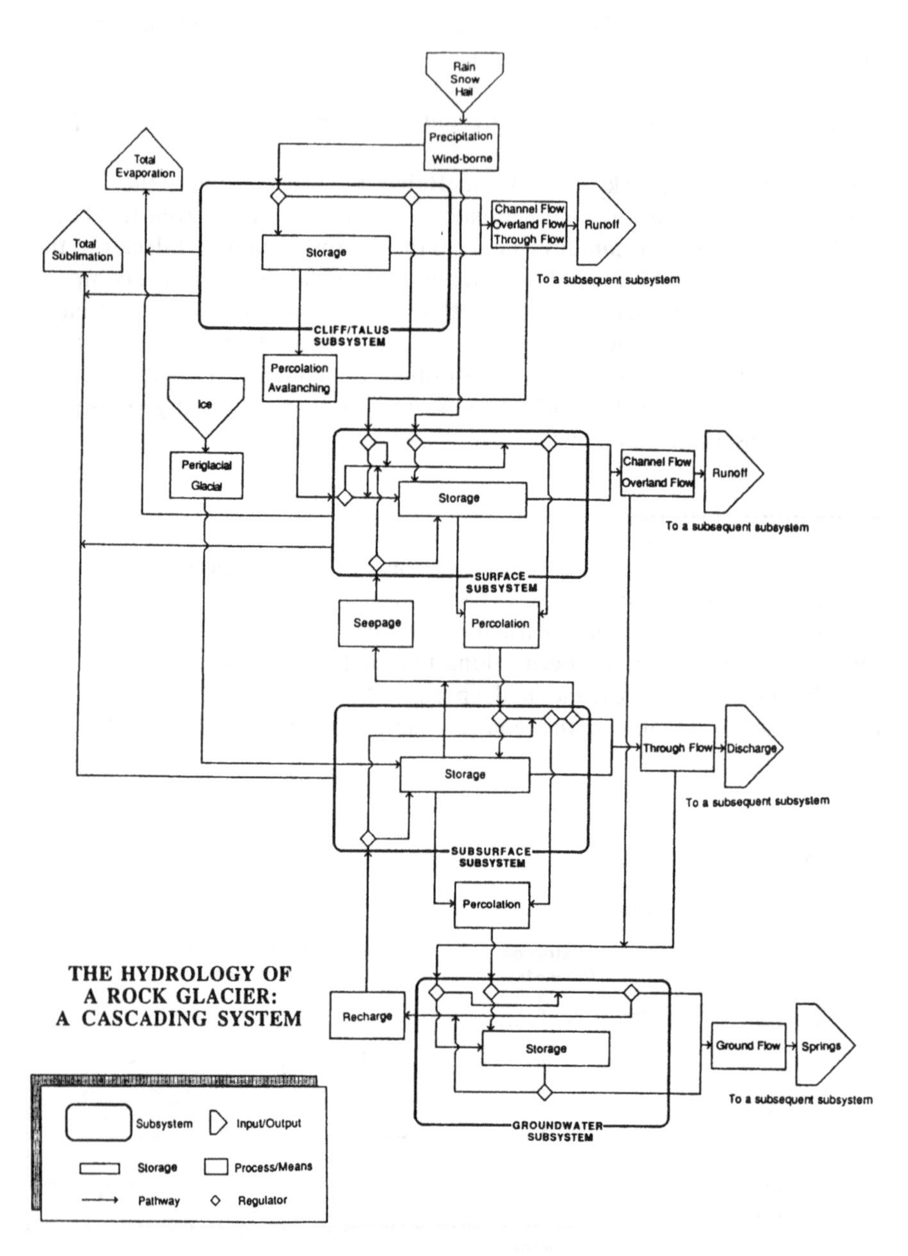

Figure 7.4. A cascading system model of water flow associated with a rock glacier

glacier. A frozen layer of ice-cemented clastics and/or ice lenses or an ice core probably separates the two pathways. The location of a frozen layer (i.e., permafrost) most likely increases and decreases in depth and vertical extent throughout the rock glacier as a function of the thickness of the rock debris over the ice (Moore, 1991), the amount of precipitation, and the internal temperature. The exposure of an internal longitudinal cross-section of a rock glacier 5 to 14 m thick as a result of an underground mine collapse revealed surficial rock debris and internal ice similar to the model (Figure 7.3) (Moore, 1991). This site could be used to test a variety of hypotheses relating to internal characteristics.

To illustrate the hydrological characteristics of a rock glacier in an alpine setting, a cascading system was utilized to define the manner in which water passed through a rock glacier (Figure 7.4). The system consists of (1) the cliff-talus subsystem, (2) the surface subsystem, (3) the subsurface subsystem, and (4) the groundwater subsystem. Water follows many complex paths through the system because the output of water from one system serves as the input of water to another subsystem. The amount of water and the rate at which it moves through the rock glacier is a function of many external and internal characteristics (e.g., input, storage, output, and pathway).

The primary inputs are direct precipitation, runoff from adjacent slopes, ice and snow from avalanching, "groundwater", and in situ glacial and/or periglacial ice. The primary outputs of a rock glacier hydrological system are surface runoff, subsurface discharge, subsurface seepage, sublimation, and evaporation. Water in liquid and solid form is capable of being stored on and below the surface of a rock glacier for extended periods of time.

Study area

To ascertain whether water quality could be used to interpret flow paths through rock glaciers, three rock glaciers were selected for study in the Blanca Massif area of the Sangre de Cristo Mountain Range of south-central Colorado, U.S.A. (Figure 7.5). Several researchers have worked in the area and provided basic characteristics of many of the rock glaciers (Parson, 1987; Morris, 1987). Olyphant (1983) assessed factors of talus formation in the region, conditions which also influence debris supply to rock glaciers. In considering extrinsic influences on the characteristics of water discharging from the rock glaciers, only geology and climate were evaluated.

Extrinsic variables

GEOLOGY

The Blanca Massif is primarily Precambrian granites, hornblende gneiss, and quartzites. Paleozoic sedimentary units, sandstones, conglomerates, limestones, and shales are present along the western and eastern sides of the Massif, and Tertiary igneous dikes and sills occur throughout. Quaternary deposits (i.e., moraines, landslides, talus cones, alluvial fans, alluvium, and rock glaciers) cover a significant portion of the bedrock and influence local processes (McCulloch, 1955; Olyphant, 1983; Parson, 1987).

CLIMATE

The climate of the Sangre de Cristo Mountains is controlled by its location in the central part of the continent and its considerable topographic relief (Reifsnyder, 1980). Summers are cool, with much of the precipitation occurring as rain during thunderstorms. Winters are cold. Unfortunately, continuous records of precipitation and temperature do not exist for the alpine region. Records from nearby Colorado stations (e.g., Alamosa, Great Sand Dunes National Monument, Red Wing, San Luis, Walsenburg, and Westcliffe) are available. These records (annual temperature, 5.1 to 10.8°C; annual precipitation, 18.11 to 41.89 cm yr^{-1}; and annual snowfall, 91 to 236 cm yr^{-1}: National Oceanic and Atmospheric Administration, 1986) do not accurately represent conditions 1000 m higher. Vitek (1973) used these regional weather station data to estimate total precipitation of the Massif area at ca. 50 cm yr^{-1}, most of which falls as snow. In general, climatic conditions on the Massif are expected to be colder and wetter than at lower stations.

Rock glaciers

Three of the 44 rock glaciers mapped within the Massif by Parson (1980, 1987) were selected for this study (Figure 7.5) based on the presence of flowing streams with measurable discharge. Table 7.1 identifies the attributes of the rock glaciers in this study. The California Rock Glacier (Figure 7.6), has a large, classic tongue-shaped form. The surface displays a typical series of ridges and furrows perpendicular to the direction of flow. Furrow faces are relatively free of lichen because of snow kill (Carroll, 1974) or surface disruption caused by internal flow mechanics. Morris (1987) used relative dating techniques to suggest an early Neoglacial age for this rock glacier.

The Lost Lake Rock Glacier (Figure 7.6) is a small, lobate-shaped form. Talus from north-facing and, to a lesser extent, northeast-facing walls of the cirque feed this rock glacier. Surficial lichen cover is not extensive, which suggests that the Lost Lake Rock Glacier could be a younger form, a more active form, and/or covered by snow for longer periods of time than the California Rock Glacier. The age of the Lost Lake Rock Glacier is early Neoglacial (Morris, 1987).

The Dutch Creek Rock Glacier is situated in a northeast-facing cirque between Amelung Peak and an unnamed peak (Figure 7.7). A series of lobes

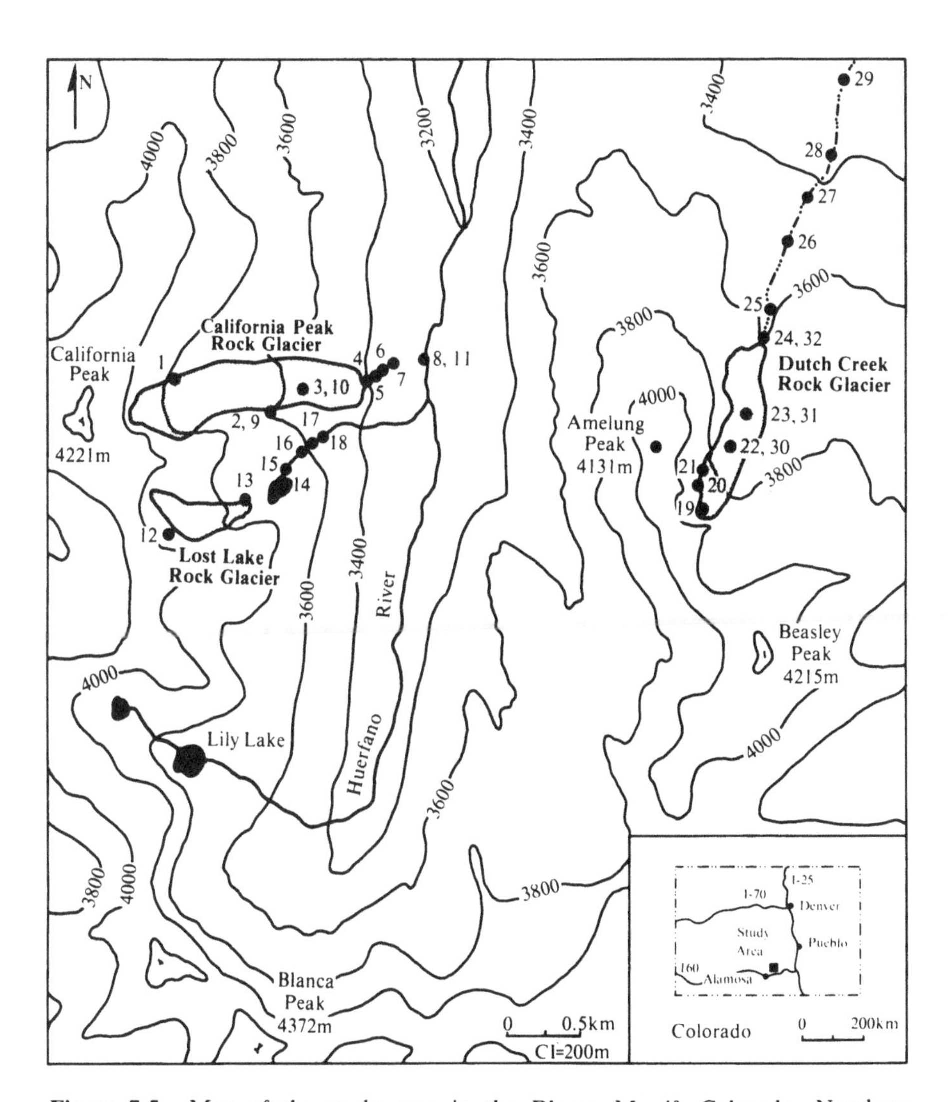

Figure 7.5. Map of the study area in the Blanca Massif, Colorado. Numbers represent points at which water samples were collected. Two numbers at one point indicate that a sample was collected at a low and a high discharge event. Water samples numbered 5–8, 11, 14–18, and 25–28 were collected in the streams that exited the rock glaciers. All other samples are rock glacier samples

trending in a northeasterly direction exhibit ridges, furrows, and several depressions or conical-shaped pits. The depressions, occurring primarily along the eastern flank, result from ablation and/or complex internal mechanics. One conical-shaped depression, ca. 30 m across, near the northeastern flank of the rock glacier, contains water for extended periods of time during

Table 7.1. Rock Glacier Attributes

Rock Glacier	Elevation (m)	Area (km^2)	Length (km^2)	Width (m)	Front Angle (degrees)
California	3775–3475	0.29	1.1	340–275	37
Lost Lake	3880–3790	0.06	0.137	550	40
Dutch Creek	3840–3560	0.54	0.433	1480	38

Figure 7.6. Aerial photograph of the California Peak Rock Glacier (right) and the Lost Lake Rock Glacier (left) in the Blanca Massif, Colorado

Figure 7.7. The Dutch Creek Rock Glacier occupies the upper one-third of this aerial view

the summer (Figure 7.2). Of the three rock glaciers studied, the greatest water discharge emerges from the Dutch Creek Rock Glacier.

Methodology

Field data, including surface morphometry, were collected during the summer and fall months of 1987 and 1988. Longitudinal profiles of the rock glaciers were mapped using a transit and stadia rod (Gardiner and Dackombe, 1982). The mineralogy of rock samples representing the major lithologies found on the rock glaciers was determined in the laboratory in an effort to understand water quality.

Water samples were collected in one-liter polyethylene bottles and examined for chemical and physical characteristics. Colorimetric and titrametric methods were used to determine total dissolved solids, silica, total hardness, alkalinity, calcium, magnesium, iron, *p*H, sulfate, dissolved oxygen, chloride, copper, nitrate, and manganese levels of the water samples.

WATER MOVEMENT: FIELD OBSERVATIONS

Field evidence was sought for water input, output, storage, and flow-through. Surface pathways associated with water inputs, through flows, and output were identified and mapped using field mapping, aerial photographs (1:5000), and topographic maps (1:24,000). Evidence for surficial water movement (e.g., mineral staining, silt and/or clay deposits) confirmed the location of various pathways.

The sound produced by subsurface movement of water through the rock glaciers confirmed subsurface flow. Shallow excavations provided the opportunity to assess the rates and intensity of these flows. Pits, less than 1.5 m in depth, were excavated by hand at sites where subsurface water flow could be heard clearly. Unfortunately, because we did not have earth-moving machinery, deeper subsurface flow could not be observed. Solid cones of red and green Rhodium-B dyes were introduced at sites of surface flow to determine internal-flow pathways, through-flow rates, and water-residence times. Water exiting and flowing near the surface of the rock glaciers was visually monitored, daily, for traces of the dye.

Surface water impoundment on the rock glaciers was investigated. Depth of water in surface depressions (i.e., ponds, pits, and holes) was measured using a tape measure that was weighted at one end. Estimates of former water levels were based upon water stains and deposition of fine sediments.

The perennial snow cover on the three rock glaciers was analyzed in terms of aspect, extent of coverage, and condition of the remaining snow. Meltwater issuing from the snow patches adjacent to and on the rock glaciers was visually followed and mapped.

Near-surface ice, identified in hand-dug holes less than 2.5 m deep, was evaluated for its condition and stratigraphy. A number of researchers have used seismic methods to analyze the ice component of rock glaciers (Muhll and Haeberli, 1990; van Tatenhove, 1990; Barsch and King, 1989; Evin, 1988; Barsch, 1987; Haeberli, 1985; Evin and Fabre, 1990). We used a four-channel recorder on one of the rock glaciers, but failed to differentiate ice locations and characteristics. Because the power source of the unit did not provide sufficient return energy, we were unable to achieve results comparable to Evin and Fabre (1990).

INPUT, STORAGE, AND OUTPUT VOLUME METHODS

To estimate water input, a network of rain gauges was placed adjacent to the three rock glaciers. Because the sites were inaccessible during winter, only summer precipitation was recorded. Estimates of the water equivalent in snow were derived using aerial photography. Estimates of water input from adjacent cirque walls and talus slopes were computed using a standard runoff calculation equation (Table 7.2).

Water storage in rock glaciers is primarily a function of ice content, the extent of which has been determined primarily by seismic methods (Barsch, 1977; Haeberli, 1985; Evin and Fabre, 1990). Using the Barsch (1977) model to determine the clastic-ice composition of a rock glacier, we estimated the

Table 7.2. Yearly Potential Runoff Volumes from Slopes Adjacent to the Three Selected Rock Glaciers

Rock Glacier	Catchment Area (m^2)	Volume of Water (liters)
California	9.25×10^5	3.57×10^8
Lost Lake	1.62×10^5	6.09×10^7
Dutch Creek	3.94×10^5	1.48×10^8

volume of internal ice for the three rock glaciers. Ice volumes were converted to water-equivalent volumes to estimate the storage component in the rock glaciers (Table 7.3).

Actual discharges of the streams issuing from the rock glaciers were measured hourly from 0600 to 1800 hours. Sampling was conducted during periods with overcast skies, clear skies, and rainstorms in an effort to understand how discharge responds to input. Stream discharge values were estimated utilizing channel cross-sections and stream velocity measurements (Gregory and Walling, 1973). Local rock was used to construct weirs. Channel profiles at the sampling sites were surveyed and velocity measurements were obtained with a current meter. Because the volume of the water in the stream issuing from the frontal slope of the Lost Lake Rock Glacier was less than the minimum level required for use of the current meter, velocity was estimated by timing the movement of a cork. The discharge data are shown in Figure 7.8.

WATER QUALITY MEASUREMENT METHODS

Water associated with rock glaciers was analyzed to assess (1) changes in water quality, and (2) internal characteristics. To assess the effect of a rock glacier on water quality, the water entering, flowing on and through, and exiting the form was examined. In addition to rain and hail samples, water was obtained from snow patches, pits, channels, and ponds on the rock glaciers. Other samples were acquired just below the surface of each rock glacier, adjacent to the side and frontal slopes of the rock glaciers, and in the streams issuing from the frontal slopes. Streams were sampled at the toe of each rock glacier and downstream approximately 300 m lower in elevation. How water quality changed in channels was assessed for comparison with changes within the rock glaciers.

Water samples were analyzed with Hach water-testing kits utilizing colorimetric and titrametric methods (Hach Chemical Company, 1973); a Devon Products conductivity meter was used. Temperature, conductivity, total dissolved solids, and dissolved oxygen levels were determined on-site within

Table 7.3. Ice and Water Volumes for the Three Selected Rock Glaciers[a]

Rock Glacier	Total Volume 100%[b]	Ice Volume		Water Volume	
		50%[b]	60%[b]	50%[b]	60%[b]
California	23.2×10^6 m^3	11.6×10^6 m^3	13.9×10^6 m^3	9.0×10^9 l	10.8×10^9 l
	8.2×10^8 ft.^{3}	4.1×10^8 ft.^{3}	4.9×10^8 ft.^{3}	2.4×10^9 gal.	2.8×10^9 gal.
Lost Lake	3.0×10^6 m^3	1.5×10^6 m^3	1.8×10^6 m^3	1.2×10^9 l	1.4×10^9 l
	1.0×10^8 ft.^{3}	5.3×10^7 ft.^{3}	6.4×10^7 ft.^{3}	3.1×10^8 gal.	3.7×10^8 gal.
Dutch Creek	43.2×10^6 m^3	21.6×10^6 m^3	25.9×10^6 m^3	16.8×10^9 l	20.2×10^9 l
	15.3×10^8 ft.^{3}	7.6×10^8 ft.^{3}	9.2×10^8 ft.^{3}	4.4×10^9 gal.	5.3×10^9 gal.

[a]Source: Barsch (1977) and Giardino (1979).
[b]Form percentage involved.

one hour of sampling. Values for *p*H, silica, alkalinity, total hardness, magnesium, iron, and copper were determined at base camp within 8 hours of collection. All tests were run following the guidelines listed in the Hach Chemical Company (1973) manual. Distilled water was used in the cleaning and the rinsing of glassware and test equipment. Tests for chloride, sulfate, calcium, and nitrate were performed on the initial samples but were discontinued after insignificant values were obtained. Additional details of sample collection and analysis are available in DeMorett (1989).

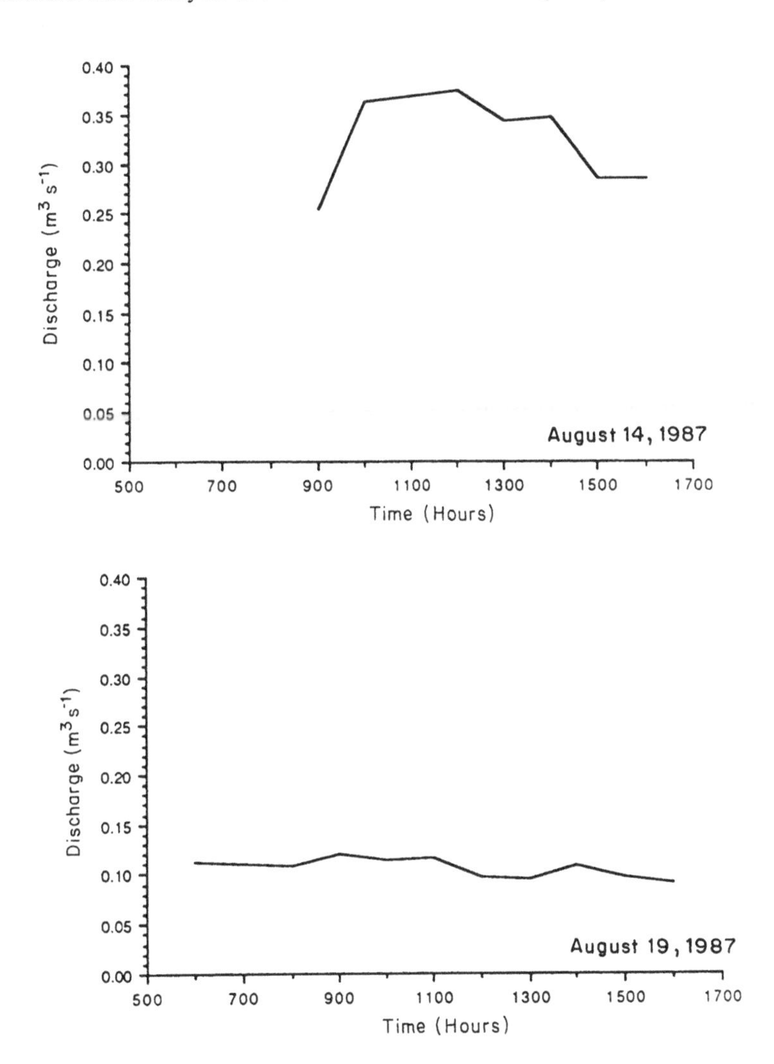

Figure 7.8. Discharge from the toe of the Dutch Creek Rock Glacier monitored on two different days

Acid rain ($pH < 5.6$) occurs in the area (U.S. Department of Interior, 1982). Acidity in water is also a function of the materials through which the water flows (Hem, 1985). The *p*H of the water samples should provide an indication of how rock glaciers affect water acidity in the Massif area. Calcium and magnesium, constituents of many of the igneous and metamorphic rocks present in the Massif, impact total hardness. Dissolved oxygen, essential for chemical processes and the survival of aquatic plant and animal life, is a function of temperature, pressure, and concentrations of other solutes. The hourly sampling procedure was designed to correspond to hourly discharge measurements. Weather conditions were recorded to assist with the explanation of variability.

Data analysis

WATER INPUTS

Water enters a rock glacier system as direct precipitation, snowmelt, and groundwater recharge. The volume of water a rock glacier receives from precipitation is highly variable. From May through July, snowmelt is one of the primary sources of water. Between June and October of 1987, ca. 14.8 cm of precipitation was recorded in rain gauges. The severe drought conditions observed throughout the alpine in early October 1987 suggest the recorded value was below normal.

Water reaching the coarse-textured surface of the rock glacier quickly infiltrates the landform. The volume of water generated by the slope walls adjacent to the three rock glaciers was estimated based upon the following assumptions: infiltration into talus is high (i.e., an infiltration coefficient of ca. 0.75), runoff from the cirque walls is high (i.e., a runoff coefficient of ca. 0.90), and approximately 50 cm of precipitation is received yearly. Using the rational equation and the above assumptions, runoff volumes from adjacent areas were determined (Table 7.2). The values suggest that runoff from adjacent cirque walls and slopes is important in the water budget of a rock glacier.

Avalanches transport large water-equivalent volumes of snow and ice to lower elevations (Clark, 1987). Several avalanche chutes terminate on the three rock glaciers and supply snow and ice which melts and is incorporated into the subsurface ice mass. The amount of this input is difficult to estimate because supply occurs during times of site inaccessibility and melts before measurements can be made.

THROUGH FLOW

Water can enter a rock glacier system above or below the freezing plane through the talus or slope debris that merge with the rock glacier. During periods of heavy rainfall, water flow was observed on the rock debris and over deposits of fine sediment in the headward zone of two of the three rock glaciers. This type of flow was not observed in the toe portions of the rock glaciers. Direct precipitation plus runoff from adjacent cirque walls contributes to saturation and subsequent overland flow. Near the cirque walls, sheet flow was observed on surfaces composed of fine clastics.

Channels on the surfaces of all three rock glaciers, generally in the troughs of the ridge and furrow structure, lack vegetation, contain an abundance of silt- and sand-sized particles, and exhibit discoloration from previous flows. Evidence from water flow at least 12 cm deep was noted. Numerous excavations (<2 m deep) on three rock glaciers revealed 1 m of filled voids and/or open voids. Rates of infiltration through the surface are controlled by clast size, texture range, depth to permafrost, and season. The surfaces of the rock glaciers studied are composed of unsorted, subangular clasts with a wide range of textures that imparts great spatial variability to infiltration rates.

Subsurface flow in the rock glaciers was detected on several occasions as a result of sound generated by the flowing water. This sound was most audible in the deeper furrows and at the headward portions of two rock glaciers suggesting structurally controlled flow. In late August, however, the sounds associated with flowing water were not detected.

STORAGE

Rock glaciers store significant volumes of water for extended periods of time (Johnson, 1981; Haeberli, 1985; Gardner and Bajewsky 1987; Bajewsky and Gardner, 1989). Several researchers describe water in the liquid and solid states in storage within and upon the surfaces of rock glaciers (e.g., Potter, 1972; Corte, 1978; Barsch, 1977; Giardino, 1983). Water collects on the surface of the rock glaciers in depressions (Figures 7.1 and 7.2). A pond (ca. 30 m in diameter) on the Dutch Creek Rock Glacier (Figure 7.2) contained water until late August. When the pond drained, the base was a matrix of fine-grained sediments cemented by ice. During the warm, dry summer, evaporation and probable lowering of the permafrost level promoted drainage.

Small (<10 m in diameter) patches of snow near cirque headwalls remained during the 1987 and 1988 field seasons. Remnant surficial ice was also visible in the headward areas where talus slopes merged with the rock glaciers. Shallow excavations (<2.5 m) near the toe of each rock glacier failed to reveal subsurface ice or fluid water. The introduction of flourocene dye into the rock glaciers was not detected in the discharge because the dye either remained within the rock glaciers for the duration of the field season, or emerged in undetectable concentrations. The volume of ice and water in these three rock glaciers (Table 7.3) was estimated using a procedure developed by Barsch (1977).

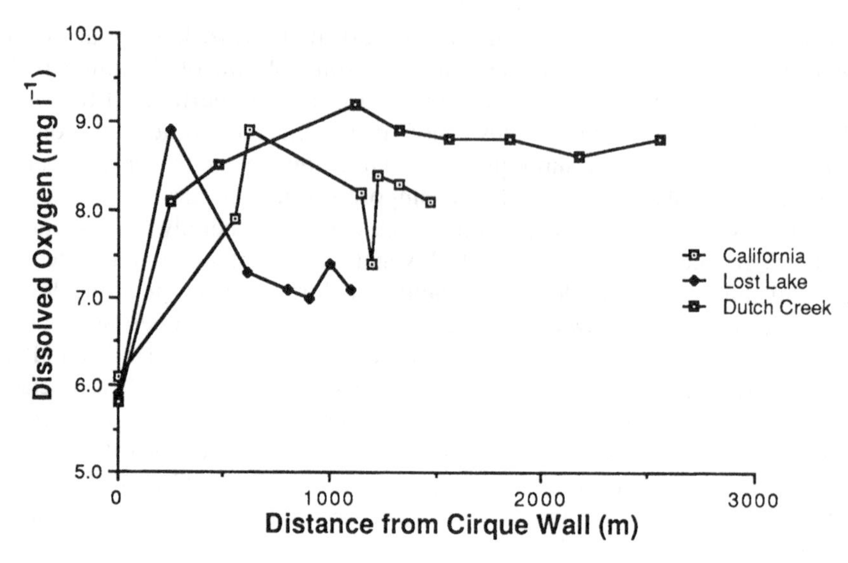

Figure 7.9. Changes in dissolved oxygen in water samples associated with three rock glaciers

OUTPUT

The changes in discharge from a rock glacier are shown with data from the stream exiting the toe of the Dutch Creek Rock Glacier (Figure 7.9). Monitored hourly from 0600 to 1800 hours on August 14 and 19, 1987, discharge varied throughout the day, with peaks occurring between 0900 and 1200 hours on both days (Figure 7.8).

On August 19, discharge values were about one-half as great as those obtained on August 14 because of the heavy rain and hail that occurred on August 13. A typical lag time between precipitation and discharge was observed. The discharge on August 19, not associated with a precipitation event, is most likely a function of ice melt and storage reduction.

WATER QUALITY ANALYSIS

The quality of water associated with rock glaciers was determined from 32 water samples (Figure 7.5) representing water input, through flow, and output sources. Eleven water samples came from the California Rock Glacier, 7 from the Lost Lake site, and 14 from the Dutch Creek site (Table 7.4).

WATER QUALITY–DISCHARGE RELATIONS

Changes in water quality can be related to discharge fluctuations. For California Rock Glacier total dissolved solids are significantly higher during low

Table 7.4. The Chemical and Physical Characteristics of Water Samples from Three Rock Glaciers and Related Streams

Sample Number	Distance from Cirque Wall (m)	Water Temperature (°C)	Dissolved Oxygen (mg l^{-1})	Conductivity (μmhos)	Total Dissolved Solids (mg l^{-1})	Silica (mg l^{-1})	Iron (mg l^{-1})	Copper (mg l^{-1})	Alkalinity (mg l^{-1} $CaCO_3$)	Total Hardness (mg l^{-1} $CaCO_3$)	*p*H
1	0	0.0	6.1	2.0	1.5	t[a]	0.1	0.0	1.3	0.0	6.4
2	556	1.1	7.9	48.0	32.0	1.6	0.0	0.0	15.8	22.3	8.7
3	640	0.0	8.9	15.0	10.0	2.0	0.0	0.0	8.5	7.1	7.1
4	1155	4.5	8.2	80.0	54.0	5.0	t	t	44.4	50.1	8.4
5[b]	1202	5.6	7.4	98.0	65.0	6.3	0.0	t	43.4	51.5	7.8
6[b]	1233	3.9	8.4	9.3	61.0	6.8	0.0	0.0	48.6	53.9	7.8
7[b]	1350	5.6	8.4	89.0	58.0	9.0	t	0.0	47.5	56.1	7.5
8[b]	1475	5.0	8.2	89.0	58.0	6.0	0.0	0.0	39.8	49.8	7.4
9	556	1.1	8.0	99.0	66.0	4.3	0.0	t	30.9	38.8	8.1
10	640	1.1	7.6	19.0	13.0	2.4	0.0	0.1	8.1	9.0	7.2
11[b]	1475	5.6	—	110.0	75.0	8.5	0.0	0.0	43.8	45.9	7.3
12	0	0.0	5.9	3.0	2.0	t	0.0	0.0	2.0	0.0	6.7
13	241	0.0	9.0	44.0	29.0	6.0	0.0	0.0	21.5	23.1	7.3
14[b]	613	11.1	7.3	26.0	17.0	1.0	t	0.0	13.2	13.4	7.9
15[b]	800	10.0	7.1	26.0	17.0	2.0	0.0	t	19.2	15.1	8.1
16[b]	914	10.0	7.0	29.0	19.5	2.1	0.0	t	18.6	16.7	7.6
17[b]	1006	8.9	7.4	31.0	20.8	2.3	0.0	t	19.7	17.3	7.3
18[b]	1094	9.4	7.1	31.0	20.9	2.3	0.0	t	20.1	17.9	7.2
19	0	0.0	5.7	9.0	6.0	t	0.0	0.0	2.7	0.0	6.9
20	0	0.0	6.0	3.0	2.0	t	0.0	0.0	1.9	0.0	6.8
21	0	0.0	8.0	5.0	3.0	0.1	0.0	0.0	2.1	0.0	7.0
22	297	−0.6	8.1	15.0	10.0	1.8	0.0	0.0	4.6	6.9	6.8
23	501	−0.6	8.5	28.0	19.0	2.1	0.0	0.0	10.0	13.1	7.2
24	1152	0.0	9.2	44.0	29.4	2.2	0.0	0.0	18.8	23.3	7.8
25[b]	1356	1.7	8.9	39.0	26.5	2.3	0.0	0.0	19.0	23.9	7.5
26[b]	1573	2.8	8.8	45.0	30.5	2.1	0.0	0.0	25.3	26.0	7.6
27[b]	1865	2.8	8.8	45.0	30.0	2.9	0.0	0.0	25.9	26.1	7.5
28[b]	2179	2.8	8.6	48.0	32.5	2.3	t	0.0	26.5	26.8	7.4
29	2551	2.8	8.8	54.0	36.0	3.3	0.0	0.0	28.6	26.8	7.3
30	297	0.0	8.5	12.0	8.0	1.4	0.0	0.0	6.7	8.6	7.3
31	501	0.0	9.7	27.0	18.4	1.9	0.0	0.0	11.7	13.5	7.8
32	1152	1.1	—	45.0	30.4	1.9	0.0	0.0	22.5	24.9	8.0

[a]t Denotes trace concentration (< 0.1 mg l^{-1}). [b]Stream sample.

flows than high flows. For example, water in the creek situated ca. 500 m downslope from the cirque headwall contained twice the concentration of dissolved solids at a low discharge than during a period of high discharge (compare samples 9 and 2 in Table 7.4); and water sampled from a depression on the surface of the rock glacier increased slightly in dissolved solids during low flows (compare samples 3 and 10). Such increases are to be expected because low flows have more time in contact with sediment and therefore acquire larger concentrations of dissolved ions. Silica, alkalinity, and total hardness increased when discharge decreased at the upper creek (compare samples 9 and 2). This inverse relationship between solute concentration and discharge is also typical of ice glaciers (Collins, 1979).

In contrast, a comparison of water quality values from sites on the Dutch Creek Rock Glacier obtained at low discharges (samples 30, 31, and 32) with values obtained at the same sites during high discharges (samples 22, 23, and 24) reveals a direct relationship between discharge and total dissolved solids. During low discharge, total dissolved ion concentrations in the surface water on the rock glacier were approximately 3 to 20% less than values at the high discharges. Although silica concentrations displayed this same trend, dissolved oxygen, alkalinity, and total hardness increased by approximately 2 to 25% when discharge decreased. Also, as discharge decreased, *p*H increased ca. 0.5.

Total ion concentration in the subsurface waters of the Dutch Creek Rock Glacier, although increasing slightly (from 29.4 to 30.4 mg l^{-1}) as discharge decreased, displayed no significant changes during discharge fluctuations. Unlike the California Rock Glacier, dilution does not occur within the Dutch Creek Rock Glacier. Either the internal characteristics of the two rock glaciers are different or discharge variation at the time of sampling was not sufficient to show this relationship.

WATER QUALITY CHANGES WITHIN THE STUDY SITES

Ion levels in all the samples were low (1.5 to 75.0 mg l^{-1}) and were consistent with the concentrations determined by Vitek, Deutch, and Parson (1981). Of the three sites sampled, the highest ion concentrations were found in the water sampled at the California Rock Glacier. A low discharge volume, flow through a moraine before emerging as California Creek, and mineralogy contribute to high ion concentrations.

At all sites water exiting rock glaciers was slightly more basic (up to 2 units of *p*H) than surface samples at the heads of the rock glaciers. Water issuing from the California, Dutch Creek, and Lost Lake rock glaciers had *p*H values of 8.4, 7.8, and 7.3, respectively. Exit water contained higher dissolved oxygen concentrations than samples taken at the heads of the rock glaciers (Figure 7.9). As water flows through rock glaciers, turbulent flow and melting ice contribute to the percent of dissolved oxygen in the water. But dissolved oxygen concentrations decrease with increasing distance from the rock glaciers as vegetation and/or aquatic life use dissolved oxygen.

The data for water quality from the California Rock Glacier and to a lesser extent the Dutch Creek Rock Glacier suggest two different flow pathways through the rock glaciers. Water emerging from these rock glaciers displayed different water quality signatures than water flowing at or just below the surface of the two rock glaciers. Water sampled on the surface with a temperature near 0°C possessed lower total dissolved ion, alkalinity, and total hardness concentrations than subsurface waters. In addition, the *p*H of the surface waters was lower than that of the subsurface waters. Although dissolved oxygen concentrations were greater in the near-surface water on the Dutch Creek Rock Glacier than in its subsurface waters, the opposite was found to be true of the California Rock Glacier.

Discussion

A model of a cascading system of the hydrology of a rock glacier provides a perspective on water movement through a complex landform. Because little direct evidence of the internal characteristics of rock glaciers exist, researchers have used seismic and water quality data as surrogates for additional internal information. Water quality data exhibit differences between surficial water and flow emerging from the subsurface. Rock glaciers physically and chemically influence the water that passes through them by acting as a concentrating rather than a filtering mechanism. Water input into the system had total dissolved loads ranging from 1.5 to 6.0 mg l^{-1}, whereas outputs had total dissolved solids ranging from 8.0 to 54.0 mg l^{-1}; silica concentrations ranged from 0 to 0.1 mg l^{-1} in inputs and from 1.9 to 6.0 mg l^{-1} in outputs; alkalinity values for inputs ranged from 1.3 to 2.7 mg l^{-1} $CaCO_3$, whereas outputs ranged from 18.1 to 44.4 mg l^{-1} $CaCO_3$; total hardness values for inputs were 0.0 mg l^{-1} $CaCO_3$, whereas outputs ranged between 21.0 and 50.1 mg l^{-1} $CaCO_3$. Water exiting the three selected rock glaciers had higher ion concentrations than water entering these landforms. Moreover, when all values are considered collectively, values recorded for water inputs were similar on all three rock glaciers, and values recorded for water exiting the three rock glaciers were also similar.

The water quality values from Colorado were compared to values obtained in other rock glacier studies. Ion values obtained in our study are lower than those monitored in other studies (Johnson, 1981; Gardner and Bajewsky, 1987; Bajewsky and Gardner, 1989). Corte (1978) found total hardness values of 204 to 208 mg l^{-1} in water from rock glaciers in the Andes. Variation in concentration levels between the studies, however, can be explained by lithologic differences between study areas. The Massif area is composed primarily of hard rocks: amphibolites, granite diorites, hornblende gneisses, and quartzites. These rocks account for the majority of the ions (e.g., Ca, Mg, SiO_2, CO_3, and H_2CO_3) present in the water on the Massif. Rock glaciers that have been subject to water quality analysis reported in the literature are located in areas of soft rock, such as limestones and sandstones. Moreover, rock glaciers reported on in the literature have been influenced by glacial ice (Johnson, 1981). Water input into the Blanca Massif rock glaciers had *p*H values ranging between 6.4 and 7.0, whereas outputs had *p*H values ranging between 7.2 and 8.5. The latter values are similar to the range of values displayed in groundwater (Hem, 1985). Although the rock glaciers in the Massif contribute near-neutral water to the streams, acidity increases with increasing distance from the rock glacier.

Although the ion concentrations measured in the waters of the rock glaciers and streams within the Massif were low, concentrations represent significant amounts of material when viewed on a long-term basis (Vitek, Deutch, and Parson, 1981). Ion values in water from rock glaciers represent a significant part of the total dissolved load further down valley.

Discharge from the three rock glaciers was influenced by temperature and precipitation events. For example, discharge increased 3 to 6 hours after a major precipitation event. These lag times are consistent with those reported by others (Evin and Assier, 1983b; Haeberli, 1985). Water in the lower flow path, however, would not be associated with major input events and would have a greater residence time in the rock glacier. Water exiting the rock glaciers of the Massif area contains little suspended sediment. The observation that solution is the important mechanism of removal of material from a rock glacier supports conclusions by Caine and Thurman (1991), Barsch and Caine (1984), and Vitek, Deutch, and Parson (1981). The rock glaciers around the Massif influence the regional hydrological regime. In over 20 years of personal observation, streams emerging from rock glaciers have continued to flow long after streams in basins without rock glaciers ceased. Rock debris covering interstitial or remnant glacial ice provides protective insulation against melting. The external variables of temperature and moisture input, however, contribute to the release of water for longer periods.

Although we were unable to trace the movement of dye through the rock glaciers under study, the effort should be repeated with short half-life radioactive substances. Being able to determine the rate of movement through a rock glacier is critical to understanding how its hydrological system responds to the input of moisture.

Conclusions

The use of a cascading-system model to depict the pathways of movement for the liquid and solid phases of water in a rock glacier accentuates the complexity of the system. Because rock glaciers can be created by periglacial or glacial processes (Giardino and Vitek, 1988), internal characteristics are considered to be variable depending upon age (the time since development) and local conditions. The cascading model of water movement provides detail to the model developed by Haeberli (1985). Specifically, the concept of through flow has been added to indicate that the permafrost zone can include liquid water as evidenced by water bursting out of the toe of a rock glacier (Giardino, 1983). Changes in form as a result of age and local conditions will impact how water moves through the system.

The hydrology of alpine basins is clearly influenced by rock glaciers. Field observations led to the discovery of overland flow in the headward reaches of rock glaciers, evidence for channel flow in furrows, and the formation of surface ponds by ice in interstitial voids. Whereas our attempt to use seismic equipment to interpret internal ice failed, other researchers have documented complex internal structures and have confirmed how complicated water pathways through a rock glacier can be.

The water quality characteristics from water on and exiting rock glaciers confirmed that residence time permits a significant dissolved load to be acquired during movement through a rock glacier. Major differences in water quality are primarily a function of local lithology. More detailed investigations of internal characteristics, assessed by bore holes or shafts through to bedrock, are necessary to provide evidence that the cascading model is a correct representation of how water flows through a rock glacier.

References

Bajewsky, I., and J. S. Gardner, Discharge and sediment-load characteristics of the Hilda rock-glacier stream, Canadian Rocky Mountains, Alberta, *Physical Geography*, **10**, 295–306, 1989.

Barsch, D., Nature and importance of mass wasting by rock glaciers in alpine permafrost environments, *Earth Surface Processes*, **2**, 231–245, 1977.

Barsch, D., The problem of the ice-cored rock glacier, in *Rock Glaciers*, edited by J. R. Giardino, J. F. Shroder, Jr., and J. D. Vitek, pp. 45–53, Allen and Unwin, Boston, 1987.

Barsch, D., and N. Caine, The nature of mountain geomorphology, *Mountain Research and Development*, *4*, 287–298, 1984.

Barsch, D., and L. King, Origin and geoelectrical resistivity of rockglaciers in semi-arid subtropical mountains (Andes of Mendoza, Argentina), *Zeitschrift für Geomorphologie*, **33**, 151–163, 1989.

Brown, W. H., A probable fossil rock glacier, *Journal of Geology*, **33**, 464–466, 1925.

Caine, N., and E. M. Thurman, Temporal and spatial variations in the solute content of an alpine stream, Colorado Front Range, *Geomorphology*, **4**, 55–72, 1991.

Capps, S. R., Jr., Rock glaciers in Alaska, *Journal of Geology*, **18**, 359–375, 1910.

Carroll, T., Relative age dating techniques and a late Quaternary chronology, Arikaree Cirque, Colorado, *Geology*, **2**, 321–325, 1974.

Clark, M. J., Geocryological inputs to the alpine sediment system in glacio-fluvial sediment transfer, in *An Alpine Perspective*, edited by A. M. Gurnell, and M. J. Clark, pp. 33–58, John Wiley, New York, 1987.

Collins, D. N., Hydrochemistry of meltwater draining from an alpine glacier, *Arctic and Alpine Research*, **11**, 307–323, 1979.

Corte, A. E., The hydrological significance of rock glaciers, *Journal of Glaciology*, **17**, 157–158, 1976a.

Corte, A. E., Rock glaciers, *Biuletyn Peryglacjalny*, **26**, 175–197, 1976b.

Corte, A. E., Rock glaciers as permafrost bodies with debris cover as an active layer, A hydrological approach, Andes of Mendoza, Argentina, *Proceedings of the Third International Conference on Permafrost*, **1**, 262–269, 1978.

DeMorett, J. L., Hydrological and water quality characteristics of three rock glaciers, Blanca Massif, Colorado, USA, M.A. thesis, 146 pp., Texas A & M University, 1989.

Domaradzki, J., Blockströme im Kanton Graubünden, Ergebnisse der wissenschiftlichen untersuchtungen des Schweiz Nationalparks, Band. 3, No. 24, *Herausgegeben von der Kommission der Schweizerischen Naturerforschenden Gesellschaft, 54*, 177–235, 1951.

Evin, M., Structure et mouvement des glaciers rocheux des Alpes du Sud, These de 3 cycle, 343 pp., Université de Grenoble, 1983.

Evin, M., Caracteristiques physico-chimiques des eaux issues des glaciers rocheux des Alpes du Sud (France), *Zeitschrift für Gletscherkunde und Glazialgeologie*, **20**, 27-40, 1984.

Evin, M., Repartition morphologie et structure interne des glaciers rocheux des Alpes du Sud en fonction de la lithologie et de la fracturation, *Bulletin du Centre de Geomorphologie du C.N.R.S., Caen*, **34**, 137–158, 1988.

Evin, M., and A. Assier, Relations hydrologiques entre glacier et glacier rocheux: l' exemple du cirque de Marinet (Haute-Ubaye, Alpes du Sud), *Journées de la S.H.F. Section de glaciologie, Grenoble*, **21–22**, 1–5, 1983a.

Evin, M., and A. Assier, Glacier et glaciers rocheux dans le haut-vallon du Loup (Haute-Ubaye, Alpes du Sud, France), *Zeitschrift für Gletscherkunde und Glazialgeologie*, **19**, 27–41, 1983b.

Evin, M., and D. Fabre, The distribution of permafrost in rock glaciers of the southern Alps (France), *Geomorphology*, **3**, 57–71, 1990.

Gardiner, V., and R. Dackombe, *Geomorphological Field Manual*, 254 pp., Allen and Unwin, London, 1982.

Gardner, J. S., and Bajewsky, I., Hilda rock glacier střam discharge and sediment load characteristics, Sunwapta Pass area, Canadian Rocky Mountains, in *Rock Glaciers*, edited by J. R. Giardino, J. F. Shroder, Jr., and J. D. Vitek, pp. 161–174, Allen and Unwin, Boston, 1987.

Giardino, J. R., Rock glacier mechanics and chronologies; Mount Mestas, Colorado. Ph.D. Dissertation, 244 pp., University of Nebraska, 1979.

Giardino, J. R., Movement of ice-cemented rock glaciers by hydrostatic pressure: An example from Mount Mestas, Colorado, *Zeitschrift für Geomorphologie*, **27**, 297–310, 1983.

Giardino, J. R., and S. G. Vick, Geologic engineering aspects of rock glaciers in *Rock Glaciers*, edited by J.R. Giardino, J. F. Shroder, Jr., and J. D. Vitek, pp. 265–287, Allen and Unwin, Boston, 1987.

Giardino, J. R., and J. D. Vitek, The significance of rock glaciers in the glacial-periglacial landscape continuum, *Journal of Quaternary Science*, **3**, 97–103, 1988.

Gregory, K. J., and D. E. Walling, *Drainage Basin Form and Process: A Geomorphological Approach*, 456 pp., Edward Arnold, London, 1973.

Hach Chemical Company, *Hach Methods Manual*, 118 pp., Hach Chemical Company, Ames, Iowa, 1973.

Haeberli, W., Creep of mountain permafrost: Internal structure and flow of alpine rock glaciers, *Mitteilungen der Versuchsanstalt für Wasserbau, Hydrologie und Glaziologie [Zurich]*, **77**, 142 pp., 1985.

Hem, J. P., Study and interpretation of the chemical characteristics of natural water, *U.S. Geological Survey Water-Supply Paper 2254*, 263 pp., 1985.

Jackson, L. E., and G. M. MacDonald, Movement of an ice-cored rock glacier, Tungsten, N.W.T., Canada, 1963–1980, *Arctic*, **33**, 842–847, 1980.

Johnson, P. G., Rock glacier types and their drainage systems, Grizzly Creek, Yukon Territory, *Canadian Journal of Earth Sciences*, **15**, 1496–1507, 1978.

Johnson P. G., The structure of a talus derived rock glacier as deduced from its hydrology, *Canadian Journal of Earth Sciences*, **18**, 1422–1430, 1981.

McCulloch, D. S., Late Cenozoic erosional history of Huerfano Park, Colorado. Ph.D. Dissertation, 158 pp., University of Michigan, 1955.

Moore, D. W., Longitudinal section of an alpine rock glacier exposed south of Berthoud Pass, Central Colorado Front Range, *Geological Society of America, Abstracts with Programs*, **23**, 50, 1991.

Morris, S. E., Regional and topoclimatic implications of rock glacier stratigraphy: Blanca Massif, Colorado, in *Rock Glaciers*, edited by J. R. Giardino, J. F. Shroder, Jr., and J. D. Vitek, pp. 107–125, Allen and Unwin, Boston, 1987.

Muhll, D. V., and W. Haeberli, Thermal characteristics of permafrost within an active rock glacier (Murtel/Corvatsch, Grison, Swiss Alps), *Journal of Glaciology*, **36**, 151–158, 1990.

National Oceanic and Atmospheric Administration, *Climatological Data; Annual Summary Colorado*, **91**, 36 pp., 1986.

Olyphant, G. A., Analysis of the factors controlling cliff burial by talus within Blanca Massif, southern Colorado, U.S.A., *Arctic and Alpine Research*, **15**, 65–75, 1983.

Parson, C. G., Rock glaciers and site characteristics on the Blanca Massif in South-Central Colorado, Ph.D. Dissertation, 212 pp., University of Iowa, 1980.

Parson, C. G., Rock glaciers and site characteristics on the Blanca Massif, Colorado, U.S.A., in *Rock Glaciers*, edited by J. R. Giardino, J. F. Shroder, Jr., and J. D. Vitek, pp. 127–143, Allen and Unwin, Boston, 1987.

Potter, N., Jr., Rock glaciers and mass-wastage in the Galena Creek area, northern Absaroka Mountains, Wyoming, Ph.D. Dissertation, 150 pp., University of Minnesota, 1969.

Potter, N., Jr., Ice-cored rock glacier, Galena Creek, northern Absaroka Mountains, Wyoming, *Bulletin of the Geological Society of America*, **83**, 3025–3058, 1972.

Reifsnyder, W. F., *Weathering the Wilderness*, 276 pp., Sierra Club Books, San Francisco, 1980.

Schweizer, G., Der Formenschatz des sät-und Postglazials in den Höhen Seealpen, *Zeitschrift für Geomorphologie, Supplement Band* **6**, 1–167, 1968.

Steenstrup, K. J. V., Bidrag til Kjendskab til Braeerne og Braeisen i Nordgronland, *Meddelelser om Gronland*, **4**, 69–112, 1883.

Tyrrell, J. B., Rock glaciers or chrystocrenes, *Journal of Geology*, **18**, 549–553, 1910.

U.S. Department of the Interior, Air Pollution and Acid Rain Report 3, *The Effects of Air Pollution and Acid Rain on Fish, Wildlife, and Their Habitats*, 181 pp., U.S. Environmental Protection Agency, Washington D.C., 1982.

van Tatenhove, F., Past and present permafrost distribution in the Turtmanntal, Wallis, Swiss Alps, *Arctic and Alpine Research*, **22**, 302–316, 1990.

Vitek, J. D., The mounds of south-central Colorado: An investigation of geographic and geomorphic characteristics, Ph.D. Dissertation, 229 pp., University of Iowa, 1973.

Vitek, J. D., A. L. Deutch, and C. G. Parson, Summer measurements of dissolved ion concentrations in alpine streams, Blanca Peak Region, Colorado, *Professional Geographer*, **33**, 436–444, 1981.

Wahrhaftig, C., and A. Cox, Rock glaciers in the Alaska Range, *Bulletin of the Geological Society of America*, **70**, 383–436, 1959.

White, S. E., Rock glacier studies in the Colorado Front Range, 1961–1968, *Arctic and Alpine Research*, **3**, 43–64, 1971.

8 Snow-avalanche Paths: Conduits from the Periglacial-Alpine to the Subalpine-Depositional Zone

David R. Butler
Department of Geography, University of Georgia

George P. Malanson
Department of Geography, University of Iowa

Stephen J. Walsh
Department of Geography, University of North Carolina

Abstract

Morphometric characteristics of over 250 snow-avalanche paths in northwestern Montana and southwestern Alberta differ with aspect and with position relative to the Continental Divide. Particularly notable is the difference between generally south-facing and north-facing paths. South-facing avalanche paths descend to lower elevations on gentler valleyside slopes and produce runout zones of greater extent. These same south-facing paths also are more prone to debris flows. Many avalanche paths debouche into subalpine lakes. Avalanche paths act as conduits that bring the hazard of the periglacial environment down to lower elevations where roads, trails, and human-use facilities are concentrated.

Introduction

Periglacial processes operative on hillslopes range in velocity from imperceptible to very rapid. Rapid mass movements occur in the periglacial landscape through various processes such as snow avalanches, debris avalanches, slush avalanches, mudflows, and combinations and gradations of these. These mass

Periglacial Geomorphology. Edited by J. C. Dixon and A. D. Abrahams

movements represent some of the most rapid and potentially dangerous processes with which human activities in the periglacial environment must deal. Because of their speed and mobility, snow avalanches and associated processes extend their influence across a wide elevation range, with avalanche paths acting as conduits for sediment and debris from the alpine periglacial zone well down into valley bottoms where human activity and structures are typically located.

This paper examines morphometric characteristics of over 250 snow-avalanche paths in a mountainous region of northwestern Montana, U.S.A., and southwestern Alberta, Canada. It describes the differences which occur in morphometric parameters amongst the studied sample of paths and identifies morphometric characteristics which typify the most hazardous sites. It also illustrates the conduit nature of snow-avalanche paths, by which snow avalanches extend the influence of periglacial hazards to relatively low elevations.

Background

Snow avalanches typically occur in well-defined locations on the landscape, known as snow-avalanche paths (Figure 8.1). (For simplicity's sake, we shall refer to these as "avalanche paths", except where they may be confused with debris avalanche processes.) Avalanche paths typically originate above upper treeline in the alpine periglacial environment and may descend well into the forested subalpine environment.

The mobility and velocity of snow avalanches provide great erosive and transportational capabilities (Rapp, 1960; Luckman, 1977; Ward, 1985), although not all snow avalanches are necessarily geomorphically effective. Frozen terrain, a stationary snow cover, and/or a flexible but armoring vegetative cover may exist underneath the moving snow and preclude effective erosional contact with the surface of the hillslope (Luckman, 1977; Gardner, 1983; Butler and Malanson, 1990). Nevertheless if a snow avalanche has sufficient mobility and velocity to reach the valley bottom it may impact with sufficient force to scour and excavate a variety of erosional forms. These include boulder holes, avalanche impact tongues, pits (also called avalanche tarns and avalanche lakes), and impact pools (Peev, 1966; Corner, 1980; Hole, 1981; Fitzharris and Owens, 1984; Ward, 1985; Butler, 1989a). Impact pressures necessary to create such landforms may range as high as 665 kN m^{-2} or larger (Fitzharris and Owens, 1984). Snow avalanches descending into valley bottoms may temporarily impound streams and rivers, creating the possibility of outburst floods analogous to jökuhlhaups (Butler, 1989b). Avalanches can also transfer eroded sediment and vegetative materials from high-elevation sites onto the frozen surfaces of lakes or directly into unfrozen lakes at the base of avalanche paths (Bujak, 1974; Luckman, 1975; Corner,

Figure 8.1. 1966 U.S. Geological Survey aerial photograph of snow-avalanche paths, northwestern Glacier National Park. North is at top of photo. Note differences in shading, vegetation, and path runout characteristics between north- (lower part of photo) and south-facing slopes (upper half of photo). A large creek flows roughly west-east across the center of the photo

1980; Fitzharris and Owens, 1984; Weirich, 1985; Butler, 1989a). In such instances, avalanche paths act as conduits for periglacial sediment transfer into the forest environment below.

Snow-avalanche paths also frequently show evidence of debris flowage and/or debris avalanching within their boundaries (Desloges and Gardner, 1981; VanDine, 1985; Kostaschuk, MacDonald, and Putnam, 1986; Sauchyn, 1986; Jackson 1987; Jackson, Kostaschuk, and MacDonald, 1987). Runout zones of these avalanche paths often display an alluvial fan-like morphology, and the deposits found there have been in some cases mistakenly mapped as alluvial fans, which are also possibly affected by debris flows (Carrara, 1990). Morphometric analyses of these fans, however, clearly distinguish them from true alluvial fans on the basis of slope and basin ruggedness parameters (Kostaschuk, MacDonald, and Putnam, 1986; Jackson, Kostaschuk, and MacDonald, 1987). Mapping debris-flow and snow-avalanche fans as alluvial fans without reference to the additional hazardous processes typical of such environments provides a major disservice to hazard planners. Snow-avalanche/debris-flow fans are also locations where avalanche damming and high avalanche impact pressures are frequently experienced; true alluvial fans are not.

The study area

This study was carried out in Glacier National Park (GNP), Montana, and Waterton Lakes National Park (WLNP), Alberta, collectively referred to as Waterton-Glacier International Peace Park (Figure 8.2). The parks are of primarily mountainous terrain, with elevations ranging from approximately 960 to 1000 m in the glacial valleys on the western flank of Glacier Park, to more than 3050 m on the Parks' highest peaks. Deeply incised glacial valleys are primarily, but not exclusively, oriented in a northeast-southwest fashion (Figure 8.3). These valleys, typically asymmetrical with steeper north-facing slopes, provide the primary corridors for transportation into and across the Parks. Aspect and elevation combine to produce a variety of microclimates. In general, the western half of GNP experiences a modified Pacific-maritime climate, with maximum precipitation levels along the Continental Divide (up to 2500 mm; Finklin, 1986). All of WLNP is east of the Continental Divide, with a climate similar to that of the eastern half of GNP. The eastern slopes are windy and relatively dry (approximately 585 mm annually), with a continental interior regime. Extreme storms are significant contributors to high-magnitude flood, mass movement, and snow-avalanche events (Butler, Oelfke, and Oelfke, 1986). Such storms can produce significant sediment mobilization in the ephemeral channels of avalanche paths above, and on, fan surfaces (Desloges and Gardner, 1984).

The periglacial environment in the two parks is currently encountered

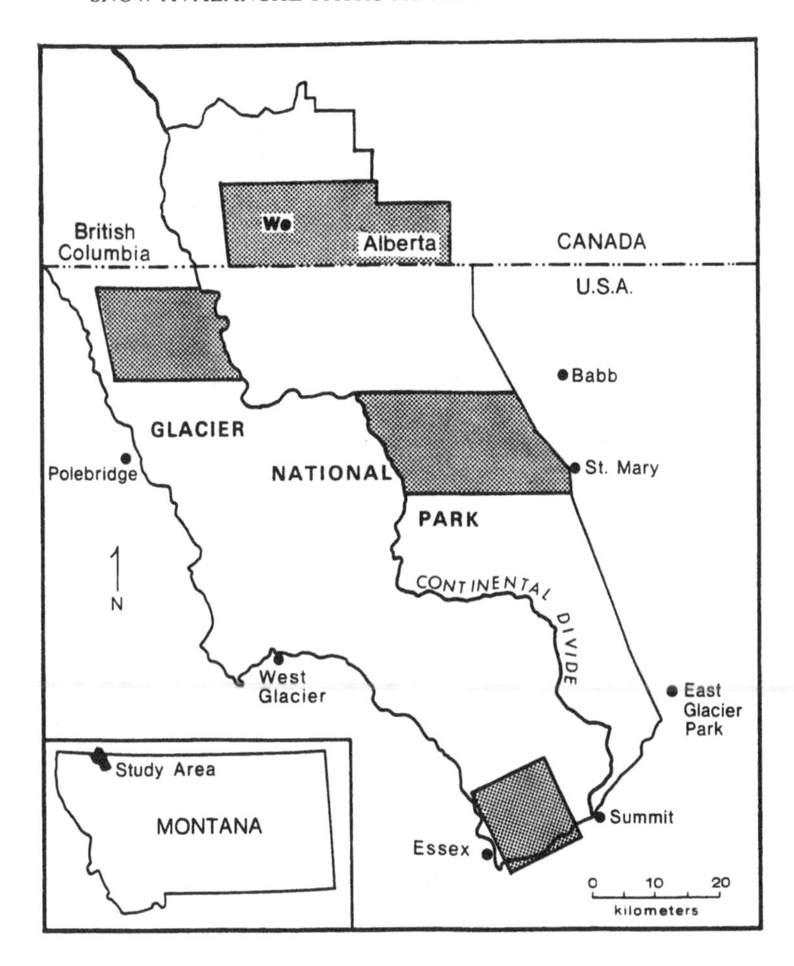

Figure 8.2. Map of the study area, Glacier National Park, Montana, and adjacent portions of Waterton Lakes National Park, Alberta. *W* indicates the townsite of Waterton Park. Enclosed areas delineate the study subsets of WLNP, northwest GNP, east-central GNP, and southern GNP

above approximately 1950 to 2000 m. Currently operative processes producing patterned ground exist above ca. 2170 m, and relict features, which were probably active during the Little Ice Age occur in the roughly 200 m of transition between 1950 and 2170 m (see Butler and Malanson, 1989, for a discussion of these features).

Snow avalanches in the two parks include both wet-snow and dry-powder types, but historical data suggest that the former is much more common, especially on the western slopes of GNP. Windblasts associated with dry-powder avalanches have been described from both sides of the Continental

Figure 8.3. SPOT image of avalanche paths in northeast-southwest trending valleys in southern GNP. The arrow points to the avalanche path illustrated in Figure 8.4, whereas the asterisk indicates the avalanche path in Figure 8.5. Ole Creek is the major valley north of asterisked path. North is at top of figure

Divide (Butler, 1986). Avalanches in the Parks can originate at either a point in the snowpack or a coherent slab of hard metamorphosed snow (Butler, 1989a). The moisture content of the wet-snow avalanches on the western slopes in particular imparts a relatively high fluidity to the avalanche, allowing densely packed wet snow to travel well into path runout zones with great impact (Butler and Malanson, 1985).

Snow-avalanche hazards of the study area

Snow avalanches are not simply of geomorphic interest in GNP and WLNP. Human-built structures in both Parks have experienced damage, and people have been injured and killed as a result of avalanche activity. The townsite of Waterton Park, home to approximately 100 year-round residents, is located on a delta at the base of Bertha Peak along the western shore of Upper Waterton Lake (Figure 8.2). Both flooding and snow avalanches create hazards here. Snow avalanches from Bertha Peak have caused major damage to cottages and other facilities in the townsite (*Hungry Horse News*, 1954a,b, 1957).

Several paved roads penetrate the mountain interior of the two Parks. Going-to-the-Sun Road crosses the Continental Divide in GNP from St. Mary to West Glacier. It is closed during the winter season, but late-fall and springtime avalanches have caused damage to vehicles along the road (Butler, 1986). Other roads that enter but do not cross the mountains include the Red Rock Canyon and the Akamina Parkways in WLNP, and the Many Glacier Road in GNP. In addition, U.S. Highway 2 and the Burlington Northern Railroad cross the Continental Divide and parallel the southern boundary of GNP from West Glacier to East Glacier, where several avalanche paths descend onto and occasionally across these transportation routes (Butler and Malanson, 1985). The Waterton roads may, depending on weather conditions, be open for vehicular travel during winter, and U.S. Highway 2 is open year-round. These roads offer access to popular cross-country skiing sites in both Parks. Avalanche paths are also numerous throughout the high mountain areas of the Parks, reached in winter only by using snowshoes or cross-country skis.

Methodology

We have mapped over 1000 avalanche paths in GNP (Butler, 1979; Butler and Malanson, 1985; Butler and Walsh, 1990; Walsh, Butler, Brown, and Bian, 1990), and over 150 have been mapped by others in WLNP (Coen and Holland, 1976). A recently published surficial geology map of GNP cannot be accurately used to locate avalanche paths; some major avalanche tracks are shown on this map, but this is not a complete inventory. Only paths designated as prominent are shown, and with approximate boundaries (Carrara, 1990). The map omits many paths which we judge to be major and hazardous.

We first mapped avalanche paths of GNP onto 1:24,000-scale U.S. Geological Survey topographic maps, after interpretation of panchromatic aerial photographs flown in 1966 (Figure 8.1). The contour interval of these maps is 80 ft (24.4 m), so mapped boundaries of each path are accurate to plus or minus 40 ft (12.2 m). Because both parks have experienced major snow-avalanche winters since the time of aerial photography and mapping (1968), particularly in 1979 when many avalanche paths experienced boundary expansion by 100-year avalanche events, we field checked the longitudinal and transverse boundaries and land cover characteristics of over 150 avalanche paths (mostly in GNP) over the course of several field seasons (primarily in 1987, 1988, and 1990, but also including 1975 and 1983). We used, and are currently using, Landsat (Walsh, Bian, Brown, Butler, and Malanson, 1989) and SPOT (Figure 8.3) multispectral digital data for gross comparisons of 1966 and post-1966 avalanche paths boundaries for an additional ca. 200 paths. The resolution of these images is 30 m × 30 m for Landsat and 20 m × 20 m for SPOT data.

AVALANCHE PATH MORPHOMETRY

We chose four subsets of avalanche paths within the two parks (Figure 8.2), two on each side of the Continental Divide, from which to take morphometric measurements. Because of the poorer quality of aerial photos and the fact that the largest-scale map is 1:50,000 in WLNP, we randomly sampled only about one-third of the avalanche paths there (Table 8.1). Data from the GNP subset east of the Continental Divide (Figure 8.2) have been previously published (Butler and Walsh, 1990). Data for the other three subsets of avalanche paths, collected following techniques described by Butler and Walsh (1990, pp. 368–369), are presented here for the first time. For each avalanche path we recorded the landcover type surrounding the path starting zone (categorized as bedrock above treeline, mixed shrubs/herbaceous, or forest). We also recorded the path highest and lowest elevations (defined following Butler, 1979), the horizontal length of the path on the map, and the slope aspect (ranging from 0 to 359°). We calculated path relief and overall path slope angle. The fire history of these four subsets was examined to determine whether the past forest fires may have influenced avalanche path characteristics (Mackenzie, 1973; Finklin, 1986). For the purposes of this

Table 8.1. Morphometric Data from Four Major Snow-Avalanche Path Zones within Waterton-Glacier Parks[a]

Group	Slope Angle (degrees)	Slope Aspect (degrees)	Top Elevation (m)	Bottom Elevation (m)	Relief (m)
Waterton Lakes (N = 54)	23	231	2067	1573	494
Northwest Glacier (N = 34)	29	167	2011	1433	578
East-Central Glacier (N = 121)	29	104	2054	1662	392
Southern Glacier (N = 80)	27	190	1959	1298	661
Mean of Means	27	212	1991	1395	596

[a]All data presented are arithmetic means.

study, we chose not to examine the effects of bedrock lithology in the starting zone or the effects of geologic structure on path morphometry; the effects of these variables have been discussed elsewhere (Butler and Walsh, 1990).

In the course of our measurements we observed a visible dichotomy between north-facing and south-facing avalanche paths in terms of size, steepness, location of debris-flow channels and levees, and erosional power at the base of paths as evidenced by avalanche-impact landforms (Figure 8.1). We therefore chose the southern subset to collect additional morphometric data. This subset was chosen because of the distinct northeast-southwest trending glacial valleys which, in turn, result in the clearest distinction between avalanche paths oriented broadly to the north and to the south (Figure 8.3). This region of GNP has also experienced many more forest fires, particularly in the early twentieth century because of cinders ejected from steam locomotives along the railroad tracks on the park's southern boundary (Figure 8.4). In this subset, approximately 50 paths were oriented either generally to the north or generally to the south (Table 8.2) in two adjacent stream valleys. From these paths we also delineated and measured the areas of the starting zone and the runout zone, following Butler and Walsh (1990).

Figure 8.4. Southeastern-facing avalanche path, Bear Creek valley, along the southern border of GNP. Note trees along ridge crest at head of avalanche path starting zones. A snowshed covers railroad tracks crossing the avalanche path

Table 8.2. Morphometric Data from the Southern Glacier Subgroup by Valley and Aspect

Group	Slope Angle (degrees)	Slope Aspect (degrees)	Top Elevation (m)	Bottom Elevation (m)	Relief (m)	Source Area (m^2)	Runout Area (m^2)
Ole Creek, North-facing (N = 14)	31 (3)[a]	329 (16)	2054 (191)	1317 (99)	737	60,896	17,179
Ole Creek, South-facing (N = 13)	24 (2)	158 (50)	2012 (100)	1282 (106)	730	147,797	32,491
Bear Creek, North-facing (N = 10)	30 (5)	0 (18)	1948 (151)	1300 (51)	648	97,336	10,617
Bear Creek, South-facing (N = 12)	26 (3)	156 (36)	1824 (93)	1294 (39)	530	81,500	27,306

[a]Standard deviation values in parentheses.

MORPHOMETRIC MEASUREMENTS OF RUNOUT ZONE FANS

Because of the importance for hazard-planning, we attempted to discern which mapped (Carrara, 1990) alluvial fans were sites subject to avalanche runout and debris flows. We used the same low-altitude panchromatic photographs from 1966 to determine these locations and the 1968 1:24,000-scale topographic maps of GNP to make morphometric measurements of site characteristics. We also consulted archival information that clearly described locations of avalanche-uprooting of trees and deposition of avalanche debris. WLNP was excluded from this portion of the study because of the lower quality and smaller scale of the aerial photos reproduced in Coen and Holland (1976).

The location of each alluvial fan was noted relative to avalanche paths and outlined from the aerial photographs onto the 1:24,000-scale topographic maps. Those fans not spatially associated with avalanche paths and debris-flow landforms were eliminated from further analysis. Data on the following variables were collected from the topographic maps for 54 fans: fan upper and lower elevation, fan length, drainage basin upper and lower elevation, fan area, and drainage basin area. These variables were chosen for ease of data extraction from the maps, for use in calculations of additional terrain variables, and because they typify the morphology of the fans in a straightforward fashion. From these data we calculated fan relief, fan slope, basin relief, and basin ruggedness, R, using Melton's (1965) equation:

$$R = H_b A_b^{-0.5} \quad (8.1)$$

where H_b is basin relief and A_b is basin area. The presence of debris-flow levees or lobes on these fans (as interpreted from aerial photographs, and field-checked on ten fans) was also recorded.

Results

AVALANCHE PATH MORPHOMETRY

The data in Table 8.1 reveal the results of the morphometric measurements from the four data subsets within the two parks. Striking similarities exist in slope angle for all but the WLNP data set. This may be attributable to the lower accuracy in path boundary placement and the greater contour interval (100 ft or 30.5 m) on the 1:50,000-scale topographic map. Field reconnaiss-ance in the two parks suggests a uniformity of angle on avalanche paths.

The highest elevations of the starting zone occur primarily above upper treeline. The major exception is in the southern Glacier data set, which is explained below. Even in southern Glacier, the upper elevations approach 2000 m. The starting zone elevations are consistently higher on the east side of the Continental Divide (2067 and 2054 m versus 2011 and 1959 m), and the lowest elevations of the runout zone are also higher there. On average, paths terminate at least 100 m and as much as 200 m higher east of the Divide.

The higher relief and lower runout zones in the west are attributed to the greater degree of past glacier activity, lower valley floor elevations, and greater snowfall (Butler and Walsh, 1990). Available data also indicate that wet-snow avalanches are more common on the western side of the divide (Butler, 1986). Greater snowfall amounts west of the Continental Divide (Finklin, 1986) also contribute; greater snowpacks can provide larger, pre-sumably more powerful avalanches with greater travel distances. These (typically wet-snow) avalanches also travel down deeply incised gullies (Figure 8.5) with limited vegetation to impede their progress. In association with the large volume potential of these avalanches, "excessive" travel distances of westside avalanches is occasionally documented (Martinelli, 1986).

Landcover categories at the upper end of each avalanche path were similar for the WLNP, northwest GNP, and east-central GNP study subsets; only the southern GNP set was distinctly different. We compare it here to the previously published data (Butler and Walsh, 1990) from the east-central study subset. There 74% of the avalanche paths originated above upper treeline on bare bedrock. Only 12% (15 paths) originated below upper treeline in forest, and 14% began in mixed shrub/herbaceous landcover. In

Figure 8.5. Steeply plunging avalanche path that frequently experiences snow avalanches which block U.S. Highway 2. The very steep side slopes and limited vegetation are typical of paths in the study area

the southern GNP study subset of 80 avalanche paths, 56% originated on bare bedrock, 14% in mixed shrub/herbaceous cover, and 30% in forest. The last group of paths typically originated at distinctly lower elevations (Figure 8.4). A closer examination of the southern data set reveals probable reasons for the lower origin of these paths.

Table 8.2 shows that the south-facing avalanche paths of Bear Creek valley originate at surprisingly low elevations (1824 m on average), over 100 m lower than in Ole Creek or on north-facing slopes of Bear Creek. We attribute this phenomenon to the forest fire history of this part of southern GNP. The early twentieth century saw numerous major forest fires sweep these slopes (Butler, Walsh, and Malanson, 1991). Removal of the stabilizing forest cover induced subsequent winter instabilities in the snowpack at lower elevations than normal, allowing avalanche paths to originate at lower elevations below climatic treeline. The avalanche path in Figure 8.4, for example, originates well below current upper treeline. Field reconnaissance and analysis of soils from the starting zone of this path (Butler and Malanson, 1990) revealed charcoal in the uppermost soil horizon; many dead snags with burned bases are still present in the starting zone as well.

Throughout both parks, and specifically in the southern study group (Table 8.2), south-facing avalanche paths extend to lower elevations on gentler valleyside slopes than their north-facing counterparts (although not all subset cases are statistically significantly different). This is not a result of any difference in up- versus down-valley location; south-facing and north-facing paths originate directly across from each other along the entire length of most major valleys. South-facing runout zones are also distinctly larger, even though no discernible relationship exists between runout zone area and starting zone area (Table 8.2). We attribute the lower runout zone elevations and larger runout zones for paths with gentler slopes to the greater fluidity of avalanches occurring on south-facing slopes. North-facing avalanches occur on well-shaded slopes (Figure 8.1), where a minimum of freeze-thaw activity and snowpack instability develops. South-facing slopes experience numerous freeze-thaw cycles. They also experience more frequent snowmelt during the avalanche season than do the north-facing paths. Slope asymmetry inherited from the Pleistocene also produces topographic settings more conducive to avalanching. These factors combine to produce avalanches of greater saturation and fluidity that are able to flow on relatively lower slopes and to fan across larger runout zones.

Lakes at the base of over 80 avalanche paths in the two parks receive snow, vegetative matter, and rock detritus from avalanches that descend into their water or onto their winter ice cover. Numerous historical examples exist to illustrate the manner in which subalpine lakes act as sinks for material brought from above. It is more common for avalanches with southern aspects to carry material into these lakes than for those with northern aspects, and the avalanche impact landforms in the parks are almost exclusively associated with avalanche paths with southern or southeasterly exposures (Butler, 1989a, Figure 4).

Under past, more severe periglacial climates of the Neoglacial period, avalanches probably were even more effective as periglacial contributors of material into subalpine lakes. Using pollen and clast analysis from cores from two lakes in WLNP, Bujak (1974) showed that an increase in avalanche detritus and a reduction of tree pollen in favor of *Alnus* sp. pollen occurred about 1630 B.P., a time correlated with the Audubon climatic cooling of the southern Rocky Mountains.

AVALANCHE RUNOUT FANS AND DEBRIS FLOWS

The microclimate difference on north- versus south-facing slopes alluded to above manifests itself in another fashion associated with the avalanche hazard of the study area. Greater efficacy of freeze-thaw activity produces a greater detrital load on south-facing avalanche paths in the study area, detrital loads that can be moved by debris flows as well as by snow avalanches. The more

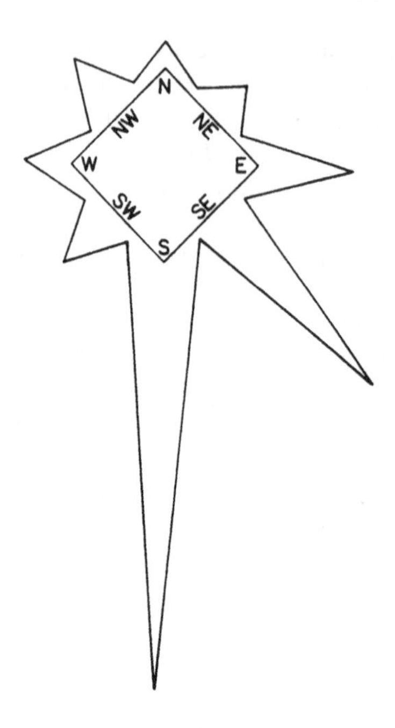

Figure 8.6. Orientation of avalanche-path runout zone fans (n = 54) which experience debris flows in GNP

xeric, less densely vegetated south-facing paths (Butler, 1979) are more susceptible to debris mobilization in the form of debris flows during heavy rain events. Categorization by slope aspect of those avalanche runout zones (avalanche "fans") which experience debris flows reveals a striking pattern that corroborates these statements. All but 11 of the 54 avalanche fans that experience debris flows possess south- or southeast-facing aspects (Figure 8.6). This aspect dichotomy is obvious on Figure 8.1, where north-facing paths do not display morphologic evidence of debris flows and do not have fan-shaped landforms in their runout zones. The south-facing paths there exhibit widespread evidence of debris flowage, and avalanche/debris flow fans are distinctly displayed. The previously mentioned slope asymmetry in the glacial valleys of the study areas also influences this pattern, with larger sediment sources on south-facing slopes.

No confusion should exist between so-called "alluvial fans" that are actually snow-avalanche/debris-flow fans and true alluvial fans. Of the 54 fans mapped by Carrara (1990) as alluvial fans but which we classify as avalanche/debris-flow fans, 28 fans displayed distinct debris flow channels and levees on the aerial photographs. These 28 fans are plotted on Figure 8.7, along with a sample of four relatively small, true alluvial fans from GNP that we examined

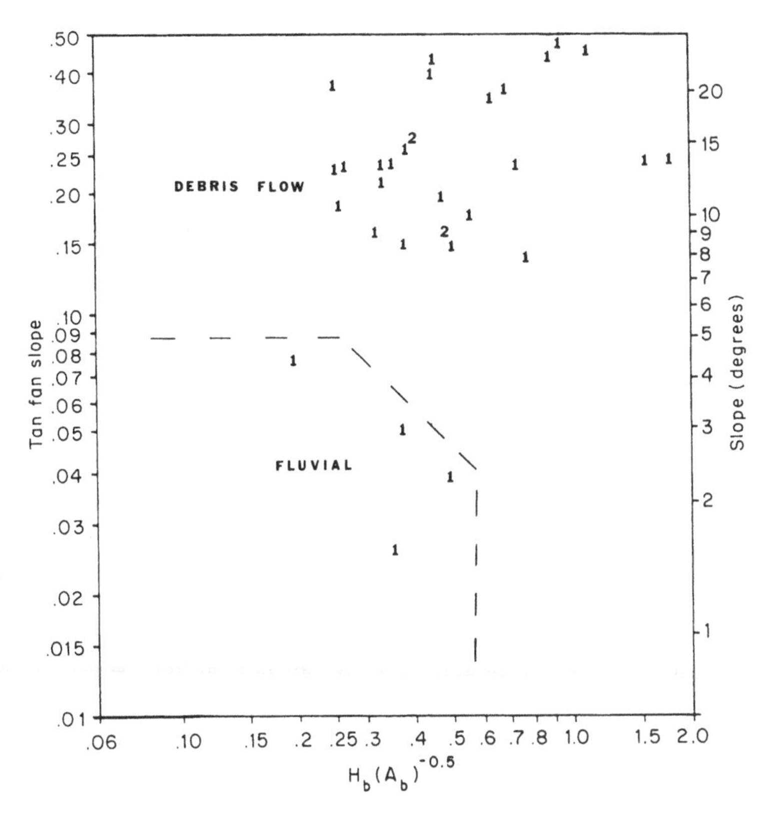

Figure 8.7. Graph of fan slope against ruggedness number showing that avalanche-path runout fans which experience debris flows and those which are largely fluvial in nature plot in distinct groups

in the field. Although the ruggedness number (Melton, 1965) of the fans shows some overlap, the fan slopes clearly distinguish the true alluvial fans with low slope angles from the steeper snow-avalanche fans which also experience debris flows. We do not imply that alluvial fans do not experience debris flows, because this phenomenon is well recognized (Jackson, 1987). The morphology and spatial coincidence with snow-avalanche paths, however, clearly differentiates avalanche/debris-flow fans from those displaying classic alluvial fan morphology. From a hazard perspective, it is irrelevant whether snow avalanches alone, debris flows, or multiple processes produce the slope-angle distinction; hazard planners need to know that alluvial fans exceeding a threshold angle are likely to experience both snow avalanching and debris flows. Future mapping of alluvial deposits in GNP should make a clear distinction between fluvially dominated alluvial fans, and fans at the base of avalanche paths which are primarily attributable to mass movements.

Conclusions

Snow avalanches are widespread periglacial phenomena in the Rocky Mountains of northwest Montana and southwest Alberta, and indeed around the world. Avalanche paths act as conduits which bring the influence of the periglacial environment down to low elevations. Climatic deteriorations of past glacial periods in the Holocene probably exacerbated this effect, delivering more sediment more frequently to low elevations. Lakes and valley bottoms at the base of avalanche paths have been the recipients of this detritus from the periglacial environment.

In our study area, valley asymmetry exists that results in south-facing avalanche paths descending to lower elevations on lower slope angles and producing runout zones of greater extent. Whether this valley asymmetry is a result of microclimatic differences on north- versus south-facing slopes or simply reflects lithologic controls exacerbated by Pleistocene glaciation is problematic. The avalanches on these paths also tend to be confined by steep channels so that they possess high energy even in the runout zone. These same runout zones experience widespread debris flowage. Proper categorization of avalanche fans, as distinct from alluvial fans, is essential for proper hazard planning in the low-slope environment where the bulk of human activity is concentrated. Although snow avalanches on any aspect can produce hazards to human occupancy and transportation in the study area, these hazards are greatest and most likely on paths with southerly exposures.

Acknowledgments

We thank officials of Glacier National Park for cooperation and collecting permits, C.M. Cowell, J. Nicholas, and J. Gao for assistance with morphometric measurements, D. Brown and L. Bian for Landsat and SPOT image processing, and the friendly personnel of Freda's for logistical support and supplies. This study was funded by grants to DRB from the University of Georgia Research Foundation, Inc., to DRB and GPM from the National Geographic Society, and to SJW from NASA (grant number NAGW-926).

References

Bujak, C.A., Recent palynology of Goat Lake and Lost Lake, Waterton Lakes National Park, M.A. thesis, University of Calgary, 1974.

Butler, D.R., Snow avalanche path terrain and vegetation, Glacier National Park, Montana, *Arctic and Alpine Research*, **11**, 17–32, 1979.

Butler, D.R., Snow-avalanche hazards in Glacier National Park, Montana: Meteorologic and climatologic aspects, *Physical Geography*, **7**, 72–87, 1986.

Butler, D.R., Canadian landform examples 11. Subalpine snow avalanche slopes, *The Canadian Geographer*, **33**, 269–273, 1989a.

Butler, D.R., Snow-avalanche dams and resultant hazards in Glacier National Park, Montana, *Northwest Science*, **63**, 109–115, 1989b.
Butler, D.R., and G.P. Malanson, A history of high-magnitude snow avalanches, southern Glacier National Park, Montana, U.S.A., *Mountain Research and Development*, **5**, 175–182, 1985.
Butler, D.R., and G.P. Malanson, Periglacial patterned ground, Waterton-Glacier International Peace Park, Canada and U.S.A., *Zeitschrift für Geomorphologie*, **33**, 43–57, 1989.
Butler, D.R., and G.P. Malanson, Non-equilibrium geomorphic processes and patterns on avalanche paths in the northern Rocky Mountains, U.S.A., *Zeitschrift für Geomorphologie*, **34**, 257–270, 1990.
Butler, D.R., J.G. Oelfke, and L.A. Oelfke, Historic rockfall avalanches, northeastern Glacier National Park, Montana, U.S.A., *Mountain Research and Development*, **6**, 261–271, 1986.
Butler, D.R., and S.J. Walsh, Lithologic, structural, and topographic controls of snow-avalanche path locations, eastern Glacier National Park, Montana, *Annals of the Association of American Geographers*, **80**, 362–378, 1990.
Butler, D.R., S.J. Walsh, and G.P. Malanson, GIS applications to the indirect effects of forest fires in mountainous terrain, U.S. Department of Agriculture Forest Service Southeastern Forest Experiment Station, *General Technical Report SE-69*, 202–211, 1991.
Carrara, P.E., Surficial geologic map of Glacier National Park, Montana, scale 1:100,000, *Miscellaneous Investigations Series Map I-1508-D*, U.S. Geological Survey, Reston, VA, 1990.
Coen, G.M., and W.D. Holland, Soils of Waterton Lakes National Park, Alberta, *Alberta Institute of Pedology S-73-33*, Information Report NOR-X-65, Edmonton, Alberta, 1976.
Corner, G.D., Avalanche impact landforms in Trøms, north Norway, *Geografiska Annaler*, **62A**, 1–10 1980.
Desloges, J.R., and J.S. Gardner, Recent chronology of an alpine alluvial fan in southwestern Alberta, *The Albertan Geographer*, **17**, 1–18, 1981.
Desloges, J.R., and J.S. Gardner, Process and discharge estimation in ephemeral channels, Canadian Rocky Mountains, *Canadian Journal of Earth Sciences*, **21**, 1050–1060, 1984.
Finklin, A.I., A climatic handbook for Glacier National Park, with data for Waterton Lakes National Park, *U.S. Department of Agriculture Forest Service General Technical Report INT-204*, 124 pp., Intermountain Research Station, Ogden, Utah, 1986.
Fitzharris, B.B., and I.F. Owens, Avalanche tarns, *Journal of Glaciology*, **30**, 308–312, 1984.
Gardner, J.S., Observations on erosion by wet snow avalanches, Mount Rae area, Alberta, Canada, *Arctic and Alpine Research*, **15**, 271–274, 1983.
Hole, J., Snow avalanche impact pits in Sunnykveb and adjacent areas in Sunnmøre, western Norway. Preliminary results, *Norsk Geografisk Tidsskrift*, **35**, 167–172, 1981.
Hungry Horse News, Slide damages Waterton cottage. Columbia Falls, Montana, p. 7, February 19, 1954a.
Hungry Horse News, Snowslides cause cottage damage. Columbia Falls, Montana, p. 10, April 9, 1954b.
Hungry Horse News, Study control of Waterton slides. Columbia Falls, Montana, p. 8, April 19, 1957.
Jackson, L.E., Jr., Debris flow hazard in the Canadian Rocky Mountains, *Geological Survey of Canada paper 86-11*, Canadian Geological Survey, Ottawa, 20 pp., 1987.

Jackson, L.E., Jr., R.A. Kostaschuk, and G.M. MacDonald, Identification of debris flow hazard on alluvial fans in the Canadian Rocky Mountains, in *Debris Flows/ Avalanches: Process, Recognition, and Mitigation*, edited by J.E. Costa and G.F. Wieczorek, pp. 115–124, Geological Society of America, Boulder, Colo., 1987.

Kostaschuk, R.A., G.M. MacDonald, and P.E. Putnam, Depositional process and alluvial fan-drainage basin morphometric relationships near Banff, Alberta, Canada, *Earth Surface Processes and Landforms*, **11**, 471–484, 1986.

Luckman, B.H., Drop stones resulting from snow-avalanche deposition on lake ice, *Journal of Glaciology*, **14**, 186–188, 1975.

Luckman, B.H., The geomorphic activity of snow avalanches, *Geografiska Annaler*, **59A**, 31–48, 1977.

Mackenzie, G., The fire ecology of the forests of Waterton Lakes National Park, M.A. thesis, University of Calgary, 1973.

Martinelli, M., Jr., A test of the avalanche runout equations developed by the Norwegian Geotechnical Institute, *Cold Regions Science and Technology*, **13**, 19–33, 1986.

Melton, M.A., The geomorphic and paleoclimatic significance of alluvial deposits in southern Arizona, *Journal of Geology*, **73**, 1–38, 1965.

Peev, C.D., Geomorphic activity of snow avalanches, *International Symposium on Snow and Ice Avalanches, International Association of Hydrological Sciences Publication*, **68**, 357–368, 1966.

Rapp, A., Recent development of mountain slopes in Kärkevagge and surroundings, northern Scandinavia, *Geografiska Annaler*, **42**, 65–200, 1960.

Sauchyn, D.J., Particle size and shape variation on alpine debris fans, Canadian Rocky Mountains, *Physical Geography*, **7**, 191–217, 1986.

VanDine, D.F., Debris flows and debris torrents in the southern Canadian Cordillera, *Canadian Geotechnical Journal*, **22**, 44–68, 1985.

Walsh, S.J., L. Bian, D.G. Brown, D.R. Butler, and G.P. Malanson, Image enhancement of Landsat Thematic Mapper digital data for terrain evaluation, Glacier National Park, Montana, U.S.A., *Geocarto International*, **4**, 55–58, 1989.

Walsh, S.J., D.R. Butler, D.G. Brown, and L. Bian, Cartographic modeling of snow avalanche path location within Glacier National Park, Montana, *Photogrammetric Engineering and Remote Sensing*, **56**, 615–621, 1990.

Ward, R.G.W., Geomorphological evidence of avalanche activity in Scotland, *Geografiska Annaler*, **67A**, 247–256, 1985.

Weirich, F.H., Sediment budget for a high energy glacial lake, *Geografiska Annaler*, **67A**, 83–99, 1985.

9 Long-term Rates of Contemporary Solifluction in the Canadian Rocky Mountains

D.J. Smith
Department of Geography, University of Saskatchewan

Abstract

This paper presents the results of a ten-year assessment of the rate of present-day solifluction movements in the Canadian Rocky Mountains. Surface displacement rates measured with the aid of inclinometer tubes show a significant range (standard deviation 0.32 cm y^{-1}), with an average value of 0.47 cm yr^{-1} recorded between 1980/1981 and 1990. Comparison with preliminary data collected between 1980/1981 and 1983 showed that short-term rates were biased by an average of 166%. As a result, it is concluded that inclinometer tubes have a 3- to 4-year stabilization period. These results demonstrate that solifluction is a moderately efficient geomorphic agency within the Canadian Rockies. The geomorphic effectiveness of this process is comparable to that of rockfalls and snow avalanches.

Introduction

Solifluction is a distinctive example of periglacial mass wasting in both permafrost and nonpermafrost settings (Harris, 1981a, 1987). It describes a particular type of slow flow-like soil movement resulting from the combined effect of frost creep and gelifluction mechanisms (French, 1976; Harris 1981a). Many attempts have been made to gauge the rate and character of contemporary solifluction movements. Nevertheless, most measurements cover only short periods (Williams, 1966; Washburn, 1967; Benedict, 1970; Price, 1973; C. Harris, 1973; Mackay, 1981; Gamper, 1983). There remain relatively few long-term records of contemporary solifluction movements

Periglacial Geomorphology. Edited by J. C. Dixon and A. D. Abrahams

(Jahn, 1978, 1981, 1989; Rapp and Strömquist, 1979; Stocker, 1984; Egginton and French, 1985). Consequently, it is rarely possible to evaluate the relative importance of solifluction as a mass wasting agent (Barsch and Caine, 1984; Caine and Swanson, 1989).

This paper presents the results of a long-term assessment of present-day solifluction movement rates in the Canadian Rocky Mountains. Alpine slopes in this region provide an important venue for monitoring the contemporary activity of a variety of mass wasting processes (Owens, 1972; S.A. Harris, 1973; Gardner, 1979; Luckman, 1988). However, until recently there was very little appreciation for the geomorphic role solifluction might play (Luckman, 1981).

Solifluction movements in the Canadian Rockies were first described by Smith (1985). Before this, the only insight into alpine solifluction in the North American cordillera came from Benedict's (1970) four-year study of slow downslope movements in the Colorado Front Range and Price's (1970, 1973) five-year examination of solifluction lobe ecology in the Yukon Ruby Range. Subsequent reports of solifluction in the Canadian Rockies have chronicled Late Holocene variations in solifluction lobe behaviour (Smith, 1987a) and stressed the importance of local site conditions to the rate of solifluction movement (Smith, 1987b, 1988). The latter records are of limited duration (1980–1983), and the present paper extends this temporal perspective. Consideration is given to the proxy merit of short- versus long-term data and to the geomorphic significance of solifluction activity in this high mountain landscape.

Characteristics of the study area

The Canadian Rocky Mountains flank the eastern edge of the North American cordillera in western Canada. This deeply sculptured landscape is a result of recent geological upheaval and extensive Pleistocene glaciation. The Rockies are characterized by two high relief structural provinces and a series of major ranges stretching 1500 km southeastward from north-central British Columbia to the international border. Along the interprovincial border the massive topography of the Eastern Main Range subprovince consists of relatively flat-lying to gently folded thrust blocks of Palaeozoic carbonates and clastics (Monger and Price, 1979). Immediately to the east are the Front Ranges consisting of strongly folded and thrust faulted Upper Cambrian to Jurassic-age bedrock (Halladay and Mathewson, 1971). The last major glacial episode to affect the area occurred during late Wisconsin time, when valley glaciers extended from the continental divide to positions near the eastern mountain front (Rutter, 1987). Recent work suggests deglaciation was underway by 13,000 years B.P., when the glaciers had receded to positions close to or inside their present limits (Osborn and Luckman, 1988). The legacy of these events is a high mountain setting profoundly accentuated by glaciation. Geomorphic changes in postglacial time have been of a much more subdued nature (Gardner, 1979; Luckman, 1981).

Numerous icefields, glaciers, and rock glaciers in the Canadian Rockies attest to alpine periglacial conditions on many of the surrounding mountain slopes (Gardner, 1978). Morphological evidence of solifluction activity is widespread in this region but restricted to a narrow 200-m altitudinal belt at or above the present-day treeline. The lower altitudinal boundary of this belt decreases in elevation northwards, from around 2400 m above sea level at 50°N latitude to 2000 m above sea level at 53°N latitude. Lobes within this solifluction zone exhibit signs of both historical (Smith, 1987a) and contemporary activity (Smith, 1987b, 1988).

Since 1980 and 1981, solifluction processes have been under continuous observation within three areas of the Canadian Rockies (Figure 9.1). Sites in the Parkers Ridge area are situated within the Main Ranges in the Columbia Icefield region, northern Banff National Park (52°11′N, 117°06′W). Study sites in the Mount Allan area (50°57′N, 115°10′W) and in the Mount Rae area in Peter Lougheed Provincial Park (50°35′N, 116°06′W) are located within the Front Ranges.

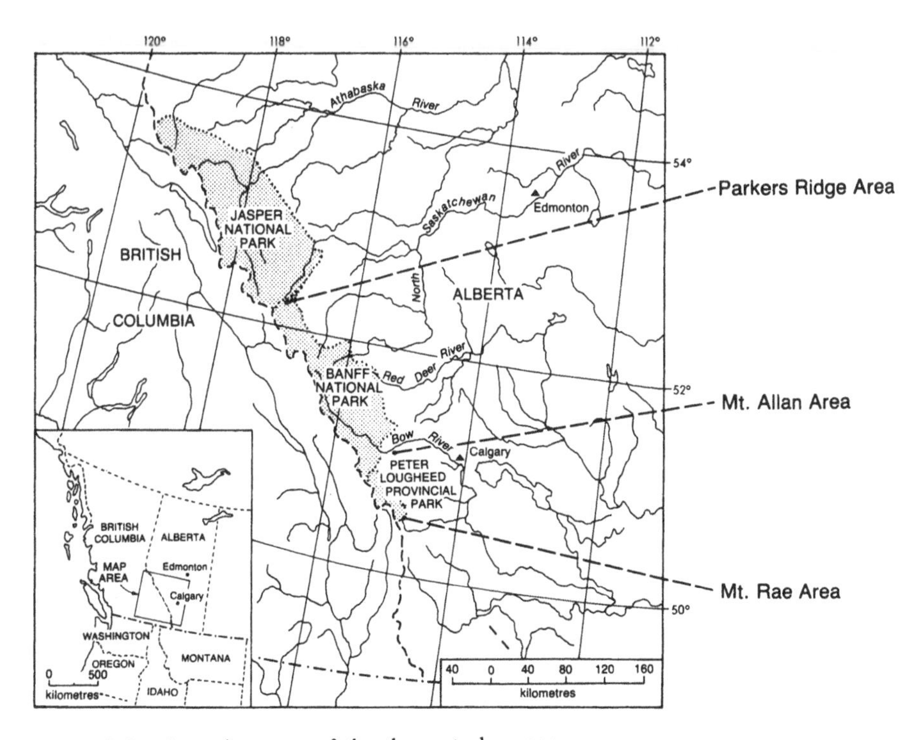

Figure 9.1. Location map of the three study areas

Methods

Forty-eight solifluction lobe sites were selected for long-term monitoring in 1980 (Smith, 1987b). All the lobes are tongue-shaped with lobate, well-vegetated frontal perimeters. Of particular interest to the present study are 60 flexible inclinometer tubes installed in 1980 and 1981. Proven useful in a number of previous investigations of solifluction (Williams, 1966; Price, 1970; Harris, 1972; Mackay, 1981), the tubes provide a depth-integrated measure of soil displacement. In this case, hollow plastic tubing was inserted into ground approximately 0.5 m behind prominent solifluction lobe snouts.

Installation followed a standard procedure. Initially a thin-walled tube auger (1.9-cm outside diameter) was hammered vertically into the ground to create an implantation hole with an overall depth close to 1 m. Thick-walled, hollow plastic tubing (1.9-cm outside diameter), heat-sealed at its bottom end, was then eased into each hole using an implantation rod. The rod was removed, and the tubes trimmed flush with the ground and sealed with cork stoppers.

Originally soil movements were monitored annually using a specially constructed inclinometer probe (Smith, 1985; pp. 86–89). Unfortunately, one-third of the original set of tubes was rendered unusable over the ten years of the survey due to disturbance, primarily by inquisitive ground squirrels. By the summer of 1989, it was clear that the magnitude of soil movement since installation at the remaining sites exceeded the capabilities of the probes. Consequently, in the second week of August, 1990, 20 of the remaining tubes were excavated to provide a final measure of soil displacement at each station. Table 9.1 provides a descriptive summary of the 20 sites.

Each excavation extended below the bottom of the tube and left it exposed along a vertical face (Figure 9.2). A vertical plumb-line was then dropped from the surface to the bottom of the tube, which was assumed not to have moved since the time of installation. The horizontal distance from this line to the leading edge of the tube was then measured to the nearest 0.1 cm. Annual movement rates were established by dividing these horizontal displacements by the elapsed time since installation. Bulk sediment samples were collected in the excavations at one-half the lobe depth. Particle size analysis of the less-than-2-mm fraction followed standard procedures (American Society of Testing and Materials, 1981, pp. D422–463), with dispersion achieved using sodium hexmetaphosphate. Determination of sand sizes (0.74 to 2 mm) was made by the hydrometer method. The index properties of these sediments were determined from samples split for grain size analysis (American Society for Testing and materials, 1981, pp. D423–466, D424–459). Bulk density estimates were obtained from multiple near-surface measurements made using a volumetric sampler (McKeague, 1978, p. 2.211).

Table 9.1. Geomorphological Characteristics at the Inclinometer Tube Sites

	Locational Characteristics			Physical Properties						
Tube Number	Elevation (m)	Aspect	Gradient (degrees)	Percent sand (%)	Percent silt (%)	Percent clay (%)	Bulk Density (kN m^3)	Liquid Limit (%)	Plastic Limit (%)	Plasticity Index
Parkers Ridge Area										
50C	2070	N	19	45	27	28	13.45	22	11	11
51	2073	N	22	51	36	13	10.68	23	12	11
52	2073	N	26	64	21	15	11.21	22	14	8
53	2071	N	20	50	38	12	12.03	21	10	11
54	2068	N	32	58	30	12	9.80	30	21	9
62B	2290	NW	12	71	24	5	14.74	25	10	15
Mount Allan Area										
5	2390	N	27							
6	2500	SW	20	64	27	9		22	20	2
7	2495	SW	17	65	23	12		23	22	NP[a]
30	2240	E	21	45	45	10		25	11	14
31	2600	W	13	69	22	9		32	32	NP
32	2492	W	20	61	38	1		27	27	NP
33	2490	W	22	57	31	12		25	15	10
37B	2389	NW	22	51	32	17		26	14	12
38	2434	NE	25	49	29	22		24	13	11
Mount Rae Area										
10	2510	NW	20	55	42	3	7.98	32	28	4
12	2509	W	23	56	32	12	7.39	29	25	4
35	2388	S	24	38	56	6		27	21	6
43	2517	W	20							
45	2393	S	24	38	60	2		24	19	5

[a]NP = nonplastic.

Figure 9.2. Magnitude of solifluction displacement at Lobe 6 between 1980 and 1990. Scale on ruler is in inches

Observations

Soil movements recorded by the inclinometer tubes at each of the 20 sites are summarized in Table 9.2. Surface movement rates show a significant range (coefficient of variation $cv = 0.69$). For example, although 11.1 cm of total displacement was observed (1980–1990) at site 31 in the Mount Allan area, only a negligible amount of surface movement was recorded during the same interval at site 7. In the majority of cases the excavations revealed surface movements closer to an average value of 0.47 cm yr^{-1} (standard deviation $sd = 0.32$ cm yr^{-1}) for the interval between 1980/1981 and 1990.

Figure 9.3 illustrates the nature of subsurface movements at the 20 sites. The majority of displacements are relatively shallow and restricted to the

Table 9.2 Summary of Long-term Solifluction Measurements

Tube Number	Interval	Tube Depth Below Surface (cm)	Maximum Depth of Movement (cm)	Annual Surface Movement (cm yr^{-1})	Annual Volumetric Movement (cm^3 cm^{-1} yr^{-1})	Shape Down-slope
5	1980–1990	104	13	0.29	5.00	Concave
6	1980–1990	96	51	0.50	21.88	Convex
7	1980–1990	86	46	0.01	3.96	Concave
10	1980–1990	88	21	0.37	3.33	Concave
12	1980–1990	86	59	0.43	6.04	Concave
30	1980–1990	88	48	0.73	17.29	Concave
31	1980–1990	102	26	1.10	105.21	Concave
32	1980–1990	90	59	0.68	24.17	Concave
33	1980–1990	87	36	0.13	4.79	Concave
35	1980–1990	97	41	0.33	6.25	Concave
37B	1980–1990	112	8	0.18	1.67	Concave
38	1980–1990	112	41	0.11	5.25	Convex
43	1980–1990	86	18	0.17	2.92	Concave
45	1980–1990	76	33	0.21	5.42	Concave
50C	1981–1990	105	36	1.17	21.88	Concave
51	1981–1990	104	32	0.36	6.25	Concave
52	1981–1990	107	15	0.52	9.17	Concave
53	1981–1990	100	35	0.75	8.96	Convex
54	1981–1990	107	15	0.46	2.92	Convex
62B	1981–1990	105	17	0.80	9.03	Concave
Average				0.47	9.33	

upper 30 to 40 cm of sediment (Table 9.2). Significantly deeper movements are visible at only a small proportion of the sites (see lobe 32, Figure 9.3).

Similar to surface movement rates, the magnitude of subsurface sediment transport ranges significantly ($cv = 0.79$). Displacements vary from a maximum of 24.17 cm^3 cm^{-1} yr^{-1} at site 31 in the Mount Allan area to a minimum of 1.67 cm^3 cm^{-1} yr^{-1} at site 37B in the same area (Table 9.2). On average 9.33 cm^3 cm^{-1} (sd = 7.36 cm^3 cm) of sediment is transported annually by solifluction.

Excavation of the tubes revealed that several styles of movement characterize solifluction displacements in this region. Most tubes exhibit a concave downslope shape (see lobe 30 and 50C, Figure 9.3). This type of velocity profile is associated with either creep or compressive shearing in solifluctING sediments (Benedict, 1970; Price, 1973; Sohma, Okazawa, and Iwata, 1979). A smaller number of tubes exhibit evidence of plug-like convex downslope profiles (see lobe 38 and 51, Figure 9.3). Recent research attributes this latter

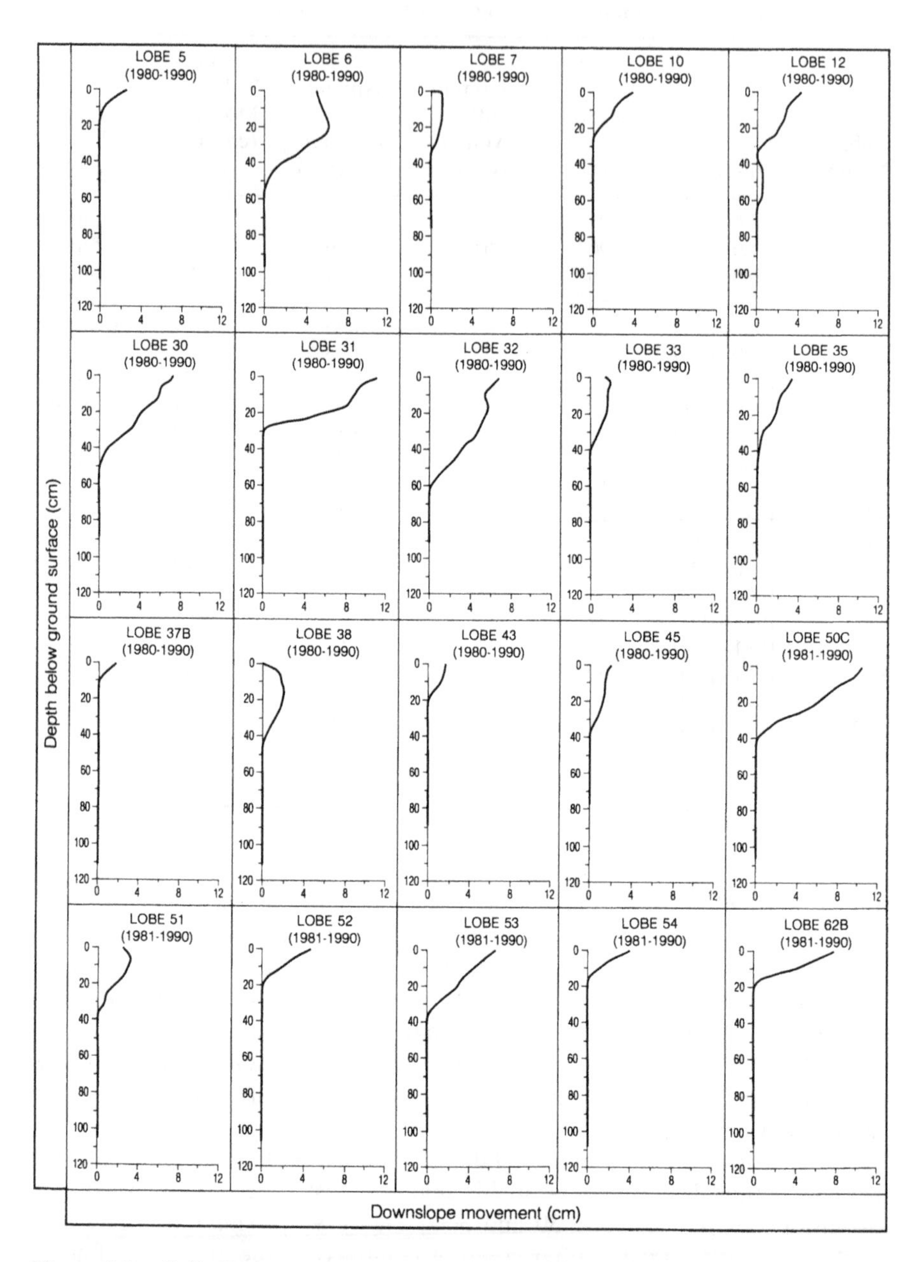

Figure 9.3. Soil displacements recorded by the inclinometer tubes. Note that the horizontal scale is exaggerated by a factor of about 6.5

type of profile to deep-seated basal sliding above an active layer–permafrost boundary (Mackay, 1981; Reanier and Ugolini, 1983). However, the lack of permafrost at these sites precludes that possibility. It seems more likely that these convex profiles develop when the overlying vegetation carpet retards the effects of creep and shallow shearing (Jahn and Cielinska, 1974; Jahn, 1978).

Discussion

These surface and subsurface measurements demonstrate that solifluction rates in the Canadian Rocky Mountains vary both spatially and temporally. For the most part this wide range in activity is due to high magnitude movements at a limited number of stations and does not reflect any significant regional variation in solifluction behaviour (Smith, 1985). Instead, the major differences between the sites are almost certainly related to local site, sedimentological, or environmental characteristics (Smith, 1987b). Unfortunately, despite the significant role assigned to local environmental characteristics in previous reports (Smith, 1988), no assessment of their long-term contribution is possible.

A significant finding of the preliminary investigations reported by Smith (1987b) was the importance of site characteristics. Despite the paucity of the current data set ($n = 20$), at least some of these earlier propositions are substantiated by Spearman rank correlation analysis. For example, there is a discernible trend of reduced surface movement with increasing elevation ($r_s = -0.308$, significant at the 90% confidence level). This observation augments preliminary findings and serves notice of the important role played by treeline in restricting the elevational extent of solifluction (e.g., Harris, 1982). Aspect was previously identified as an important determinant of solifluction rates (Smith, 1987b, pp. 316–317). Unfortunately, any influence that aspect has on the long-term rates is masked by the limited size and distribution of the data set. On the other hand, both surface and subsurface displacement rates are shown to be strongly influenced by the local gradient ($r_s = -0.376$; $r_s = -0.412$, significant at the 95% confidence level). Field observations suggest this decrease in displacement with increasing slope is largely influenced by the rapid drainage of soil water on steeper slopes (e.g., Strömquist, 1983). Drainage reduces the potential for solifluction by decreasing the likelihood for saturated gelifluction flows (Smith, 1988).

Sediments at the 20 sites share many physical attributes (Table 9.1). They are best described as silty to stony loams. The less-than-2-mm fraction is dominated by either sand (mean 56% ± 9%) or silt (mean 33% ± 10%), with only minor amounts of clay present (11% ± 7%). No significant relationship exists between these textural parameters and the long-term rates of movement. Specifically, this analysis indicates that the differences in movement recorded in Table 9.2 are not due solely to grain size considerations.

The liquid limits of the site sediments are low (average 26%), and most have a narrow plasticity range (Table 9.1). In general, these index properties suggest sediments sensitive to liquid or plastic failure at relatively low field moisture conditions (e.g., Harris, 1987). There is, however, no indication that differences in these properties provides an explanation for spatial variations in the observed rates of solifluction.

In summary, the long-term data demonstrate that spatial variations in solifluction are not solely the result of one particular locational or physical property. However, the fact that elevation and gradient are shown to affect the rate of flow means that solifluction activity is, to some extent at least, dependent upon its spatial setting.

COMPARISON OF SHORT-TERM AND LONG-TERM MOVEMENT

Estimates of contemporary geomorphic activity rates from short-term records are frequently challenged by the findings of more prolonged research efforts. Unfortunately, there remains little awareness for this problem in mass wasting research where long-term studies remain the exception (Saunders and Young, 1983). Clearly, however, long-term records do provide a more adequate appraisal of average geomorphic activity (Gardner, 1979; Egginton and French, 1985) and serve to reduce measurement bias, which can be as high as 300% in the short-term (Caine, 1981).

The short- and long-term data (Table 9.3) permit examination of the impact of time on the absolute annual measure of both surface and subsurface solifluction rates. Annual surveys at the 20 sites between 1980 and 1983 illustrate the pervasiveness of year-to-year rate changes (Table 9.3). In addition, both measures of solifluction record a decrease in magnitude over the short-term (1980 to 1983). This inclination is also reflected in the coefficient of variation estimates, which consistently decrease over the short-term (Table 9.3).

Figure 9.4 shows that both short-term mean movement rates approach the long-term mean rates with the passage of time. In both cases a 3- to 4-year period is needed before the short-term rate approximates the long-term rate. In effect, this discovery demonstrates that short-term surface rates are biased by an average of 166% (range 149 to 200%). Volumetric movement rates are similarly distorted, with short-term measurements overestimating displacements by an average of 125% (range 107 to 149%).

Caine (1981) illustrated a similar bias associated with surface marker stones and attributed it to the time needed for tracer material to be incorporated into the soil. The fact that a corresponding trend is evident here suggests that inclinometer tubes require an analogous stabilization period. Many factors

Table 9.3. Comparison of Average Short- and Long-term Solifluction Movement Rates

		Rate of Movement		
Interval	Number of Measurements	Mean[a]	Coefficient of Variation	Bias[b] (%)
Surface Movements				
1980–1981	14	0.94	0.581	200
1981–1982	19	0.71	0.511	151
1982–1983	19	0.70	0.548	149
1980/81–1990	20	0.47	0.658	—
Volumetric Movements				
1980–1981	14	13.89	0.778	149
1981–1982	19	11.05	0.641	118
1982–1983	19	9.96	0.511	107
1980/81–1990	20	9.33	0.789	—

[a]Surface movements in cm yr^{-1} and volumetric movements in cm^3 cm^{-1} yr^{-1}.
[b]Bias is expresed as percent difference in the short-term mean from the long-term mean.

may contribute to this bias, as several methodological problems are inherent to the technique (Anderson and Finlayson, 1975; Jahn, 1978). First, it seems inevitable that installation of the tubes results in substantial disturbance to the surrounding sediment. In view of the distinct structures identified in solifluction soils (Harris, 1981b), the impacts of installation almost certainly continue to influence solifluction for an indefinite period. In a related discussion, Jahn (1989, p. 162) indicated that several years are likely to pass before peg columns will be fully incorporated into soil. Second, there has been little consideration for how effectively plastic tubes measure solifluction. For example, one question that arises is the necessity for ensuring that tube and soil densities are similar. If they are not, the tubes may resist deformation and thereby generate spurious information about soil displacement (Williams, 1957). Similarly, most tube material has a tendency to curl with time (e.g. Price, 1970). If this is a serious problem, it can obviously lead to indeterminate or anomalous measurement errors (Price, 1970, pp. 108–111). Finally, there remains the possibility that any relationship displayed by inclinometer tube measurements is a result of natural variability in solifluction behaviour.

In the present case, it could also be argued that the short-term bias displayed in Figure 9.4 is actually either an artifact of calibration errors associated with the inclinometer probes or a result of the decision to excavate the tubes in 1990. The latter concern is of particular note, as the short-term data (relative measurement using inclinometer probes) were collected in a

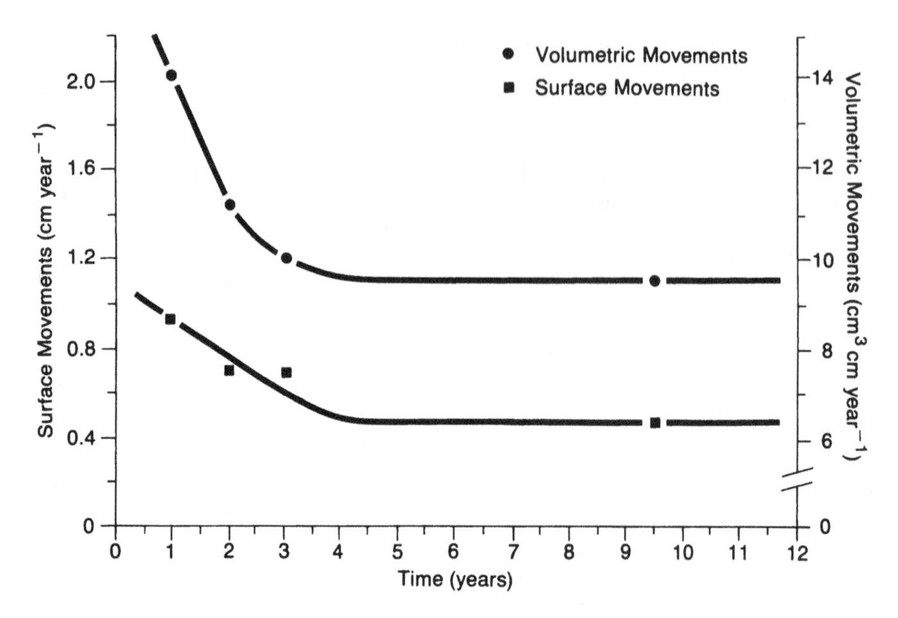

Figure 9.4. Graph of mean surface movement and mean volumetric movement against length of a record

different fashion to the long-term data (absolute measurement in excavations).

As a simple check of these concerns, two tests were undertaken. In 1981 and 1982 several of the tubes measured with the inclinometer probes were later excavated to substantiate the accuracy of the procedure. In all cases, no significant differences emerged between the actual and computed movement values. Thus, it is assumed that the long-term measurements provide a comparable measure of solifluction. As a second test, the short-term tube measurements were compared to replicate data gleaned from adjacent dowel pillar columns installed in 1980 (Smith, 1987b). The columns consist of 2.5-cm segments of 1.3-cm-wide wooden dowelling stacked in vertical boreholes. All were excavated and measured in 1982 using the same technique designed to assess long-term tube displacements in 1990. An empirical comparison of the short-term displacements recorded by the pillars and tubes demonstrated that both provide a complementary measure of solifluction. In the case of neighboring installations, significant surface (r^2 = 0.62) and volumetric (r^2 = 0.68) correlations were noted between the tube and pillar data. Both tests support the contention that the short-term decline in solifluction described in Table 9.3 is genuine.

Thus it is concluded that the approach over a 3- to 4-year period of both short-term measures of solifluction to their respective long-term means is real. This conclusion suggests that inclinometer tubes have a stabilization period that extends beyond the duration of most of the mass wasting studies in which they have been employed (Saunders and Young, 1983). Furthermore, it seems likely that many of the techniques used by process geomorphologists in hillslope studies face similar methodological problems.

GEOMORPHIC SIGNIFICANCE OF SOLIFLUCTION

As one of a suite of mass wasting processes operating on slopes in the Canadian Rocky Mountains (Slaymaker, 1990), the geomorphic significance of solifluction can be evaluated in relation to other slope processes. Although general indices provide a crude measure of the effectiveness of the different geomorphic processes (Slaymaker, 1974), the results of these analyses are not easily compared. One solution to this problem is to consider sediment transport as a form of physical or geomorphic work (Embelton and Whalley, 1979). Since it requires a reduction in the potential energy of the landscape for sediment to be transported, any change in the potential energy can be used to express the geomorphic work done as a measure of erosional intensity (Caine, 1976). In the case of solifluction transport the change in potential energy ΔE can be defined as $\Delta E = v\varrho_s g(d\sin\theta)$, where v is the volume of sediment transported on a yearly basis per unit of contour (m^3), ϱ_s is the bulk density of the sediment ($kg\ m^{-3}$), g is the gravitation acceleration ($m\ s^{-2}$), d is the mean velocity ($m\ yr^{-1}$), and θ is the slope angle (Caine, 1976). The dimensions of work accomplished are joules ($J = kg\ m^2\ s^{-2}$) per year, which are a measure of the magnitude of work.

The average work done by solifluction per year within each of the study areas is summarized in Table 9.4. These estimates range from a minimum value of 8.05 $J\ yr^{-1}$ in the Mount Rae area to a maximum of 36.27 $J\ yr^{-1}$ in the Parkers Ridge area. Interestingly, the amount of geomorphic work attributable to solifluction in this region is similar to that at a variety of permafrost and nonpermafrost sites (Table 9.5). Thus what appears to be a wide spatial range of geomorphic work in the Rockies may simply epitomize solifluction behavior in general.

These results demonstrate that solifluction is a moderately efficient geomorphic agent within the Canadian Rockies. However, the relative effectiveness of this activity can only be expressed with reference to the work accomplished by other contemporaneously operating geomorphic processes. At present, this is only possible within the Mount Rae area, where the long-term rates of a number of different process sets are reasonably well-established.

The Mount Rae area is described in detail by Gardner, Smith, and Desloges (1983). Suffice it to say that quantitative classification of the land surface of the study area (150 km^2) indicates that 26% percent is covered by subalpine

Table 9.4. Average Annual Geomorphic Work Done by Solifluction per Unit Area of Solifluction

Area	Number of Observa- tions	Volume[a] (m^3)	Bulk Density ($kg\ m^{-3}$)	Mean[b] Velocity ($m\ yr^{-1}$)	Slope (sine)	Work ($J\ yr^{-1}$)
Parkers Ridge	6	0.3	1227	0.0271	0.371	36.27
Mount Allan	9	0.364	760	0.0117	0.352	11.17
Mount Rae	5	0.344	754	0.0088	0.360	8.05

[a]Volume (m^3) = area (m^2) times average thickness of movement (m).
[b]Mean velocity ($m\ yr^{-1}$) = average transport distance ($m^2\ yr^{-1}$) divided by average thickness of movement (m).

forests, 21% by alpine tundra, 19% by steep debris and talus slopes, 2% by perennial snowpacks or glacier ice, and 32% by rock slopes and bedrock free-faces. Continuing investigations (1974 to 1990) have provided a quantitative perspective of sediment transfer rates and magnitudes in this area. For example, the amount of soil transported downslope by ground squirrels is estimated at $1.24\ t\ ha^{-1}\ yr^{-1}$ by Smith and Gardner (1985). Likewise, estimates of rockfall and snow avalanche accretion rates indicate that $1.38\ mm\ yr^{-1}$ of sediment is added to debris slopes within this area (Gardner, 1979, 1983). Hydrochemical surveys indicate that $2.63 \times 10^6\ kg\ yr^{-1}$ of

Table 9.5. Comparison of the Average Geomorphic Work Done by Solifluction per Unit Area at a Variety of Sites in Permafrost and Nonpermafrost Sites

Location	Number of Observations	Work ($J\ yr^{-1}$)	Source
Caledonian Mountains, Northern Sweden	2	5.24	Rapp (1960
Okistindan Mountains, Northern Norway	5	7.66	Harris (1977)
Mount Rae area, Alberta	5	8.05	This study
Mount Allan area, Alberta	9	11.17	This study
Rocky Mountains, Colorado	1	26.19	Benedict (1970)
Schefferville, Quebec	6	28.20	Williams (1966)
Parkers Ridge area, Alberta	6	36.27	This study
Mesters Vig, Greenland	6	44.56	Washburn (1967)

Table 9.6. Estimates of Geomorphic Work in the Mount Rae Area

Process set	Work (10^6 J km^2 yr^{-1})	Percent of Total Work	Source
Ground squirrel burrowing	0.07	0.01	Smith and Gardner (1985)
Solifluction	1.69	0.2	This study
Rockfalls & avalanches	4.81	1.6	Gardner (1979, 1983)
Solute transport	860.60	99.2	Gardner, Smith, and Desloges (1983)

sediment is removed from the area as part of the local solute loads (Gardner, Smith, and Desloges, 1983).

An estimate of the amount of work performed by the various process sets is reasonable if one extrapolates the average rate values to the landform areas described above. On this basis, the work done by solifluction in the Mount Rae area is estimated at 1.69×10^6 J km^2 yr^{-1}. The data demonstrate that geochemical transfers (8.61×10^8 J km^2 yr^{-1}) are the most geomorphologically efficient process at the present time. On the other hand, the geomorphic work done by solifluction and rockfalls/avalanches (4.81×10^6 J km^2 yr^{-1}) is of the same order of magnitude but of only intermediate significance. A relatively insignificant amount of work is done by such biological actors as ground squirrels (6.587×10^4 J km^2 yr^{-1}) (Table 9.6).

The applicability of these estimates to the Parkers Ridge and Mount Allan areas is difficult to assess, as there is an almost complete lack of complementary process studies in those areas. Certainly, variations in the relative significance of solifluction are expected due to a variety of intangibles associated with the different spatial and temporal scales of investigation (Barsch and Caine, 1984).

Conclusions

The morphological effects of solifluction are only subtly manifest in the Canadian Rocky Mountains. Nonetheless, the results of this long-term research program firmly establish that solifluction is a highly active process in this mountain landscape.

The principal results of the study are as follows:

1. Long-term rates of solifluction movement in the Canadian Rocky Mountains average 0.47 cm yr^{-1} at the ground surface. Over the study period these displacements annually result in the downslope transfer of 9.33 cm^3 cm^{-1} of sediment. The preponderance of concave downslope velocity profiles indicates that most of this activity occurs in response to either creep or shallow shearing forces.

2. The movement data are not indicative of any regional variation in solifluction behavior. Instead, differences in magnitudes between the sites are almost certainly a consequence of local site characteristics. Similar to the findings reported previously (Smith, 1987b, 1988), elevation and local gradient play an important role in determining the magnitude of displacement.
3. A comparison of the short- and long-term data shows that inclinometer tubes have a lengthy stabilization period. Both surface and volumetric measurements record a bias of up to 200% if the tubes are measured within 3 to 4 years of installation. As a consequence of these and other findings (Caine 1981; Egginton and French, 1985), further research is obviously required to appraise the short-term utility of most hillslope measurement techniques.
4. Solifluction is recognized as a moderately competent geomorphic agent in the Canadian Rockies. Estimates of the average work done by solifluction range from a minimum value of 8.05 J yr^{-1} to a maximum of 36.27 J yr^{-1}. The geomorphic effectiveness of this activity in the Mount Rae area is comparable to that of rockfalls and snow avalanches.

Acknowledgments

The field investigations described in this paper were supported by the Boreal Institute for Northern Studies at the University of Alberta and the Natural Sciences and Engineering Research Council of Canada (A1930). Alberta Recreation and Parks granted permission to work in Peter Lougheed Provincial Park and in the adjoining Kananaskis Country. Parks Canada and the chief warden of Banff National Park permitted the monitoring of solifluction activity at Parkers Ridge.

A number of friends and colleagues have contributed to this study in the decade since it was initiated. Thanks are due in particular to Jamie Steel who was an essential ingredient in 1980; to Johanna Andersson who helped with the excavations in 1990; and to Keith Bigelow for his assistance in preparing the illustrations.

References

American Society for Testing and Materials, *1981 Annual Book of ASTM Standards. Part 19: Natural Building Stones; Soil and Rock, Philadelphia, 1981.*

Anderson, E.W., and B. Finlayson, Instruments for Measuring Soil Creep, Geomorphological Research Group, Technical Bulletin 16, 51 pp. GeoAbstracts, Norwich, 1975.

Barsch, D. and N. Caine, The nature of mountain geomorphology, *Mountain Research and Development*, **4**, 287–298, 1984.

Benedict, J.B., Downslope soil movement in a Colorado alpine region: Rates, processes and climatic significance, *Arctic and Alpine Research*, **2**, 165–226, 1970.
Caine, N., A uniform measure of subaerial erosion, *Bulletin of the Geological Society of America*, **87**, 137–140, 1976.
Caine, N., A source of bias in rates of surface soil movement as estimated from marked particles, *Earth Surface Processes and Landforms*, **6**, 69–75, 1981.
Caine, N., and F.J. Swanson, Geomorphic coupling of hillslope and channel systems in two small mountain basins, *Zeitschrift für Geomorphologie*, **33**, 189–203, 1989.
Egginton, P.A., and H.M. French, Solifluction and related processes, eastern Banks Island, N.W.T., *Canadian Journal of Earth Sciences*, **22**, 1671–1678, 1985.
Embelton, C., and B. Whalley, Energy, forces, resistances and responses, in *Process in Geomorphology*, edited by C. Embelton and J. Thornes, pp. 11–38, Edward Arnold, London, 1979.
French, H.M., *The Periglacial Environment*, 309 pp., Longman, London, 1976.
Gamper, M., Controls and rates of movement of solifluction lobes in the eastern Swiss Alps, *Proceedings of the Fourth International Conference on Permafrost*, 328–333, 1983.
Gardner, J.S., Wenckchemna: Glacial, periglacial and permafrost conditions in the Valley of the Ten Peaks, Alberta, Canada, *Guide book to the Third International Conference on Permafrost: Appendix A*, National Research Council of Canada, Ottawa, 1978.
Gardner, J.S., The movement of material on debris slopes in the Canadian Rocky Mountains, *Zeitschrift für Geomorphologie*, **23**, 45–57, 1979.
Gardner, J.S., Accretion rates on some debris slopes in the Mt. Rae area, Canadian Rocky Mountains, *Earth Surface Processes and Landforms*, **8**, 347–355, 1983.
Gardner, J.S., D.J. Smith, and J.R. Desloges, *The Dynamic Geomorphology of the Mt. Rae Area: A High Mountain Region in Southwestern Alberta*, 237 pp., University of Waterloo, Department of Geography Publication Series No. 19, 1983.
Halladay, I.A.R., and D.H. Mathewson, *A Guide to the Geology of the Eastern Cordillera Along the Trans Canada Highway Between Calgary, Alberta and Revelstoke, British Columbia*, 123 pp., Alberta Society of Petroleum Geologists, 1971.
Harris, C., Processes of soil movement in turf-banked solifluction lobes, Okistindan, Northern Norway, in *Polar Geomorphology*, edited by R.J. Price and R.B. Parry, pp. 155–174, Institute of British Geographers Special Publication 4, 1972.
Harris, C., Some factors affecting the rates and processes of periglacial mass movements, *Geografiska Annaler*, **55A**, 24–28, 1973.
Harris, C., Engineering properties, groundwater conditions, and the nature of soil movement on a solifluction slope in North Norway, *Quarterly Journal of Engineering Geology*, **10**, 27–43, 1977.
Harris, C. *Periglacial Mass Wasting: A Review of Research*, British Geomorphological Research Group, Research Monograph 4, 204 pp., GeoAbstracts, Norwich, 1981a.
Harris, C., Microstructures in solifluction sediments from south Wales and north Norway, *Biuletyn Peryglacjalny*, **28**, 221–226, 1981b.
Harris, C., The distribution and altitudinal zonation of periglacial landforms, Okistindan, Norway, *Zeitschrift für Geomorphologie*, **26**, 283–304, 1982.
Harris, C., Mechanisms of mass movement in periglacial environments, in *Slope Stability*, edited by M.G. Anderson and K.S. Richards, pp. 531–560, John Wiley, Chichester, England, 1987.
Harris, S.A., Studies of soil creep, western Alberta, 1970–1972, *Arctic and Alpine Research*, **5**, A171–A180, 1973.
Jahn, A., Mass wasting in permafrost and non-permafrost environments, *Proceedings of the Third International Conference on Permafrost*, **1**, 296–300, 1978.

Jahn, A., Some regularities of soil movement on the slope as exemplified by the observations in the Sudety Mts., *Transactions of the Japanese Geomorphological Union*, **2**, 321–328, 1981.

Jahn, A., The soil creep on slopes in different altitudinal and ecological zones of Sudetes Mountains, *Geografiska Annaler*, **71A**, 161–170, 1989.

Jahn, A., and M. Cielinska, The rate of soil movement in the Sudety Mountains, *Abhandlungen der Akademie der Wissenschaften in Göttingen, Mathematisch-Physikalische Klasse*, **29**, 86–101, 1974.

Luckman, B.H., The geomorphology of the Alberta Rocky Mountains: A review and commentary, *Zeitschrift für Geomorphologie Supplement Band*, **36**, 91–119, 1981.

Luckman, B.H., Debris accumulation patterns on talus slopes in Surprise Valley, Alberta, *Geographie Physique et Quaternaire*, **42**, 247–278, 1988.

Mackay, J.R., Active layer slope movement in a continuous permafrost environment, Garry Island, Northwest Territories, Canada, *Canadian Journal of Earth Science*, **18**, 1666–1680, 1981.

McKeague, J.A., *Manual on Soil Sampling and Methods of Analysis*, 2nd edition, 212 pp., Canadian Society of Soil Science, Ottawa, 1978.

Monger, J.W.H., and R.A. Price, Geodynamic evolution of the Canadian cordillera, *Canadian Journal of Earth Science*, **16**, 770–791, 1979.

Osborn, G.D., and B.H. Luckman, Holocene glacier fluctuations in the Canadian Cordillera (Alberta and British Columbia), *Quaternary Science Reviews*, **7**, 115–128, 1988.

Owens, I.F., Morphological characteristics of alpine mudflows in the Nigel Pass area, Canadian Rocky Mountains, in *Mountain Geomorphology*, edited by H.O. Slaymaker and H.J. McPherson, pp. 93–100, Tantalus Press, Vancouver, B.C., 1972.

Price, L.W., Morphology and ecology of solifluction lobe development—Ruby Range, Yukon Territory, Ph.D. dissertation, 325 pp., University of Illinois, Urbana, 1970.

Price, L.W., Rates of mass wasting in the Ruby Range, Yukon Territory, *North American Contribution to the Second International Conference on Permafrost*, 235–246, 1973.

Rapp, A., Recent development of mountain slopes in Karkevagge and surroundings, northern Scandinavia, Geografiska Annaler, **42**, 65–200, 1960.

Rapp, A., and L. Strömquist, Field experiments on mass movements in the Scandinavian mountains, with special reference to Karkevagge, Swedish Lappland, *Studia Geomorphologica Carpatho-Balancia*, **8**, 23–40, 1979.

Reanier, R.E., and F.C. Ugolini, Gelifluction deposits as sources of paleoenvironmental information, *Proceedings of the Fourth International Conference on Permafrost*, 1042–1047, 1983.

Rutter, N.W., Glacial processes in the central Canadian Rocky Mountains, in *Geomorphic Systems of North America*, edited by W.L. Graf, pp. 228–238, Geological Society of America, Centennial Special Volume 2, Boulder, Colo., 1987.

Saunders, I., and A. Young, Rates of surface processes on slopes, slope retreat and denudation, *Earth Surface Processes and Landforms*, **8**, 473–501, 1983.

Slaymaker, H.O., Rates of operation of geomorphological processes in the Canadian Cordillera, *Abhandlungen der Akademie der Wissenschaften in Göttingen, Mathematisch-Physikalische Klasse*, **29**, 319–332, 1974.

Slaymaker, H.O. Climate change and erosion processes in mountain regions of western Canada, *Mountain Research and Development*, **10**, 183–195, 1990.

Smith D.J., Turf-banked solifluction lobe geomorphology in the Alberta Rocky Mountains, Canada, Ph.D dissertation, 300 pp., University of Alberta, 1985.

Smith, D.J., Late Holocene solifluction lobe activity in the Mount Rae area, southern Canadian Rocky Mountains, *Canadian Journal of Earth Science*, **24**, 1634–1642, 1987a.
Smith, D.J., Solifluction in the Canadian Rockies, *The Canadian Geographer*, **31**, 309–318, 1987b.
Smith, D.J., Rates and controls of soil movement on a solifluction slope in the Mount Rae area, Canadian Rocky Mountains, *Zeitschrift für Geomorphologie Supplement Band*, **71**, 25–44, 1988.
Smith, D.J., and J.S. Gardner, Geomorphic effect of ground squirrels in the Mt. Rae area, Canadian Rocky Mountains, *Arctic and Alpine Research*, **17**, 205–210, 1985.
Sohma, H., S. Okazawa, and S. Iwata, Slow mass movement processes in the alpine region of Mt. Shirouma Dake, the Japan Alps, *Geographical Review of Japan*, **52**, 562–579, 1979.
Stocker, E., Ergebnisse elfjähringer Messungen der Bodenbewgung in der alpinen stufe der Kreuzeckgruppe (Kärten), *Wiener Geographische Schiften*, **59/60**, 27–35, 1984.
Strümquist, L., Gelifluction and surface wash, their importance and interactions on a periglacial slope, *Geografiska Annaler*, **38**, 303–317, 1983.
Washburn, A.L., Instrumental observations of mass wasting in the Mesters Vig District, northeast Greenland, *Meddelelser om Grønland*, **166**, 1–297, 1967.
Williams, P.J., Some investigations into certain solifluction and patterned ground features in Norway, *Geographical Journal*, **123**, 49–58, 1957.
Williams, P.J., Downslope movement at a subarctic location with regards to variations with depth, *Canadian Geotechnical Journal*, **3**, 191–203, 1966.

10 Factors Influencing the Distribution and Initiation of Active-layer Detachment Slides on Ellesmere Island, Arctic Canada

Antoni G. Lewkowicz
Department of Geography, Erindale College, University of Toronto

Abstract

Active-layer detachment slides were mapped at three locations on the Fosheim Peninsula. Most occur on the upper and middle parts of slopes within plan form concavities, and generally they appear unrelated to undercutting by streams or to geological structure. The failure histories of individual slides range from simple, involving the movement of a single slab, to complex, with progressive failure from the base upslope as blocks become unsupported. At all sites, the process involves sliding of relatively rigid, strong soil masses over a thawing layer with minimal shear strength.

The average rate of activity in an area of 5.3 km^2 within the valley of lower Black Top Creek is 3.5 failures per year (1950–1990), but 75 slides were initiated during July and August 1988. During this time thaw depths at nearby Hot Weather Creek were 56 to 65 cm, and the same levels were attained in the following two years, confirming that failure took place within the active layer where sampling showed ice contents and liquid limits were moderate. The importance of thaw rates was demonstrated by values for the basal part of the active layer in 1988, which were approximately double those in 1989 and 1990 when no failures occurred. The high density of active-layer detachment slides on the Fosheim Peninsula compared to most of the Canadian Arctic Archipelago is linked to the enhanced probability of substantial

Periglacial Geomorphology. Edited by J. C. Dixon and A. D. Abrahams

surface energy inputs in late summer when the thaw front is encountering segregated ice near the base of the active layer.

Introduction

Active-layer failures are shallow landslides that develop in permafrost areas in response to particularly high air temperatures (Edlund, Alt, and Young, 1989), summer rainfall events (Carter and Galloway, 1981; Cogley and McCann, 1976; Hodgson, 1977), rapid melting of a snowbank upslope (Mathewson and Mayer-Cole, 1984), or surface disturbances such as fire (Hardy and Morrison, 1972; Harry and MacInnes, 1988). The failure process involves a reduction in effective stress and strength at the contact between the thawing overburden and the underlying frozen material. Movement downslope takes place over a distinct failure plane that roughly parallels the preexisting surface topography. Depending on the degree of saturation, the soil type, and the velocity of the movement, the detached mass may move by sliding with little internal deformation (Mackay and Mathews, 1973), flowing (McRoberts and Morgenstern, 1974), or a combination of the two.

Active-layer failures are widely referred to in the North American literature and have been of particular interest in the Mackenzie Valley (e.g., Hughes, Veillette, Pilon, Hanley, and van Everdingen, 1973; Mackay and Mathews, 1973; McRoberts and Morgenstern, 1973) where they represent potential hazards to pipelines. Their initiation has been linked to excess porewater pressures induced during thaw consolidation (e.g., McRoberts and Morgenstern, 1974) or the development of a particulate soil structure in which frozen soil lumps are surrounded by thawing ice veins and lenses (Vallejo, 1980; Vallejo and Edil, 1981). In the Canadian Arctic Archipelago, detachment failures have been mentioned as occurring on Banks (French, 1976, p. 146), Ellesmere (Cogley and McCann, 1976; Edlund, Alt, and Young, 1989) and Ellef Ringnes Islands (Hodgson, 1977). More detailed studies of particular failure sites have been undertaken on Bathurst Island (Mathewson and Mayer-Cole, 1984) and Melville Island (Stangl, Roggensack, and Hayley, 1982).

In an earlier publication, the morphometry of active-layer detachment slides on the Fosheim Peninsula, Ellesmere Island, was described and the geomorphic significance of several hundred of these features was assessed (Lewkowicz, 1990). It was shown that in areas with high slide concentrations, denudation rates due to active-layer detachment exceed probable values for slow mass movement. The aim of the present paper is to describe the distribution of failures at the same locations in order to examine medium-scale factors influencing slope instability. In addition, surface and subsurface properties at three failure sites and ground temperature data are presented in an assessment of soil conditions at the time of failure.

Terminology

Although a considerable number of terms have been used to describe shallow landslides over permafrost (e.g., skinflows, Hughes et al., 1973, earth flows, Holmes and Lewis, 1965, and active layer glides, Mackay and Mathews, 1973), the *Glossary of Permafrost and Related Ground-Ice Terms* (Permafrost Subcommittee, 1988) recommends 'active-layer failures' or 'detachment failures'. In this paper, these general terms are used to refer to forms where it is unclear whether the major motion of the failed mass was slide or flow. The more specific term active-layer detachment slide is employed for the features studied on the Fosheim Peninsula, all of which involve the sliding downslope of a relatively rigid, dry active layer. The distinction is made between sliding and other motions of the detached mass because different external initiating factors may be responsible for failure of the slope (e.g., high temperatures for slides or heavy rainfall for flows). This is relevant when considering the potential increase in frequency of this type of mass movement in response to climatic change (Lewkowicz, 1990).

Study area

The study was undertaken in the vicinity of Eureka, Ellesmere Island (Figure 10.1A), during the summers of 1988 to 1990. Two valleys were selected for detailed investigation in 1988: the lower part of Black Top Creek (79°58′N, 85°40′W) and the central part of an unnamed valley, hereafter referred to as Big Slide Creek (unofficial name) (79°42′N, 84°23′W) (Figure 10.1B). These sites were revisited in June 1989 and a third area, the lower portion of Hot Weather Creek (79°58′N, 84°28′W), was examined in the light of the numerous new failures which had occurred there and elsewhere in late July and August 1988 (Edlund, Alt, and Young, 1989) (Figure 10.2). All three locations were visited again in the summer of 1990.

Climate records for the Fosheim Peninsula from Eureka (1951–1980) show mean air temperature and mean annual precipitation to be −19.7°C and 64 mm, respectively (Atmospheric Environment Service, 1984). Average June and July air temperatures (1.8°C and 5.4°C, respectively) are unusually high for this latitude within the Queen Elizabeth Islands. The anomaly is due to the intermontane position of the Fosheim Peninsula relative to the prevailing northwesterly flow from the Arctic Ocean (Edlund and Alt, 1989). High temperatures are significant, as active-layer detachment slides can result from periods of particularly warm weather.

Recent measurements indicate that summer temperatures inland may substantially exceed those recorded at the Eureka weather station (Edlund, Alt, and Young, 1989). For example, comparisons of hourly temperatures at Eureka with those at Hot Weather Creek and those at a site located 1.5 km north of Slidre Fiord and half-way between Eureka and Black Top Creek show that significant differences exist (Figure 10.3A). There is considerable scatter, but more than 90% of the points lie above the 1:1 line and the greatest temperature gradients develop when Eureka values are 2° to 6°C.

The mean temperatures at the three sites over the measurement period (July 5–10, 1988) exhibit a logarithmic relationship relative to distance from the edge of Slidre Fiord (Figure 10.3B). This indicates that Eureka temperatures may be representative only of terrain within a few hundred metres of the

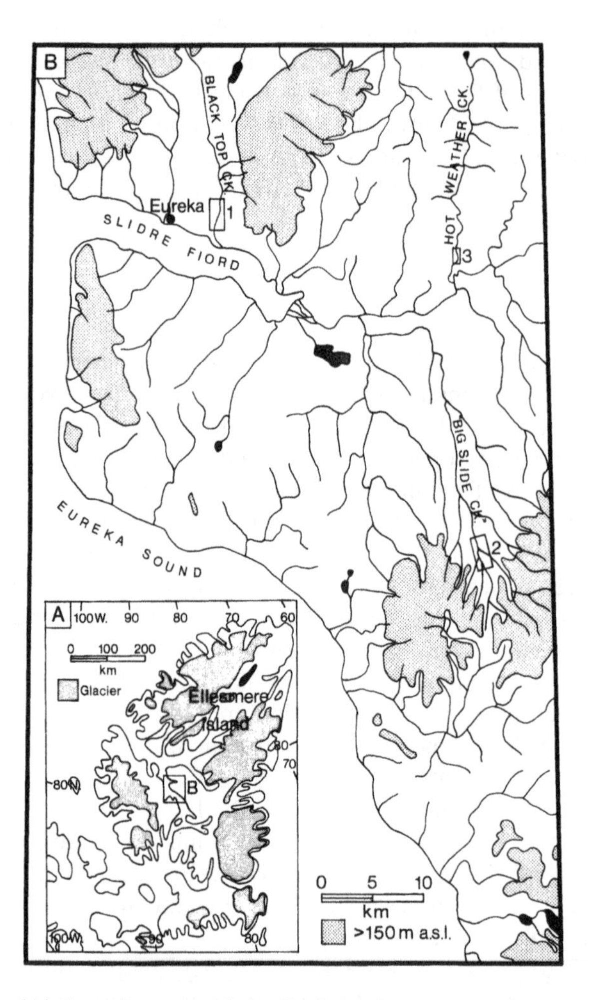

Figure 10.1. (A) Location map of the Fosheim Peninsula study area on Ellesmere Island, Northwest Territories. (B) Study sites on the Fosheim Peninsula: 1 = Black Top Creek, 2 = Big Slide Creek, 3 = Hot Weather Creek

Figure 10.2. Change in active-layer detachment slide distribution in lower Black Top Creek. (A) Photo taken on July 11, 1988, showing site 1, a failure that had occurred in 1987. (B) Photo of the same part of the slope in June 1989 after the activity in late July and August of the previous year; arrows indicate new failures

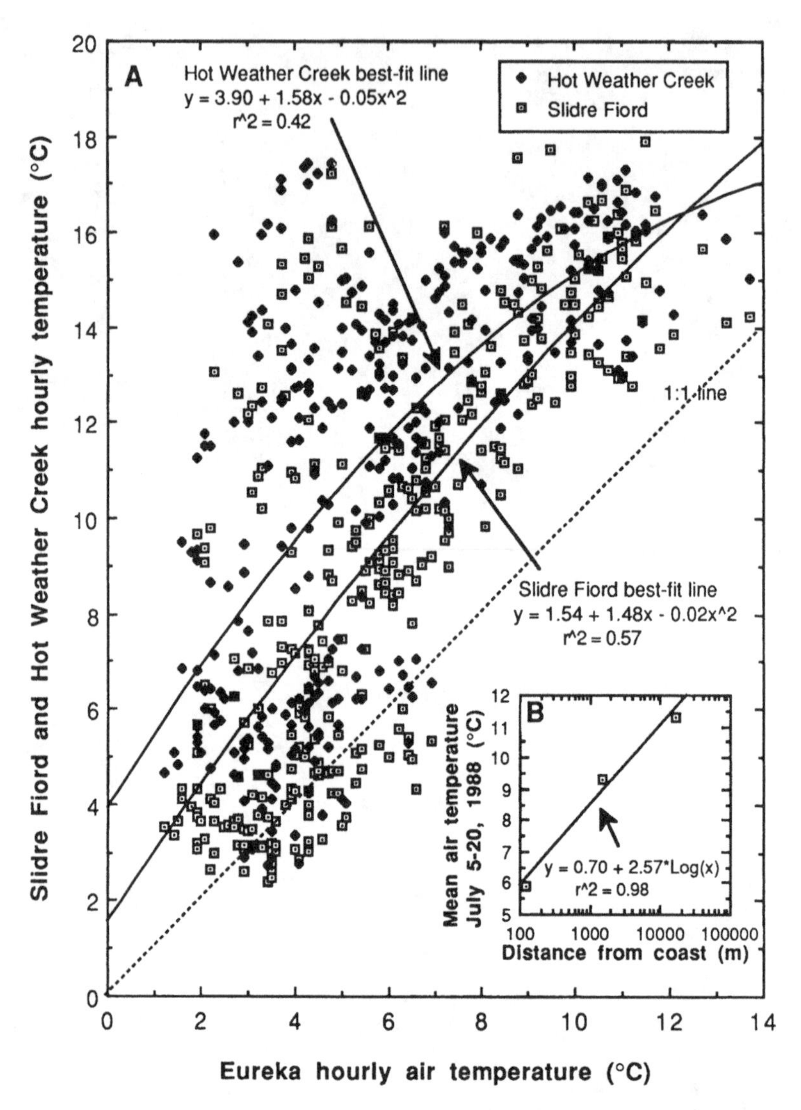

Figure 10.3. Air temperature relationships, Fosheim Peninsula, July 5–20, 1988. (A) Comparison of hourly data for Eureka weather station with site 1.5 km north of Slidre Fiord and Hot Weather Creek (best-fit lines are second-order polynomials). (B) Influence of distance to the coast on mean air temperatures that for the period. Note that averages do not include 2300 to 2400 hours CDT, as data are not available from Eureka. Data were obtained from weather station records (Eureka), Atmospheric Environment Service automatic weather station (Hot Weather Creek), and field measurements (Slidre Fiord)

coast, whereas the Hot Weather Creek values represent much of the central Fosheim Peninsula, where distances from the coast range from 10 to 30 km. Although these observations were made for only two weeks, monitoring at Hot Weather Creek in 1989 and 1990 confirms that inland temperatures in summer are persistently higher than Eureka values (e.g., Woo, Edlund, and Young, 1991).

Borehole temperature measurements from a site 10 km east of Hot Weather Creek indicate that permafrost is about 500 m thick (Taylor, Burgess, Judge, and Allen, 1982). Mean ground temperatures at the depth of zero annual amplitude are approximately −17°C, well below the threshold required for two-sided freezing of the active layer during the autumn. Thus segregated ice can develop at both the base and the top of the active layer in frost-susceptible soils (see Mackay, 1981). Ice wedges are present in all three areas and there are exposures of massive ice within retrogressive thaw slumps in the valley of Station Creek (5 km to the east of Black Top Creek) and at one site within the Hot Weather Creek basin.

Bedrock affects active-layer detachment sliding where surficial deposits are thin or absent. In the Black Top Creek area, bedrock consists of poorly lithified shales, siltstones, sandstones, and mudstones belonging to the Awingak, Deer Bay, and Isachsen Formations of Mesozoic age (Geological Survey of Canada, 1971a). The beds dip to the west at 16° in the valley bottom and decrease in dip towards the Weather Station Syncline lying to the west. The Big Slide Creek and Hot Weather Creek areas are underlain by Tertiary sandstones, siltstones, shales, and coal of the Eureka Sound Formation (Geological Survey of Canada, 1971b). Dips are about 10° towards the east at Big Slide Creek and close to zero at Hot Weather Creek. Surface exposures of bedrock are common in the scar floors of active-layer detachment slides at Big Slide Creek but are less frequent at the other two sites.

There is considerable uncertainty about the Quaternary history of the Fosheim Peninsula (e.g., England, 1987; Hodgson, 1985), and no surficial geology map has been compiled. Fine-grained surface deposits, apparently of marine origin, are present in parts of the Black Top Creek and Hot Weather Creek study areas, but Big Slide Creek is located above the Holocene marine limit of 140 m (Hodgson, 1985). The majority of slopes at all three sites are covered by a veneer of colluvium or residual material derived from the weathering of bedrock. Vegetation cover on slopes is usually less than 20% but is strongly dependent on water supply. At the sites of late-lying snow-banks or in poorly drained swales, cover in sedge-willow communities can be more than 50%. The main effect of vegetation on active-layer detachment slides is that roots increase the shear strength of near-surface soil layers (Mathewson and Mayer-Cole, 1984), potentially inhibiting movement down-slope.

Results

DISTRIBUTION AND FREQUENCY

Airborne observations made within a 50 km radius of Eureka showed that detachment failures are common landscape features, occurring singly or in clusters of up to 100 or more. Based on the literature and personal observations made over the past 15 years, it is clear that the density of active-layer detachment slides on the Fosheim Peninsula is anomalously high compared to most parts of the Canadian Arctic Archipelago.

At the three study locations, the distribution of active-layer detachment slides was studied using a combination of vertical aerial photography, ground survey, and low-level oblique aerial photography. The resulting maps (Figures 10.4–10.6) allow an assessment of the influence of terrain on the failure process. Ground surveys and sequential aerial photographs also enable some slope failures to be dated, either to a particular year or a range of years.

Black Top Creek

The best available time control on failure activity is for Black Top Creek, as the area appears on five sets of vertical aerial photographs (taken 1950, 1959, 1974, 1982, and 1986). New active-layer detachment slides occurred in each period between photography, showing that the conditions required for slope failure occurred repetitively. As the activity is not a regular, seasonal phenomenon, these slopes belong within Classes II (recurrence interval ≤5 years) or III (recurrence interval >5 years) of Crozier's stability classification based on frequency and potential (Crozier, 1984). The dominant episode in the last 40 years was during the summer of 1988 when 75 new failures were initiated (Lewkowicz, 1990), compared to an average rate of 3.5 per year (1950–1990). As in earlier years, three-quarters of the 1988 landslides occurred on the long, relatively rectilinear slope that composes the western side of the valley. They are not uniformly distributed, however, and are virtually absent from two parts of the slope, each about 300 m wide (Figure 10.4). No new failures were observed in Black Top Creek in 1989 or 1990.

Mass movements can be caused by several external or internal changes in stability conditions (e.g., Brunsden, 1979; Terzaghi, 1950). The 1988 active-layer detachment slides occurred during a period without precipitation or seismic activity, leaving the only potential triggering factors as loss of shear strength due to undercutting or a change in pore water pressure affecting the effective shear stress. Fewer than 5% of such slides were at sites that could have been affected by undercutting of the bank by Black Top Creek. Indeed, most of the slides did not reach the floor of the valley so that the initial

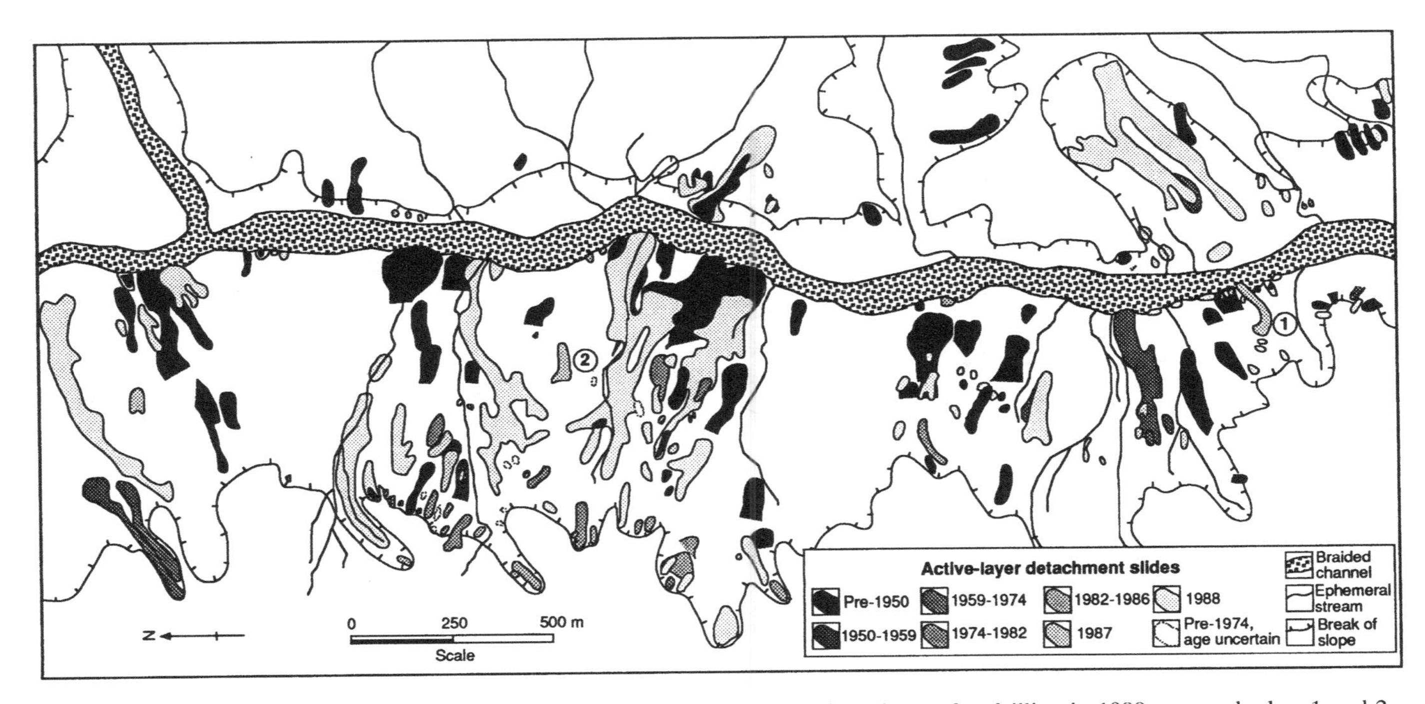

Figure 10.4. Active-layer detachment slides in lower Black Top Creek. Sites chosen for drilling in 1988 are marked as 1 and 2. Map was originally drawn using photos at a scale of 1:7500, and slide ages were determined from aerial photo interpretation and field observations. Map shows only the outlines of failures that could positively be identified on the ground; those shown as 'age uncertain' are visible on 1974 photographs but are too small to be identified in earlier sets. More recent failures are generally shown superimposed on older ones except where this would completely obscure the earlier landslide

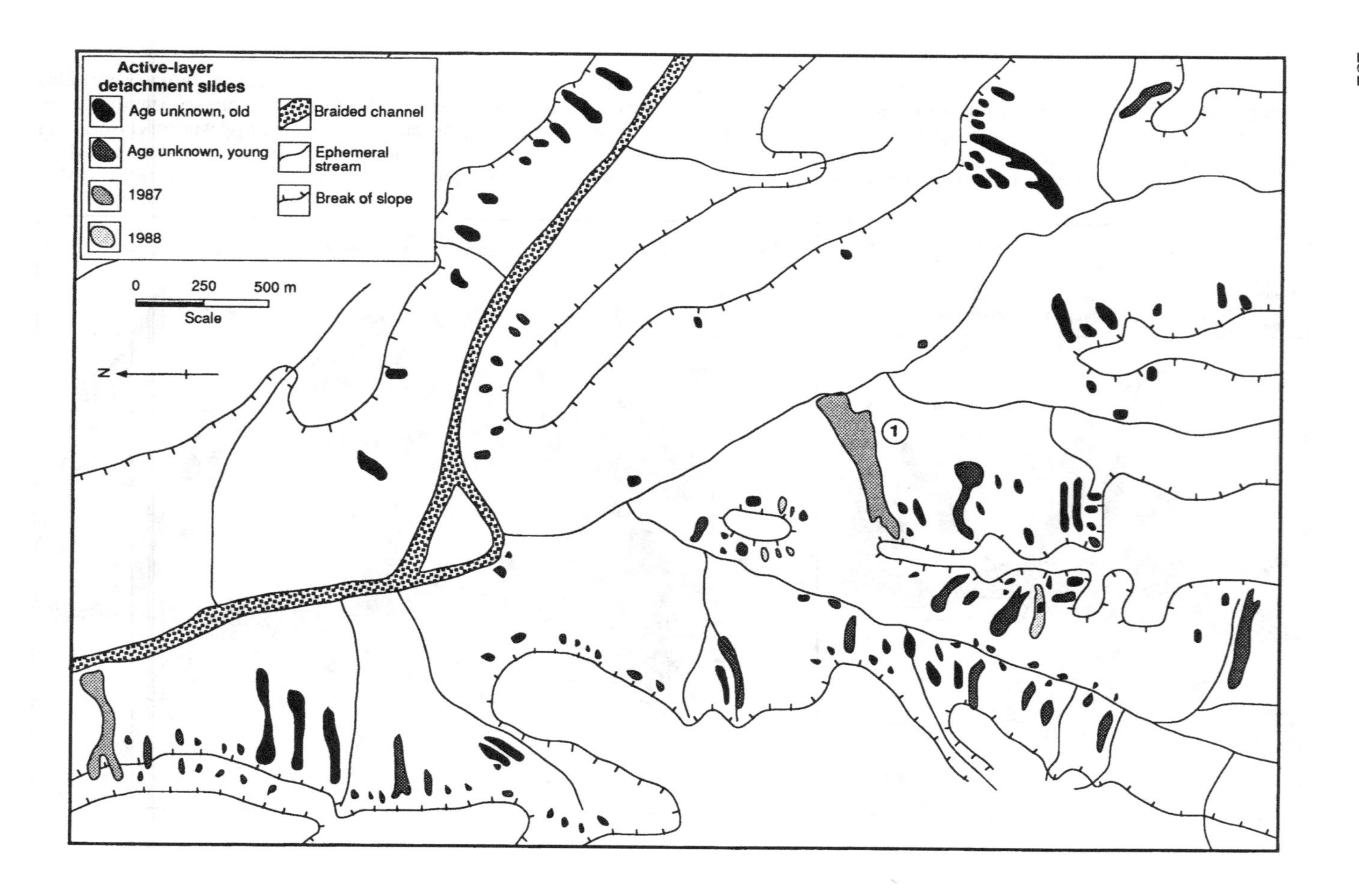
Active-layer
detachment slides
Age unknown, old
Age unknown, young
1987
1988
Braided channel
Ephemeral stream
Break of slope
0
250
500 m
Scale
N
1

sediment transfer was within the slope system. The probable cause for the 1988 active-layer detachment slides is the generation of excess pore water pressures associated with thaw consolidation (McRoberts and Morgenstern, 1974). The numerous small failures adjoining the plateau edge where the gradient is up to 5° greater than the remainder of the slope indicate the importance of slope angle in increasing the local shear stress. Slides also frequently occurred within local cross-slope concavities where supra-permafrost groundwater could have concentrated (e.g., Woo and Steer, 1983). In contrast, the two sections of the slope where slides are absent appear in plan form as convexities where downslope drainage of the active layer would diverge rather than converge. At some sites, however, there is no obvious indication in the topography why one part of the slope failed and an adjoining section did not.

Big Slide Creek

This area was photographed only in 1950 and 1959 and both sets of photos are at too small a scale to enable positive identification of active-layer detachment slides. The area was mapped in the field in 1988 prior to the failures that took place that year and remapped in the following two years (Figure 10.5). The two largest landslides are inferred to have occurred in 1987. Retrogressive thaw slumps developed over ice wedges exposed in the sidewalls of the scars of these slides. The distances that the slump headwalls retreated outside the margins of the slides then doubled between 1988 and 1989 (Figure 10.7). Although no time control is available for the remaining active-layer detachment slides, they can be divided into two subjective categories of young and old. Young failures have unvegetated scar areas, well-defined lateral berms, and transported blocks with near-vertical sides. Visual comparisons with the Black Top Creek area where ages are known suggest that the young category includes failures from about 1975 or later. Old failures have partially revegetated scar and track zones along with lateral berms and blocks that have been weathered and eroded into less distinct forms. This category represents failures that occurred prior to about 1975. The surface expression of an active-layer detachment slide becomes progressively less clear with time as slopewash, solifluction, and freeze-thaw processes act to break down individual blocks and smooth the scar and toe zones. Thus the oldest failures and mapped are probably between 50 and 100 years old (Lewkowicz, 1990).

Figure 10.5. Active-layer detachment slides in Big Slide Creek. Site chosen for drilling in 1988 is marked as 1. Map was originally drawn using aerial photos at a scale of 1:30,000; see text for criteria used for young and old categories. Map shows only the failures that could be positively identified on the ground; owing to map scale, shapes are approximate

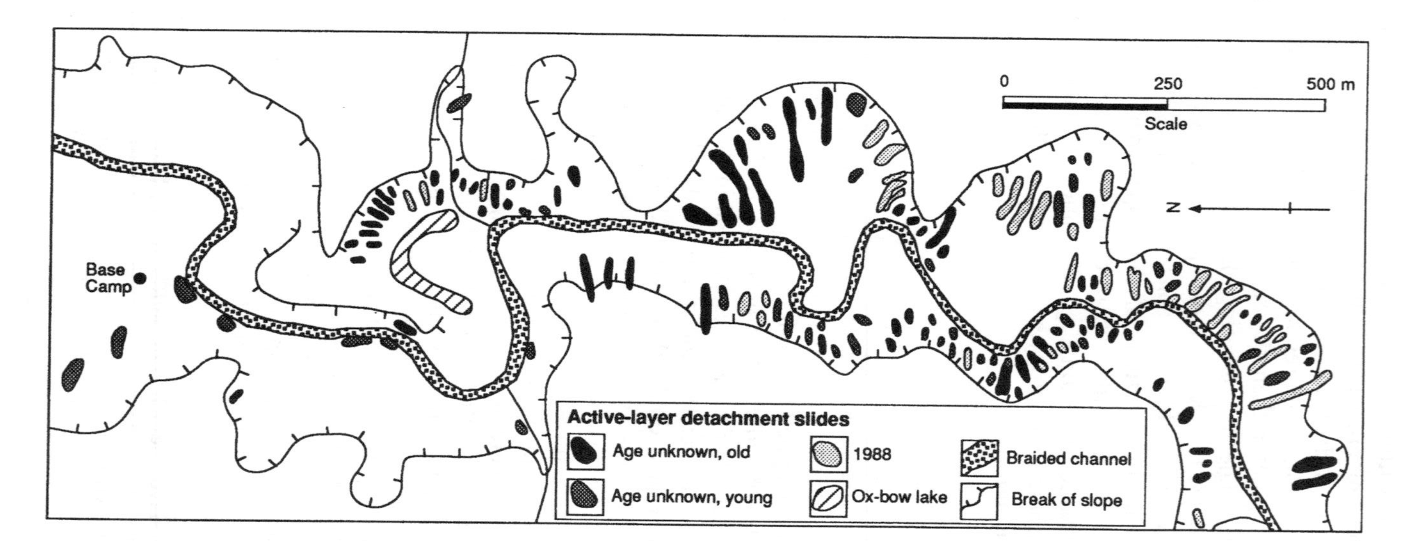

Figure 10.6. Active-layer detachment slides in lower Hot Weather Creek. Map was originally drawn using photos at a scale of 1:14,500; see text for criteria used for young and old categories. Map shows only the failures that could be positively identified on the ground; owing to map scale, shapes are approximate

Of the active-layer detachment slides mapped in July 1988, 44% are in the young category. Only six new failures developed in the area later in 1988 and none in the following two summers. The different response in 1988 compared to Black Top Creek (and Hot Weather Creek, see below) is thought to be a consequence of the numerous failures in the preceding few years, most of which probably occurred in 1987. Although higher air temperatures produced conditions more favorable for slope failure in 1988, few suitable slopes

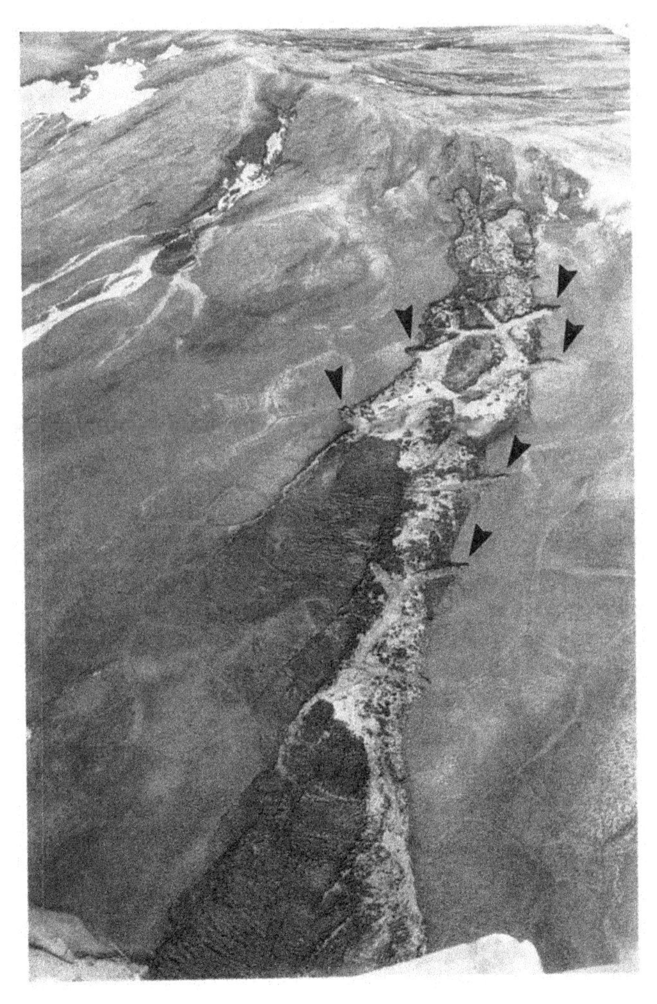

Figure 10.7. Photo of site 1, one of the two large active-layer detachment slides in Big Slide Creek that occurred in 1987. The failure is 680 m long, up to 150 m wide, and crosses a network of ice wedge polygons. Postfailure thermokarst led to the formation of small retrogressive thaw slumps (arrowed), which by June 1989 had retreated up to 16 m outside the margins of the slide track

remained available. Analysis of overlaps between active-layer detachments at Black Top Creek indicates that it takes at least 6 years and often more than 15 years before a slope will fail in the same place again (Lewkowicz, 1990).

Big Slide Creek has a lower areal density of active-layer detachment slides than the other two locations. Only 10% of the features adjoin streams, whereas many are located on slope segments that are concave in plan form. At the northern end of the study area, numerous failures are present on relatively steep slopes that reflect the outcropping of more resistant beds of the Eureka Sound Formation. Failures are more common on east-facing slopes parallel to the local dip, but significant numbers also occur on west-facing slopes (see Figure 10.5) where they cut across the rock structure.

Hot Weather Creek

This area was mapped in 1989, but on-site records from Geological Survey of Canada scientists (S. Edlund, personal communication, 1989) permit failures that took place in 1988 to be identified. Pre-1988 failures are divided into old and young using the same criteria as in the Big Slide Creek area (Figure 10.6). Of the pre-1988 slides, slightly less than half are classified as young. As at Black Top Creek, the summer of 1988 produced an exceptional response on the slopes and there was a 32% increase in the total number of slides present in the area.

Since Hot Weather Creek is steeply incised, active-layer detachment slides here are shorter and scar slope angles are greater than in the other two areas (Lewkowicz, 1990). Dips are close to zero, so all the failures either cut across the underlying bedrock (Figure 10.8) or are in marine deposits. More active-layer detachment slides have taken place on the lower-angled west-facing slopes (60%) than on the steeper east-facing ones (40%). Many of the latter are being actively undercut by the river and fail by blockfall rather than by sliding. Fewer than 10% of the active-layer detachments could have been influenced by fluvial erosion of the toe of the slope.

SITE INVESTIGATIONS

Three active-layer detachment slides were chosen for detailed examination in 1988, two at Black Top Creek and one at Big Slide Creek (see Figures 10.4 and 10.5). The purpose of the investigations was to obtain soil samples and to assess the processes active during slope failure, so the most recent failures available (those which had occurred in 1987) were selected for study.

Sites 1 and 2 in Black Top Creek have widths and failure angles typical of the total distribution in the area, but exceed 80% of the failures there in terms of length. The site at Big Slide Creek is unusual, as only one other failure out of 145 in the area is of comparable length. The slope angle in the scar zone of

Figure 10.8. Steep active-layer detachment slides (arrowed) occurring in weathered bedrock around an ox-bow lake at Hot Weather Creek. Note the exposure of near-horizontal beds of Eureka Sound Formation on the right

12° is within the modal class, but the very low angles of the toe zone (<3°) are atypical.

Boreholes were drilled in and around the three landslides using a modified CRREL auger attached to a Stihl power-head. Continuous core 50 mm in diameter was taken from the frost table to maximum depths of 1.6 m. The thawed soil, in which the auger performed poorly, was first removed with a shovel. Moisture contents were determined for all samples, and portions of the core were selected for grain size and Atterberg limit analyses.

Black Top Creek, Site 1

This is a complex failure 140 m long and up to 28 m wide (Figures 10.2 and 10.9) that extends to the floor of Black Top Creek, partially blocking one channel of the stream in 1988. The upslope scar zone is mainly smooth and lies at an angle of 8°. It is surrounded by tension cracks attributable to the loss of support downslope. On the northern flank is an intact transported block that formed a berm 1.8 m high on collision with a stable section of the slope.

Slickensides on the failure plane in the central part of the failure confirm that the detached material moved by sliding. This area apparently acted as

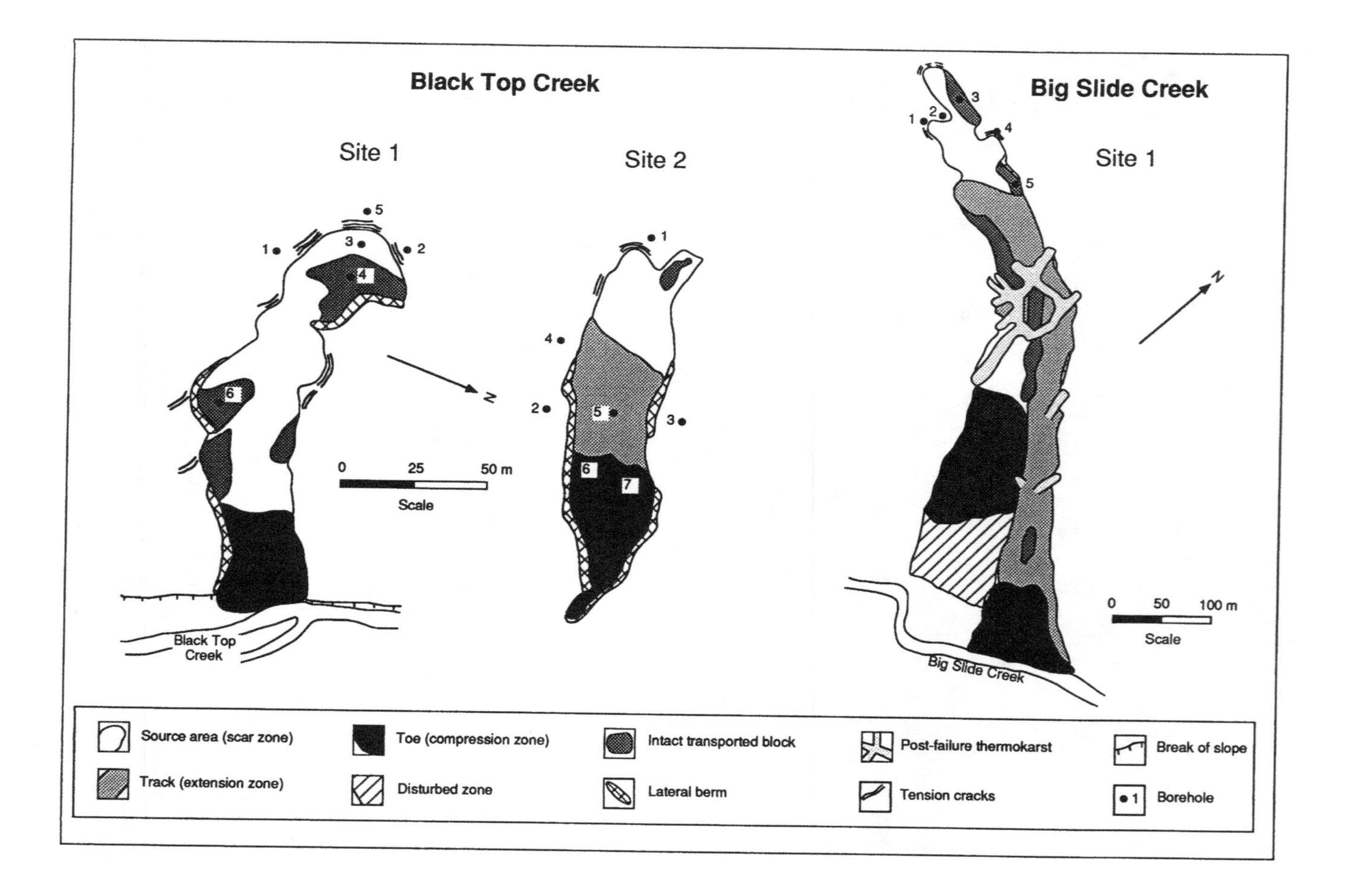
Black Top Creek
Site 1
Site 2
N
0 25 50 m
Scale
Black Top Creek
Big Slide Creek
Site 1
N
0 50 100 m
Scale
Big Slide Creek
Source area (scar zone)
Track (extension zone)
Toe (compression zone)
Disturbed zone
Intact transported block
Lateral berm
Post-failure thermokarst
Tension cracks
Break of slope
Borehole

both a source zone and a track for the movement of blocks downslope. Rills incised in the floor were produced by runoff generated during ground ice thaw immediately following detachment (Holmes and Lewis, 1965) or by snowmelt in the spring of 1988. Three transported blocks remain in this zone, one of which has raised a berm in the surrounding slope. The blocks have partly disaggregated into smaller 'block units,' typically 50 cm in diameter separated by surface cracks 10 to 15 cm wide. These units stand upright except at the edges of major blocks, and their vertical heights of 0.4 to 0.6 m indicate the range of thaw depths at the time of failure. In the lower portion of the slide, failed material is piled up and compressed. Discontinuous ridges up to 1.5 m high traverse the toe and the surface microtopography is very rough, consisting of broken-up and tilted block units.

A possible failure history for this site involves an initial movement near the toe where the slope steepens above the edge of the flood plain. Once the support of this material was lost, upslope blocks failed progressively, and most of these came to rest in the basal segment, causing considerable overthrusting and compression. The blocks that remain intact within the scar and track zones were probably the last to move and did so with the lowest velocities.

Black Top Creek, site 2

This 135-m-long landslide has a relatively simple morphology (see Figure 10.9). The upslope section is a bare scar zone at an angle of 11°, partially surrounded by tension cracks and containing a small intact transported block. The central part of the failure is an extension zone 25 m wide containing many individual upright block units that become more concentrated spatially downslope. There is no sharp demarcation between the extension and compression zones. In the lower part of the failure, a single slide block becomes progressively more compressed towards the toe. Transverse ridges are 0.3 m high in the upslope part of this zone and become 1 to 2 m high at the distal end. Prominent lateral berms surround the extension and compression zones, increasing in height from less than 1 m at the upper end of the extension zone to 2.5 m high at the slide toe.

Failure at site 2 apparently took place in one major movement. As the slab traveled downslope it pushed up lateral berms, became compressed at the toe, and underwent extension upslope, breaking down into block units. The

---

Figure 10.9. Geomorphic maps of the three active-layer detachments studied in detail in 1988. Note the scale and orientation change between the Black Top Creek and Big Slide Creek sites. The distances between the sites on the diagram are not to scale

greater than usual compression experienced in the toe is the result of narrowing of the slide caused by confinement within a preexisting topographic depression. The movement of the small block in the scar zone took place following the main failure. As this site is 200 m from Black Top Creek, the instability was probably due to loss of effective stress during thaw consolidation. The intact blocks show that the surface was dry and strong at the time of failure; friction at the failure plane must have been minimal.

Big Slide Creek, site 1

This site is the largest and most complex of the three studied (see Figure 10.7 and 10.9). The scar zone at the upslope end of the slide is similar to those at the other two sites with failure angles of 10 to 12° but a number of scarps attest to multiple block movement. Several large rills are incised into the scar floor as a result of the concentration of runoff within the basin. Downslope, an extended track runs along the northern flank of the failure at angles of 3 to 5°. Within the track there are numerous discrete block units at variable concentrations, mostly standing upright with heights of 0.35 to 0.60 m. This expanded range compared to the Black Top Creek sites is due to the wide variation in vegetation cover at this site which affected the depth of thaw at the time of movement. In the central portion of the landslide permafrost degradation occurred after failure and the underlying network of ice wedges thawed.

There are two main zones that experienced compression. These are separated by vertical shear planes and offset by more than 100 m. The material on the northern flank extends further downslope and is more compressed, with transverse ridges up to 1.5 m high. On the southern flank, the detached mass pushed against a stable portion of the slope, forming a disturbed zone with ridges up to 0.3 m high that diminish in height downslope. In both the toe and disturbed zones, the surface is deformed but generally unbroken, probably due to additional shear strength provided by the the vegetation (50–80% cover). Unlike the Black Top Creek sites, the lateral berms are low and discontinuous either because of the failure shape, which widens downslope, or as a result of the lower block velocities on the gentle slope.

The failure history of this site is uncertain because of the disintegration of large blocks within the track and the development of postfailure thermokarst. Two main possibilities exist.

First, failure could have started within the steep upper source area where the shear stress would have been greatest. The relatively small blocks could not have pushed the much larger masses downslope but could have contributed to progressive overloading, as was suggested for a detachment failure on Bathurst Island (Mathewson and Mayer-Cole, 1984). However, the overthrusting that is likely to result from this process was not observed. Moreover, soil moisture conditions at the Big Slide Creek site are the opposite of those studied by Mathewson and Mayer-Cole (1984), increasing rather than decreasing downslope. During periods without precipitation in 1988 and 1990, the surface alongside the upper part of the scar was dry, while near-saturated conditions prevailed in the toe, suggesting the presence of significant amounts of groundwater derived from thawing of ground ice.

Alternatively, failure could have started at the toe with upslope blocks moving as they became unsupported. The overall shape of the slide supports this hypothesis, as does the presence of a mainly bare scar area extending 50 m upslope of the block on the southern flank of the slide. This evidence suggests that the slope was not overloaded. The main arguments against this scenario are that it requires the initial movement to have occurred on a slope of only 2 to 3° and that the surface shear strength would have been greater in the lower part of the slope because of the binding action of plant roots, particularly *Salix arctica*.

GRAIN SIZE AND ATTERBERG LIMITS

Boreholes were drilled within and around the failures at the study sites in early July 1988 (see Figure 10.9). Grain size analyses were conducted on five core samples from each of sites 1 and 2 at Black Top Creek and from site 1 at Big Slide Creek, and on six samples from the surface of failures at Hot Weather Creek. There was considerable variability in the granulometry, both within and among sites (Figure 10.10). The samples are mainly poorly sorted clays and silts with minor percentages of sand. The least variability was at site 1 in Black Top Creek (less than 15% variation in sand, silt, or clay fractions), while the greatest heterogeneity was at Hot Weather Creek (up to a 40% variation in clay content). The latter result is not surprising, as the samples were obtained from different failure sites.

Core samples from the upper 0.8 m of the boreholes are mainly clays of low to intermediate plasticity, but all fall close to the A line (Figure 10.11). A minority of the samples are low plasticity silts, and there is one intermediate plasticity silt and one highly plastic clay. The variability reflects the granulometry, as there is a positive correlation between percentage clay and liquid limit. The grain sizes in turn are heterogeneous owing to the diverse origins of the surficial materials as weathered bedrock, colluvium, or marine deposits. Plastic limits in samples from Black Top Creek site 1 range from 17 to 23%, liquid limits from 22 to 38%, and plasticity indices from 5 to 19%. Samples from site 2 tend to have higher liquid limits with a range of 26 to 52% while plastic limits are 15 to 27%, and plasticity indices are 6 to 26%. Samples from Big Slide Creek show the highest liquid limits and the greatest range of values (see Figure 10.11).

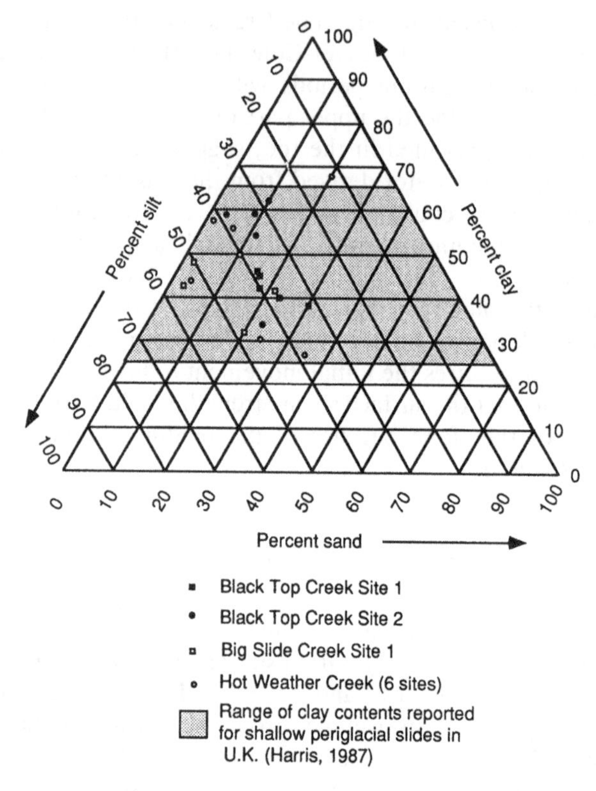

Figure 10.10. Textural properties of samples from active-layer detachments in the study areas

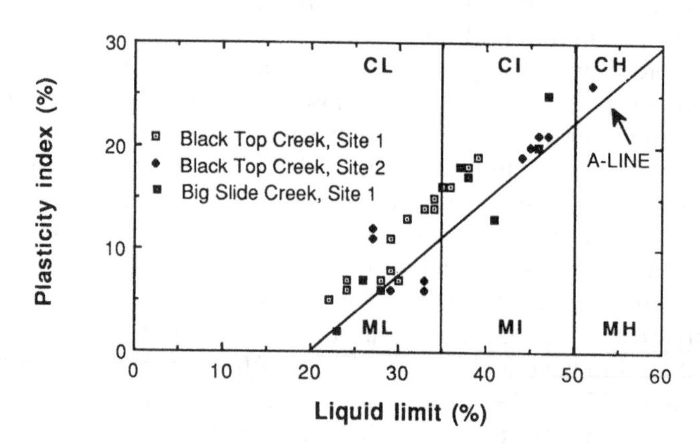

Figure 10.11. Plasticity chart for core samples from the upper 0.8 m of boreholes

GROUND ICE

Ground ice contents were studied in samples obtained from the shallow boreholes drilled in 1988. At Black Top Creek samples collected from the failure depths (0.4–0.6 m) exhibited low to moderate moisture contents, from 18 to 45% (Figure 10.12) with excess ice contents of 0 to 15%. With limited exceptions, moisture contents remained moderate to 0.8 m, although they tended to be slightly higher within permafrost than within the active layer. Moisture contents below 0.8 m were generally greater than in the upper parts of the profile, but these could not have affected the stability of the slope. Similar trends were observed for the boreholes at Big Slide Creek.

Samples obtained one year after a detachment failure has taken place give some idea of the prefailure ground ice conditions but do not preserve them. Holes drilled within the margins of the slide are in material that was unstable, but thaw subsequent to the movement will have destroyed the in situ ice content and ice fabric. Beneath any exposed portions of the failure plane, for example, a new active layer will have formed in what was previously permafrost. Destruction of the cryostratigraphy will also have occurred to a lesser degree beneath intact slide blocks. Although the movement of the block may have covered the thawing failure plane and thereby insulated it from direct energy exchange with the atmosphere, further advance of the thaw plane into the underlying sediments would be expected unless failure occurred immediately prior to freeze-up. Holes drilled outside the margin of the slide are the least disturbed in terms of sediments, but are in parts of the slope that did not fail. Moreover, even in these areas the thaw plane likely would have advanced beyond the depth it attained at the time of failure.

Despite the problems of postfailure changes, the boreholes do provide an indication of conditions at the time of active-layer detachment. First, overall ice contents were probably moderate. This is suggested by the absence of a major increase in moisture contents from the level of failure to 0.8 m, a depth that would certainly have been beneath the previous year's active layer.

Second, visible ice within the cores occurred mainly as lenses, randomly oriented veins 2 to 20 mm thick, or in a reticulate structure. About half of the samples with visible ice yielded no supernatant water upon thaw, so sediment between the lenses and veins must have been desiccated. No massive ice was encountered, showing that active-layer detachment slide distribution is unrelated to massive ice bodies.

Finally, a comparison of in situ moisture contents with Atterberg limits at the approximate depth of failure shows that for many samples, thaw could cause the liquid limit to be exceeded (Figure 10.12). The lengths of the samples were greater than the thickness of ice layers. Consequently, even where the average moisture content appears less than the liquid limit in Figure 10.12, the latter could still be exceeded within parts of the sample.

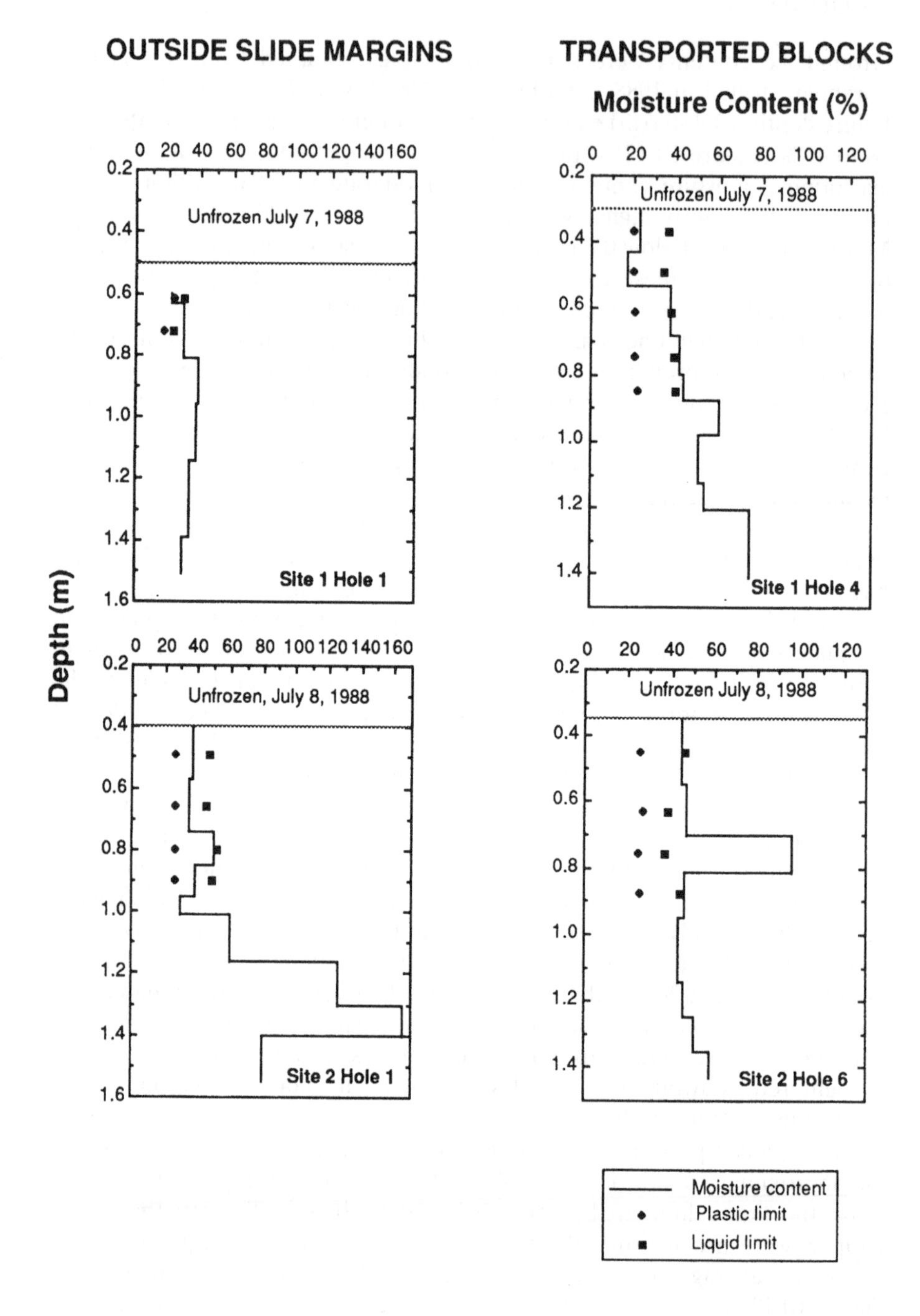
OUTSIDE SLIDE MARGINS
TRANSPORTED BLOCKS
Moisture Content (%)
Depth (m)
0 20 40 60 80 100 120 140 160
Unfrozen July 7, 1988
Site 1 Hole 1
Unfrozen, July 8, 1988
Site 2 Hole 1
0 20 40 60 80 100 120
Site 1 Hole 4
Unfrozen July 8, 1988
Site 2 Hole 6
0.2 0.4 0.6 0.8 1.0 1.2 1.4 1.6
Moisture content
Plastic limit
Liquid limit

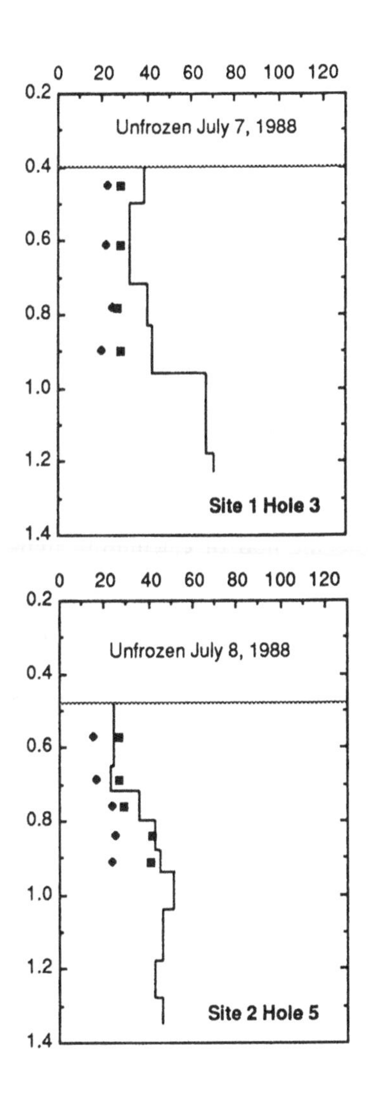

Figure 10.12. Moisture content and Atterberg limits against depth for selected boreholes at Black Top Creek sites 1 and 2. Failure depths beneath transported blocks at both sites are approximately 0.5 m. For location of boreholes, see Figure 10.9

Given the layering of ice within the soil and maximum rates of thaw plane advance of about 10 mm d^{-1} at the failure depth (see below), the thaw plane could be completely contained within individual ice lenses for periods of 1 to 2 days or more. High rates of thaw promote an increase in the thaw consolidation ratio, which in turn increases the excess pore water pressures and reduces the effective stress (McRoberts and Morgenstern, 1974). During these periods, if consolidation was sufficiently slow, the base of the thawed layer would have little or no shear strength and failure even on low-angled slopes could take place.

GROUND TEMPERATURES

Ground temperatures can be used to assess depths and rates of thaw of the active layer in relation to active-layer detachment events. Records are available from July 1988 to August 1990 from an Atmospheric Environment Service automatic weather station located on the plateau above the Hot Weather Creek base camp (Figure 10.13). Thaw depth at this site is not identical to that beneath slopes in the vicinity (e.g., Woo, Young, and Edlund, 1990, Figure 8) but can be considered to be an indication of thaw plane position and movement. A moving average is used to suppress short-term fluctuations in thaw rates caused by the method of calculating thaw depth.

Substantial differences in thaw depths and rates are present among the three summers. The deepest thaw and highest rates were in 1990, while 1989 represented the lowest thaw rates and depths. The summer of 1988 produced neither the greatest depth of thaw nor the fastest rate of thaw, yet on July 23 active-layer detachment slides began at Hot Weather Creek. Figure 10.13 shows that the depth of thaw at that time was 56 cm and that rates of thaw were 0.9 to 1.1 cm d^{-1}. Although the same depth of thaw was attained in the other two summers, rates of thaw at this depth were 0.4 to 0.6 cm d^{-1}, approximately half those of 1988.

Several points are indicated by the ground temperature data. First, a critical factor in the initiation of active-layer detachment slides in the area is a rapid advance of the thaw plane while it encounters segregated ice at depth. Second, thaw depths from the following two years confirm that during slope failures in 1988, the thaw plane was contained within the lower part of the active layer, not at the top of the permafrost. Consequently, the ground ice must have been produced by two-sided freezing during the previous autumn. This observation rules out the possibility that a build-up of ice over several years is required before failure can occur. Finally, Figure 10.13 shows that had rapid rates of thaw continued for a further week in 1990, ground temperature conditions at least would have been suitable for the initiation of additional active-layer detachment slides at Hot Weather Creek.

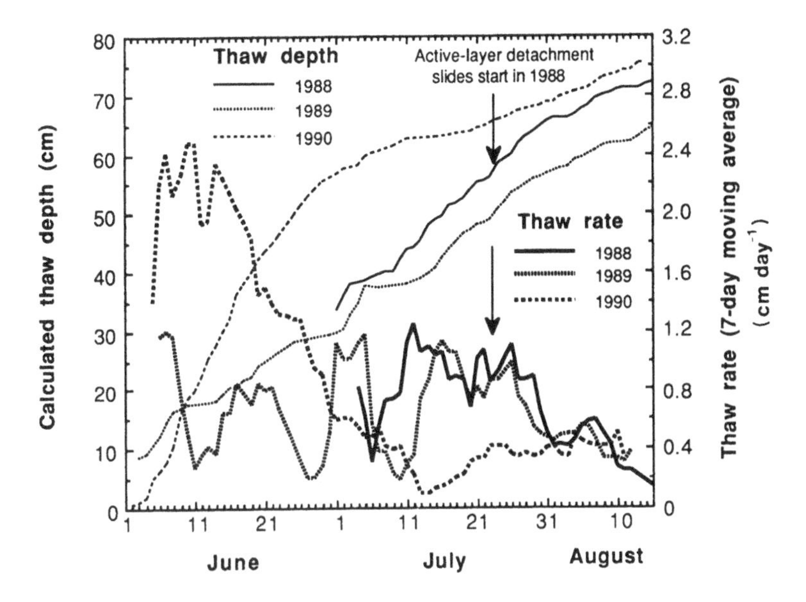

Figure 10.13. Thaw depths and rate of thaw beneath the plateau at Hot Weather Creek. Thaw depths calculated by fitting third-order polynomial curves to mean daily temperatures at depths of 0, 10, 20, 50, and 100 cm. Rates of thaw are seven-day moving averages derived from the calculated thaw depths

Conclusions

First, most active-layer detachment slides in the study area occur in the upper or central parts of slopes, often within segments that are concave in plan form; fewer than 10% border the stream channel where they could be affected by fluvial undercutting. Their spatial distribution is largely unrelated to bedrock structure. Detailed studies at three sites show that failure histories can be simple, involving the movement of a single slab, or more complex, with progressive failure from the base upslope as blocks become unsupported. In all cases, the process involves sliding of relatively rigid, strong soil masses over a thawing layer with minimal shear strength.

Second, the most important series of active-layer detachment events in the last 40 years at Hot Weather Creek and Black Top Creek took place in 1988. In the latter location, 75 failures occurred over an area of 5.3 km^2, compared to an average activity rate of 3.5 per year (1950–1990). In contrast, the most important recent active-layer detachment events in the Big Slide Creek area occurred in 1987, and only six failures were initiated in 1988. The most likely explanation for this lack of response is that few slopes with suitable drainage conditions and sufficient thickness of colluvium or weathered bedrock remained to fail.

Third, analysis of summer ground temperatures for 1988–1990 provides support for thaw consolidation as the mechanism for initiating slope failure. Data from the plateau at Hot Weather Creek show that prior to the failures in 1988, thaw near the base of the active layer (56–65 cm) took place at a rate of about 1 cm d^{-1}. Although seasonal thaw reached the same depths in the following two years, rates within these layers were approximately half those in 1988 and no new failures occurred. These data confirm that failure takes place within the active layer where ice contents and liquid limits are moderate, and not within the upper part of the permafrost.

Finally, thaw consolidation theory explains why active-layer detachment slides are more common here than in most of the Canadian Arctic Archipelago. The prevailing high summer air temperatures and low cloud covers recorded at Eureka (Edlund and Alt, 1989) are further enhanced inland, so the potential for rapid advance of the thaw plane within segregated ice in the basal part of the active layer is high wherever frost-susceptible soils are present on the Fosheim Peninsula.

Acknowledgments

This project was funded by an operating grant from the Natural Science and Engineering Research Council of Canada (A2643) and by a Research Agreement with the Department of Energy, Mines and Resources Canada (No. 206). S. Edlund, Geological Survey of Canada, supplied hospitality at Hot Weather Creek in 1989 and 1990 and recorded which active-layer detachment slides occurred in 1988. B. Alt, Geological Survey of Canada, kindly supplied the Hot Weather Creek ground temperatures. The author is grateful for the assistance of S. Campo, M. Santry, C. Spencer, K. Stoker, H. Taylor, and T. Watkins in undertaking the field research and laboratory analyses.

References

Atmospheric Environment Service, Eureka, *Principal Station Data*, vol. 79, Atmospheric Environment Service, Environment Canada, Downsview, Ontario, 1984.

Brunsden, D., Mass movements, in *Progress in Geomorphology*, edited by C. E. Embleton and J. B. Thornes, pp. 130–186, Edward Arnold, London, 1979.

Carter, L. D., and J. P. Galloway, Earthflows along Henry Creek, northern Alaska, *Arctic*, **34**, 325–328, 1981.

Cogley, J. G., and S. B. McCann, An exceptional storm and its effects in the Canadian high arctic, *Arctic and Alpine Research*, **8**, 105–110, 1976.

Crozier, M. J., Field assessment of slope instability, in *Slope Instability*, edited by D. Brunsden and D. B. Prior, pp. 103–142, John Wiley, Chichester, England, 1984.

Edlund, S. A., and B. T. Alt, Regional congruence of vegetation and summer climate patterns in the Queen Elizabeth Islands, Northwest Territories, Canada, *Arctic*, **42**, 3–23, 1989.

Edlund, S. A., B. T. Alt, and K. Young, Interaction of climate, vegetation and soil hydrology at Hot Weather Creek, Fosheim Peninsula, Ellesmere Island, Northwest Territories, *Current Research, Part D, Geological Survey of Canada Paper, 89-1D*, 125–133, 1989.

England, J., Glaciation and the evolution of the Canadian high arctic landscape. *Geology*, **15**, 419–424, 1987.

French, H.M., *The Periglacial Environment*, 309 pp., Longmans, London, 1976.

Geological Survey of Canada, Geology of Slidre Fiord, District of Franklin, scale 1:50,000, *Map 1298A*, Geological Survey of Canada, Ottawa, 1971a.

Geological Survey of Canada, Geology of Eureka Sound North, District of Franklin, scale 1:250,000, *Map 1302A*, Geological Survey of Canada, Ottawa, 1971b.

Hardy, R. M., and H. A. Morrison, Slope stability and drainage considerations for arctic pipelines, in *Proceedings, Canadian Northern Pipelines Research Conference*, pp. 249–266, Ottawa, National Research Council of Canada Technical Memorandum, 104, 1972.

Harris, C., Solifluction and related periglacial deposits in England and Wales, in *Periglacial Processes and Landforms in Britain and Ireland*, edited by J. Boardman, pp. 209–223, Cambridge University Press, Cambridge, 1987.

Harry, D. G., and K. L. MacInnes, The effect of forest fires on permafrost terrain stability, Little Chicago–Travaillant Lake area, Mackenzie Valley, N.W.T., *Current Research, Part D, Geological Survey of Canada Paper, 88-1D*, 91–94, 1988.

Hodgson, D. A., A preliminary account of surficial materials, geomorphological processes, terrain sensitivity, and Quaternary history of King Christian and Southern Ellef Ringnes Islands, District of Franklin, *Report of Activities Part A, Geological Survey of Canada Paper, 77-1A*, 485–493, 1977.

Hodgson, D. A., The last glaciation of west-central Ellesmere Island, Arctic Archipelago, *Canadian Journal of Earth Sciences*, **22**, 347–368, 1985.

Holmes, G. W., and C. R. Lewis, Quaternary geology of the Mount Chamberlain area, Brooks Range, Alaska, *U.S. Geological Survey Bulletin, 1201-B*, 32 pp., 1965.

Hughes, O. L., J. J. Veillette, J. Pilon, P. T. Hanley, and R. O. van Everdingen, Terrain evaluation with respect to pipeline construction, Mackenzie Transportation Corridor, central part, lat. 64° to 68°N., *ESCOM Report 73–27*, 74 pp., Environmental-Social Committee, Northern Pipelines, Task Force on Northern Oil Development, Information Canada, Ottawa, 1973.

Lewkowicz, A. G., Morphology, frequency and magnitude of active-layer detachment slides, Fosheim Peninsula, Ellesmere Island, N.W.T., in *Permafrost—Canada: Proceedings of the Fifth Canadian Permafrost Conference*, pp. 111–118, 1990.

Mackay, J. R., Active layer slope movement in a continuous permafrost environment, Garry Island, Northwest Territories, Canada, *Canadian Journal of Earth Sciences*, **18**, 1666–1680, 1981.

Mackay, J. R., and W. H. Mathews, Geomorphology and Quaternary history of the Mackenzie River Valley near Fort Good Hope, N.W.T., Canada, *Canadian Journal of Earth Sciences*, **10**, 26–41, 1973.

Mathewson, C. C., and T. A. Mayer-Cole, Development and runout of a detachment slide, Bracebridge Inlet, Bathurst Island, Northwest Territories, Canada, *Bulletin of the Association of Engineering Geologists*, **21**, 407–424, 1984.

McRoberts, E. C., and N. R. Morgenstern, A study of landslides in the vicinity of the Mackenzie River, Mile 205–660, *ESCOM Report 73–35*, 96 pp., Environmental-Social Committee, Northern Pipelines, Task Force on Northern Oil Development, Information Canada, Ottawa, 1973.

McRoberts, E. C., and N. R. Morgenstern, Stability of thawing slopes, *Canadian Geotechnical Journal*, **11**, 447–469, 1974.

Permafrost Subcommittee, Associate Committee on Geotechnical Research, National Research Council of Canada, *Glossary of Permafrost and Related Ground-Ice Terms*, National Research Council of Canada, Technical Memorandum no. 142, 156 pp., 1988.

Stangl, K. O., W. D. Roggensack, and D. W. Hayley, Engineering geology of surficial soils, eastern Melville Island, in *Proceedings of the Fourth Canadian Permafrost Conference*, edited by H. M. French, pp. 136–147, National Research Council of Canada, Ottawa, 1982.

Taylor, A. E., M. Burgess, A. S. Judge, and V. S. Allen, Canadian Geothermal Data Collection—Northern Wells 1981, *Geothermal Series Number 13*, Geothermal Service of Canada, Ottawa, 1982.

Terzaghi, K., Mechanisms of landslides, in *Application of Geology to Engineering Practice*, Berkey Volume, Geological Society of America, pp. 83–123, 1950.

Vallejo, L. E., A new approach to the stability analysis of thawing slopes, *Canadian Geotechnical Journal*, **17**, 607–612, 1980.

Vallejo, L. E., and T. B. Edil, Stability of thawing slopes: Field and theoretical investigations, in *Soil Mechanics and Foundation Engineering. Proceedings of the 10th International Conference*, **3**, pp. 545–548, Balkema, Rotterdam, 1981.

Woo, M.-K., S. A. Edlund, and K. L. Young, Occurrence of early snow-free zones on Fosheim Peninsula, Ellesmere Island, Northwest Territories, *Current Research, Part B, Geological Survey of Canada Paper, 91-1B*, 9–14, 1991.

Woo, M.-K., and P. Steer, Slope hydrology as influenced by thawing of the active layer, Resolute, N.W.T., *Canadian Journal of Earth Sciences*, **20**, 978–986, 1983.

Woo, M.-K., K. L. Young, and S. A. Edlund, 1989 observations of soil, vegetation and microclimate, and effects on slope hydrology, Hot Weather Creek basin, Ellesmere Island, Northwest Territories, *Current Research, Part D, Geological Survey of Canada Paper, 90-1D*, 85–93, 1990.

11 Buoyancy Forces Induced by Freeze-thaw in the Active Layer: Implications for Diapirism and Soil Circulation

Bernard Hallet and Edwin D. Waddington
University of Washington

Abstract

Indications of diapirism and soil circulation are common in periglacial areas, but governing mechanisms remain unclear. We explore the possibility that such motion may be driven, at least in part, by buoyancy forces that arise seasonally in thawing ice-rich soil. Because thawing proceeds downward from the ground surface, the duration of the thaw phase and, hence, the time available for progressive soil compaction decrease with depth. This leads to greater compaction and higher soil density near the ground surface but is partially offset by an increase in the rate of compaction and decrease in segregation ice volume with depth in the upper part of the active layer.

A theoretical model of thaw consolidation that parallels recent geophysical analyses of buoyancy-driven segregation of relatively light fluid from a viscous porous matrix provides quantitative insight into the soil compaction process. The model indicates that the density profile in a soil layer above the thaw front can be gravitationally unstable for much of the thaw season. Buoyancy can be a very effective driving force for diapirism where low permeability soil occurs at the base of the active layer. This is consistent with field evidence indicating that fine-grained soil commonly ascends to the ground surface in periglacial areas, particularly where groundwater conditions are favorable. On the other hand, buoyancy is generally incapable of driving wholesale soil circulation in unpatterned active layers but may incrementally contribute to long-term circulatory soil motion in established sorted circles.

Periglacial Geomorphology. Edited by J. C. Dixon and A. D. Abrahams

Introduction

Differential vertical displacements of soil in the active layer are characteristic of a number of common periglacial features. These range from diapirs, tongue-like masses of soil that ascend into surrounding soil, through involutions, in which adjacent soil layers interpenetrate in complex ways, to circulatory soil motion, which is often inferred in sorted circles and other forms of patterned ground. Cryoturbation and frost churning are widely used terms describing differential soil motion manifested in the ubiquitous overturn, mixing, and in some cases, homogenization of soil in areas subjected to recurrent freezing conditions.

This spectrum of features and processes undoubtedly reflects a wide range of soil conditions and may involve a variety of driving mechanisms. For example, Washburn (1980) cited six distinct mechanisms for, and a multitude of references on, differential soil motion due to frost action, which he terms "mass displacements." Our paper focuses on how patterned ground formation and maintenance may be governed or affected by buoyancy forces due to vertical gradients in soil bulk density, which have long been recognized as potentially important in this context (Sørensen, 1935; Wasiutynski, 1946). In water-saturated soils, such gradients arise from differences in porosity because mineral particles and water can be regarded as incompressible. Little is known, however, about the precise origin and magnitude of the buoyancy forces. We present a quantitative analysis of the evolution of the vertical profile of soil bulk density in thawing permafrost. We examine a range of representative soil conditions and assess whether buoyancy is likely to play a significant role in inducing periglacial soil deformation or in driving circulatory soil motion. The theoretical framework we develop permits more precise understanding of diverse known characteristics of periglacial involutions and structures reflecting soil circulation. Our focus is on soil deformation and not on the role of thermally induced buoyancy of pore water in possibly driving water convection through the active layer, as studied by Krantz (1990) and coworkers (e.g., Gleason, Krantz, and Caine, 1988).

Periglacial involutions closely resemble deformation structures in unfrozen water-saturated sediments. Conditions in the active layer during the thaw period are particularly favorable for the formation of such structures, making them ubiquitous in periglacial areas. Mortensen (1932) and Sørensen (1935) recognized that thawing of ice-rich soil liberates considerable water that cannot percolate downward because of the underlying frost table, and that with continued water production due to thawing, the resulting water-rich sediments are prone to differential motion. Our model quantifies these observations and provides a theoretical framework suitable for interpreting results of experimental (e.g., Dzulynski, 1963, 1966; Butrym, Cegla, Dzulynski, and Nakonieczny, 1964) and field studies of involutions (e.g., Vandenberghe and Van Den Broeck, 1982).

Our model also sheds new light on the long-term circulatory motion of soil evident in numerous patterned ground areas (e.g., Nicholson, 1976; Mackay, 1980; Hallet, Anderson, Stubbs, and Gregory, 1988). Along with Van Vliet-Lanoe (1988), we favor the term circulation over convection because convection may convey a parallelism with convecting fluids that is too strict, and it may imply a particular driving mechanism. The different connotations of the term convection have been problematic in studies of patterned ground from one of the earliest (Nordenskjold, 1907) to some of the most recent (Van Vliet-Lanoe, 1988; Washburn, 1989). Circulation is used here to describe the circulatory or churning pattern of soil motion (its kinematics). The pattern comprises a general ascent of fine-grained soil in the center of sorted circles, a descent of soil at the periphery of fine-grained soil domains, and appropriate soil recirculation at depths to sustain the observed near-surface soil motion. It does not imply (1) that the soil is fluidized, (2) that motion is driven by buoyancy, or (3) that the subsurface motion is closely approximated by circular particle trajectories characteristic of Bénard convection cells, as has been proposed (e.g., Hallet and Prestrud, 1986) and subsequently questioned (Pissart, 1990). Hallet et al. (1988) interpreted surface soil displacements they measured in sorted circles as reflecting intermittent soil circulation, a kinematic model consistent with the observed size, geometry, microrelief, vegetation cover, and subsurface distribution of organic carbon of sorted circles. According to Hallet et al. (1988, p. 775), "theoretical considerations suggest that intermittent [buoyancy-driven] free convection is plausible in thawed fine-grained soil, and that the requisite bulk density decrease with depth arises naturally during the thawing and consolidation of ice-rich soil." However, their observation that "the actual driving mechanism remains to be established" (p. 775) provides one of the incentives for the present paper.

We wish to stress that we do not view buoyancy as being essential or even dominant in the development of many types of patterned ground. Other forces, especially those associated with differential frost heaving, are clearly important in their development due to the large gradients in heaving stresses in soil resulting from local variations in soil texture or moisture (Pissart, 1982; Van Vliet-Lanoe, 1988; Washburn, 1991). In essence, differential frost heaving results from preferential soil heaving in domains where moisture and textural conditions are particularly favorable for segregation ice growth. A critical element of this mechanism is that frost heave tends not to be fully reversible upon thawing. As a consequence, cyclic heave-thaw generally leads to permanent upward motion of soil where recurrent frost heaving is most active. The factors responsible for this irreversibility and the resultant residual long-term motion of the soil remain somewhat of a mystery. We propose that buoyancy can provide a significant and persistent upward force on frost-susceptible soil at depth through most of the thaw season, exactly what is needed to translate cyclic motion induced by freeze/thaw into long-term net soil motion.

Evolution of the soil density profile: thaw consolidation

The well-known settling of the soil surface during the thaw phase reflects an increase in the average soil bulk density in the active layer. Surface settling can exceed 0.1 m (Jahn, 1963; Hallet and Prestrud, 1986), approximately 10% of the typical active layer depth. The phase transformation of ice to water in soil pores could only account for less than half of this because the reduction of soil volume upon thawing would be 9% of the porosity, say 0.3 to 0.4, yielding a maximum settling of 0.04 m for a 1-m-thick saturated active layer. Much of the settling must involve the progressive loss of water that is in excess of the thawed porosity. The evolution of the vertical profile of soil bulk density through the thaw season and the attendant surface settling can be investigated by analyzing the consolidation process.

SOIL DENSITY AND HEAVING STRAINS

For unfrozen water-saturated soils the bulk density is:

$$\rho = (1 - \phi)\,\rho_m + \phi\,\rho_w \tag{11.1}$$

where ϕ is the porosity and ρ_m and ρ_w are the densities of mineral particles and water, respectively. Settling during the thaw period will be accompanied by increasing soil bulk density ρ, as can be seen from the conservation statements for water and for the soil matrix. Assuming that soil and water displacements are strictly one-dimensional and that mineral particles and water are incompressible, the conservation equations are:

$$\frac{\partial \phi}{\partial t} + \frac{\partial}{\partial z}(\phi\, V_w) = 0 \tag{11.2}$$

$$\frac{\partial \tau}{\partial t} - \frac{\partial}{\partial z}\left[(1 - \phi)\, V_s\right] = 0 \tag{11.3}$$

where t is time, z is the vertical distance directed upward, V_w is the mean water velocity, and V_s is the soil matrix velocity. Differentiating (11.1) with respect to time and inserting (11.3) yields:

$$\frac{\partial \rho}{\partial t} = (\rho_w - \rho_m)\,\frac{\partial \phi}{\partial t}$$

$$= (\rho_w - \rho_m)\left[(1 - \phi)\,\frac{\partial V_s}{\partial z} - V_s\,\frac{\partial \phi}{\partial z}\right] \tag{11.4}$$

If changes in soil bulk density are small in a soil layer with a uniform initial porosity, the advection of porosity differences can be neglected, and (11.4) reduces to:

$$\frac{\partial \rho}{\partial t} = (\rho_w - \rho_m)(1 - \phi)\frac{\partial V_s}{\partial z} \tag{11.5}$$

Thus the rate of change in bulk density is simply proportional to the vertical compressive strain rate. Integrating (11.5) with respect to time and substituting in (11.1) yields the following relation between the instantaneous local density change and vertical strain ϵ_z (tensile strain being positive):

$$\Delta\rho = (\rho_w - \rho)\,\epsilon_z \tag{11.6}$$

The surface settling during the thaw period, typically amounting to 10% of the active layer thickness, reflects a vertically averaged compressive strain of 10%. This strain corresponds to an average increase in soil bulk density of 5%, given that soil bulk densities in the active layer are characteristically about 2×10^3 kg m^{-3} (Hallet and Prestrud, 1986).

CONSOLIDATION DYNAMICS

Considerable progress has been made in the development of thaw consolidation models based on extensions of Terzaghi's (1943) original soil consolidation theory. Morgenstern and Nixon (1971) provided a solution to a moving boundary thaw consolidation problem. One-dimensional closed-form solutions of practical interest have been outlined for cases when the thaw front advances at a rate proportional to the square root of time. Numerical solutions were presented later for problems with arbitrary advance rates of the thaw line (Nixon and Morgenstern, 1973). In these models, soil in the thawed region is treated as a compressible, linear elastic, porous material. The volumetric strain of the soil is assumed to respond instantaneously to any change in effective stress, and the time-dependent nature of the consolidation arises only from the temporal element inherent in the motion of pore water. A more general model for a thawing unsaturated porous soil that includes coupled heat and mass transport was presented by Corapcioglu (1983). Although he assumes a time-independent linear elastic stress-strain relation for the medium in much of his analysis, he explores the possibility of modeling thawed soil as a temperature-dependent visco-elastic material.

As mentioned above, the average volumetric strain of thawing soil in sorted circle areas is on the order of 0.1, which is generally larger than the elastic strain expected to result from variations in effective stress. A maximum increase in effective stress from zero to the overburden, which is typically 10^4 Pa (for 0.5 m of soil with a density of 2×10^3 kg m^{-3}), would produce strains on the order of 10^{-3} to 10^{-1}, based on a representative range of compressibility values for frost-susceptible soils (10^{-7} to 10^{-5} Pa^{-1} from Morgenstern and Smith, 1973). This suggests that modeling thaw consolidation for frost-susceptible soils with considerable ice should include the nonelastic part of the deformation of the soil skeleton, as well as the pore fluid migration.

Recent mathematical formulations for the buoyancy-driven segregation of a relatively light fluid from a deformable matrix (McKenzie, 1984; Richter and McKenzie, 1984; Scott and Stevenson, 1984, 1986) can provide a novel formalism for, and interesting new insights into, thaw consolidation. In these formulations, derived in the geophysical context of melt-segregation, the matrix and the pore fluid are both treated as linear viscous materials. The success of rate process modeling in analyzing soil consolidation (see results by Christensen and Wu, 1964, as illustrated by Lun and Parkin, 1985, p. 164) provides some justification for approximating the consolidating matrix as a linear viscous fluid. This assertion stems from the form of the stress-strain rate relation emanating from rate-process theory; to first order this relation is linear at low strain rates (Mitchell, 1976, p. 294).

We emphasize, along with McKenzie (1984), that compaction of a layer of porous material is controlled either by the properties of both the matrix and the pore fluid or by those of the matrix alone. No solution exists in which the behavior is controlled by pore fluid properties alone. It is therefore essential to take into account the deformation of the soil matrix when discussing thaw consolidation. This is in marked contrast with existing models in which the time dependence in the consolidation process is assumed to arise solely from the fluid motion (e.g. Nixon and Morgenstern, 1973).

The mass conservation equations (11.1) and (11.2) have already been presented. Following Richter and McKenzie (1984), Darcy's Law is assumed to govern pore fluid motion:

$$\frac{\partial p}{\partial z} + \frac{\mu}{k} \phi \, (V_w - V_s) = 0 \tag{11.7}$$

where p is the pressure in excess of "hydrostatic" $\rho_w g z$, k is the permeability, and μ is the pore fluid (i.e., water) viscosity. The matrix deforms at a rate dependent on the imbalance between the fluid pressure and the "lithostatic pressure." The balance of forces for the matrix involves the pressure gradient, a buoyancy term, and the matrix deformation:

$$\frac{\partial p}{\partial z} + (1 - \phi) \, \Delta\rho \, g - \left(\xi + \frac{4\eta}{3} \right) \frac{\partial^2 V_s}{\partial z^2} = 0 \tag{11.8}$$

The fluid-matrix density contrast is $\Delta\rho = \rho_m - \rho_w$, and the effective bulk and shear viscosities of the matrix are ξ and η, respectively.

The permeability $k(z)$ depends on the porosity of the soil $\phi(z,t)$ and soil grain radius a. Following McKenzie (1984) we use the Blake–Kozeny–Carman relation:

$$k = \frac{\phi^3 a^2}{K(1-\phi)^2} \tag{11.9}$$

K is an adjustable constant in the range 45–1000. We use $K = 100$ because it yields realistic permeability values using the mean porosity and grain size of fine-grained soils from a sorted circle area.

An important natural length scale emerges from the analysis:

$$\delta = \left[\frac{k}{\mu} \left(\xi + \frac{4\eta}{3} \right) \right]^{1/2} \tag{11.10}$$

δ is the effective thickness of the compacting boundary layer that is characteristic of many solutions. It scales with soil permeability and with the ratio of viscosity of the matrix to that of the pore fluid.

Useful insight can be gained by considering the initial one-dimensional compaction of a constant porosity layer resting on an impermeable horizontal surface. With the boundary conditions:

$$V_s(0) = V_w(0) = 0 \tag{11.11}$$

for a rigid, impermeable lower boundary at $z = 0$, the solution of equations (11.2), (11.3), (11.7), and (11.8) is (McKenzie, 1984, equation B9):

$$\frac{-\partial V_s}{\partial z} = \frac{1}{\phi} \frac{\partial \phi}{\partial t} = \frac{-V_w^*}{\delta} (1-\phi) \exp\left(-\frac{z}{\delta} \right) \tag{11.12}$$

where V_w^* is a characteristic pore fluid velocity given by:

$$V_w^* = \frac{k}{\mu} \left(\frac{1-\phi}{\phi} \right) \Delta\rho\, g \tag{11.13}$$

Expression (11.12) is valid for time scales that are short compared with τ_0 given by (McKenzie, 1984, equations B11 and B9):

$$\tau_0 = \frac{\mu\, \delta\, \phi}{k\, \Delta\rho g\, (1-\phi)^2} \tag{11.14}$$

We use this criterion later to assess the validity of our results when applying the McKenzie theory to the active layer.

We now preview the gross characteristics of consolidation in a thawing active layer in two complementary ways that illustrate the magnitude of gravitational instabilities, and the spatial and temporal extent of zones of instability: (1) profiles of soil bulk density ρ or porosity ϕ as a function of depth at sequential times, and (2) contours of constant porosity (or density) as a function of both depth and time. An initially uniform layer thawed instantly (Figure 11.1A) is neutrally buoyant at all depths at the instant of thaw, t_0 (Figure 11.1D). At a later time t_1, net consolidation at any one depth (Figure 11.1C) will be the product of the time-averaged consolidation rate, which decreases with height z away from the base of the consolidating porous medium, as is evident from (11.12), and the time available for consolidation, which is the time elapsed t^* since soil thawed at that depth (Figure 11.1A).

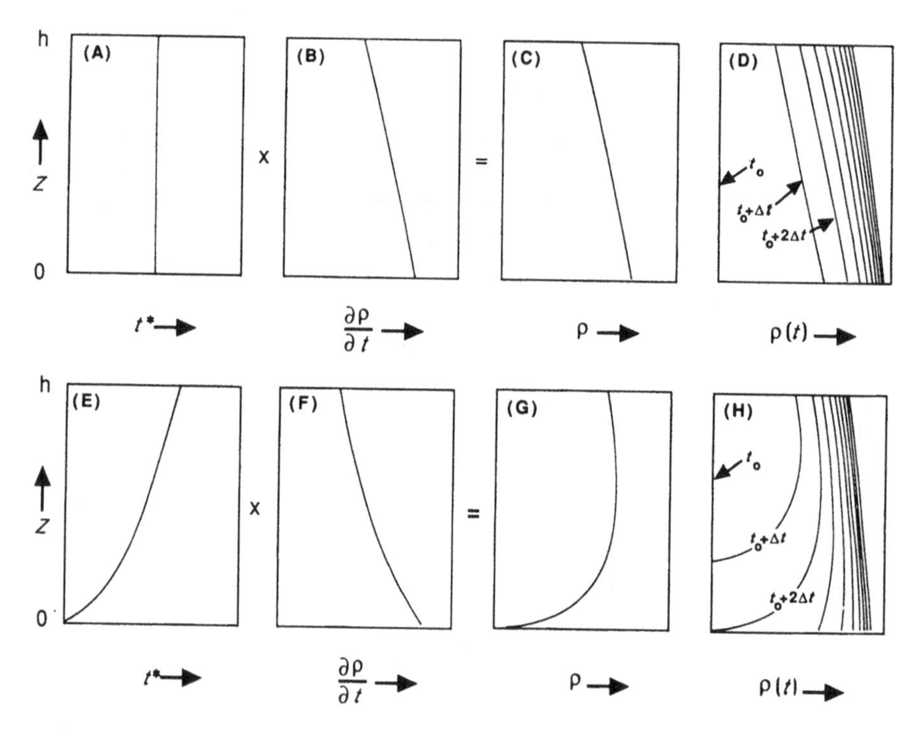

Figure 11.1 Schematic diagram illustrating the consolidation of an initially uniform saturated soil layer.
(A)–(D) Consolidation following instantaneous thaw at time t_0.
(E)–(H) Consolidation following realistic thaw penetration.
Each sequence shows: (A) time available for consolidation t^* at each depth z at some time t_1 later than t_0, (B) average rate of densification at each depth from (11.12), (C) net density profile at time t_1, given by the product of the two preceding curves, and (D) resulting density profile evolution at sequential times following onset of thaw. The reduced time available for consolidation near the thaw front in the case (E)–(H) leads to a zone of lower density at depth. Soil is gravitationally unstable where density decreases with depth in lower zones of (G) and (H)

However, when the thaw front takes a finite time to penetrate the active layer (Figure 11.1E), consolidation must vanish at the thaw front, because soil is frozen below this level. Far from the thaw front, where $z > \delta$, the net consolidation must diminish because the long time available for consolidation is compensated by slow consolidation rates. Maximum consolidation is therefore expected at a distance approaching δ above the thaw front. Assuming a uniform soil bulk density in the initial frozen state, the resulting density profile shows a maximum at that distance (Figure 11.1G). The resulting decrease in density with depth directly above the thaw is particularly noteworthy, as it renders the soil gravitationally unstable.

Alternatively, we can look at the same consolidation process of a layer of uniform initial porosity using a depth-time plot. Figure 11.2A illustrates the density evolution when thaw is instantaneous. The vertical left margin ($t = 0$) reflects the initial uniform density. The bulk density contours are concentrated near the base where consolidation is most rapid. Contours sloping up with time reflect stably stratified soil.

Consolidation of the active layer, in which thaw penetration is progressive, can give the pattern shown in Figure 11.2B. The leftmost density contour shows the penetration of the thaw front with time through a frozen soil of constant density. Contours of higher density are also distorted from their positions in Figure 11.2A; because consolidation starts later at depth, each density contour crosses a given depth later in Figure 11.2B than in Figure

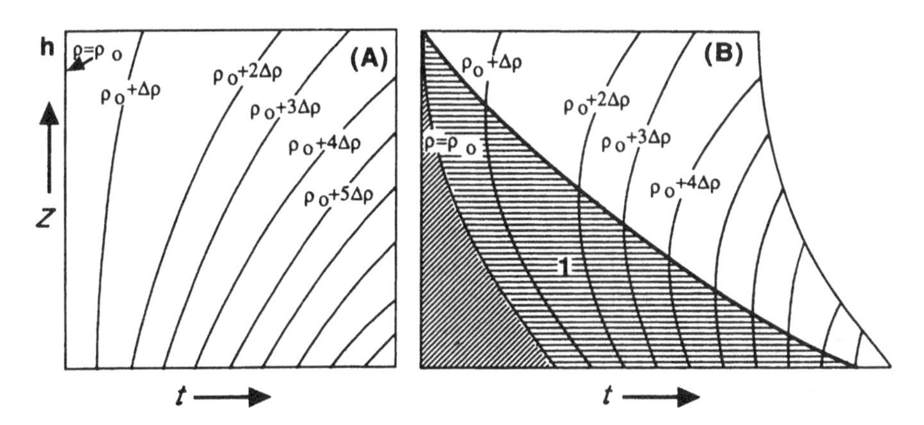

Figure 11.2. Consolidation of an initially uniform saturated layer viewed in a space-time plot. (A) Instantaneous thaw at time t_0. Consolidation proceeds faster at depth, leading to a permanently stable density distribution. (B) Delayed thaw penetration, as in Figure 11.1E. Oblique hatching represents time when soil is still frozen. Throughout zone 1 density decreases with depth, indicating gravitational instability. Density contours in (B) can be obtained by shifting those in (A) to later times by the amount $T(z)$, the time taken for the thaw front to reach depth z

11.2A. In fact, as a first approximation, we can derive Figure 11.2B from Figure 11.2A by simply time-shifting the contour pattern at each depth by the time thawing reaches that depth. This operation can be represented as a linear filter. Note that this time-shifting filter operation results in a region where contours have a negative slope and, consequently, where soil is gravitationally unstable.

SOIL PARAMETER EVALUATION

Before proceeding further in the analysis of thaw consolidation, it is useful to review what is known about the magnitude of the controlling parameters based on measurements made in an extensively studied sorted circle site in western Spitsbergen (Hallet and Prestrud, 1986; Hallet et al., 1988). All data presented below are for the sandy silt with pebbles that comprises the center of sorted circles at this site.

We have used various techniques to estimate the effective viscosity of the soil, including the rate at which a loaded sphere penetrates into thawed soil, the rate at which a decimeter-thick soil layer moves down slope, and the rate of relaxation of a cylindrical hole in the active layer. The results reflect extreme variations in effective viscosity both spatially and temporally. That the effective viscosity can change by orders of magnitude over a period of a few weeks is obvious to anyone who travels over a patterned ground area in the early thaw season and revisits the site a few weeks later; whereas deep footprints would form in many areas of wet fine-grained soil early in the thaw season, sole marks might not even be apparent later in the thaw season. Our measurements reflect this inherent variability and lead us to estimate the effective matrix viscosity $[\xi + (4\eta/3)]$ as ranging from 10^6 to 10^{10} Pa s in the first month of the thaw season. The effective bulk and shear viscosities of soil are assumed to be similar. Recognizing that the low values reflect the easily deformed material within a few days of thaw, the most representative estimate of effective soil viscosity for a month-long period directly after thaw appears to be 10^8 to 10^{10} Pa s. The viscosity of water μ is 2×10^{-3} Pa s at 0° C.

The permeability k can be evaluated directly from simple slug tests (Freeze and Cherry, 1979, pp. 339–342) in which a vertical pipe is filled with water and the rate at which it drains to a background value is monitored. A series of tests yielded permeabilities ranging from 2×10^{-15} m^2 to 2×10^{-16} m^2. These values are thought to represent the local permeability 0.3 to 0.6 m below the ground surface in the immediate vicinity of the bottom of the 40 mm-diameter pipe used in the slug test. For larger soil domains k can be estimated from measurements of the hydraulic diffusivity: $C_v = k/(\mu\ m_v)$ (Morgenstern and Smith, 1973), where m_v is the effective coefficient of compressibility of the soil (C_v is also known as the coefficient of consolidation). Representative m_v values for frost-susceptible soils range from 10^{-7} to 10^{-5} Pa^{-1} (Morgenstern and Smith, 1973). This range of m_v, together with the range of C_v (from 4×10^{-5} to 5×10^{-3} m^2 s^{-1}) deduced from our field pore pressure relaxation tests, leads to a plausible range of permeabilities of 10^{-10} to 10^{-14} m^2. That the higher values exceed typical permeabilities of silty soils (e.g. Mitchell, 1976) reflects a highly permeable soil microstructure produced by ice lens formation and subsequent thaw (Chamberlain and Gow, 1978; Van Vliet-Lanoe, 1988). We have used a plausible range of 10^{-13} to 10^{-15} m^2 in our calculations to reflect results of both techniques of estimating soil permeability. With our use of $K = 100$ in equation (11.9), these permeabilities correspond to soils comprised of grains with a characteristic radius of 13.4 and 1.3 μm, respectively. The soil from sorted circle centers is poorly sorted with a mode in the grain size distribution around 20 μm, which renders the higher permeability values more realistic, as will be confirmed below.

Although the viscosity and permeability values are poorly constrained, considerable guidance for defining realistic parameter values comes from comparisons of model predictions with field measurements of surface settling that reflect the cumulative effect of active layer consolidation. Numerous observations suggest that 50 to 100 mm of surface settling occurs as the thaw front descends through a meter-thick active layer in about one month.

THAW CONSOLIDATION OF UNIFORM SOIL LAYER

An instructive approximation of the consolidation of a thawing soil layer can be obtained by integrating (11.12) and taking into account that (1) the time available for consolidation decreases with proximity to the thaw front because no significant volumetric change is assumed to precede thawing, and (2) position z is measured relative to the impermeable base at the thaw front, which is a moving boundary. The thaw front advance $T(z)$ can be reasonably well approximated by (McRoberts, 1975)

$$\begin{aligned} (h - z) &= (\alpha\, T)^{1/2} \qquad & T \leqslant h^2/\alpha \\ z &= 0 \qquad & T > h^2/\alpha \end{aligned} \tag{11.15}$$

when thawing proceeds downward from the surface at $z = h$, α is a constant dependent on soil thermal conductivity, volumetric latent heat of frozen soil, and surface temperature. The value of α can be obtained empirically by measuring thaw depth versus time. For example, McRoberts (1975) reported a relatively narrow range of 5.8×10^{-2} to 4.5×10^{-1} mm^2 s^{-1}. Integrating (11.12) and inserting (11.9) and (11.13) yields

$$\int_{\phi_0}^{\phi} \frac{\mathrm{d}\phi}{\phi^3} = \frac{-\Delta\rho\, g\, a^2}{\mu\, \delta\, K} \int_{T}^{t} \exp\left(\frac{-z'}{\delta}\right) \mathrm{d}t \tag{11.16}$$

where $z'(t)$, the height above the thaw front, is

$$\begin{aligned} z'(t) &= z - [h - (\alpha t)^{1/2}] \qquad && t \leqslant h^2/\alpha \\ &= z && t > h^2/\alpha \end{aligned} \tag{11.17}$$

Solving for $T(z)$ in (11.15) and substituting into (11.16) leads to an approximate solution for the incipient stage of consolidation in the thawed material.

For $t \leqslant h^2/\alpha$

$$\frac{1}{\phi(z,t)^2} = \frac{1}{{\phi_0}^2} - R\,\delta \exp\left(\frac{h-z}{\delta}\right)\Big[(\ell_0 + 1) \exp(-\ell_0) - (\ell + 1)\exp(-\ell)\Big]$$

and for $t > h^2/\alpha$

$$\frac{1}{{\phi_0}^2} - R\,\delta \exp\left(\frac{h-z}{\delta}\right)\left\{(\ell_0 + 1)\exp(-\ell_0) - \left[\frac{h}{\delta} + 1 + \frac{(h^2 - \alpha t)}{2\,\delta^2}\right] \exp\left(\frac{h}{\delta}\right)\right\} \tag{11.18}$$

where

$$R = \frac{2\,k\,g\,\Delta\rho\,a^2}{(\mu\,\alpha\,K)}, \quad \ell_0 = \frac{(\alpha T)^{1/2}}{\delta}, \text{ and } \ell = \frac{(\alpha t)^{1/2}}{\delta} \tag{11.19}$$

To obtain the analytical solution (11.18), we assumed that the compaction length δ does not vary appreciably as the porosity ϕ and hence the permeability $k(\phi)$ decrease over time. This approximation should be valid over a time comparable to τ_0 given by (11.14). Note that even though (11.18) may not be quantitatively valid after time τ_0 for several reasons, (11.18) still has the correct asymptomatic behavior at longer times – that is, porosity approaches zero at a decreasing rate.

Equation (11.18) together with (11.1) can be used to compute the variation in soil bulk density ρ with depth at sequential times. Figure 11.3 shows results assuming that at the onset of thaw ρ is uniform. In this and subsequent figures, all soil densities are normalized by this initial value of ρ, the initial porosity ϕ_0 is taken to be 0.3, and the density ρ_m of soil particles is 2700 kg m^{-3}. The thaw penetration rate parameter α is 0.264 $mm^2\ s^{-1}$—that is, thaw reaches 1 m, the bottom of the active layer, in 1 month. Figure 11.3 explores expected ranges of effective soil viscosity $[\xi + (4\eta/3)]$ and particle radius a (which controls permeability) based on field measurements. In the top 3 panels, particle size is 13.4 μm, typical of a silty soil; the corresponding initial permeability is $10^{-13}\ m^2$ according to (11.9). The lower 3 panels show results for 1.3 μm particles typical of clayey silt with an initial permeability of $10^{-15}\ m^2$. The effective soil viscosity decreases from left to right in the sequence 10^{12}, 10^{10}, and 10^8 Pa s.

Note that in all cases a layer develops above the thaw front where the soil bulk density ρ_s decreases with depth because limited time has been available for matrix deformation and water percolation. The thickness of this layer scales with δ, the compaction depth, which can be evaluated using (11.10). δ decreases by a factor of 10 in each panel moving from left to right.

In Figure 11.3A, the whole active layer is buoyant, and this condition persists throughout the thaw season. As the effective soil viscosity decreases (Figures 11.3B, C), the matrix within the boundary layer of decreasing thickness δ compacts relatively rapidly, thereby accelerating stabilization of the density profile. Figure 11.3B shows stability by 2.5 months after thaw initiation, while Figure 11.3C is stable everywhere by 1.25 months.

The same trends are visible for the clayey silt in Figures 11.3D–F, although δ is an order of magnitude smaller for each corresponding effective soil viscosity, and buoyant soil conditions are less persistent in each case. When the compaction length δ is much smaller than the active layer depth (Figures 11.3D–F), there is almost no compaction at heights greater than δ above the base. Most of the compaction higher in the soil occurs rapidly as the thaw front passes that depth. At later times the gravitational forces inducing compaction of the matrix are balanced by pressure gradient (seepage) forces associated with upward water percolation.

Our solution (11.18) neglects advection of porosity differences by soil settling. According to (11.14), the results in Figure 11.3 are valid for times τ_0 (11.14) that decrease from 20 months at the lower left of Figure 11.3 (panel D) to 0.6 days at the upper right (panel C). At times greater than τ_0, the patterns are qualitatively correct but should not be interpreted quantitatively.

A useful general assessment of the model performance can be made by comparing the calculated surface settling, which results from the consolidation integrated throughout the active layer, with a sample of our measurements of settling in a sorted circles area, western Spitsbergen (Figure 11.4). The comparison is encouraging in view of the fact that this model result was not "tuned" to simulate this particular thaw settling event. Both the magni-

Table 11.1. Calculated Surface Settling after 0.15 Years

Panel in Figure 11.3	Particle Radius (μm)	Permeability (m^2)	Effective Soil Viscosity (Pa s)	Compaction Length (m)	τ_0 (months)	Surface Settling (mm)
A	13.4	10^{-13}	10^{12}	7.07	20	52
B	13.4	10^{-13}	10^{10}	0.707	0.2	118
C	13.4	10^{-13}	10^{8}	0.071	0.02	108
D	1.3	10^{-15}	10^{12}	0.707	20	5.4
E	1.3	10^{-15}	10^{10}	0.071	2.0	6.5
F	1.3	10^{-15}	10^{8}	0.007	0.2	6.5

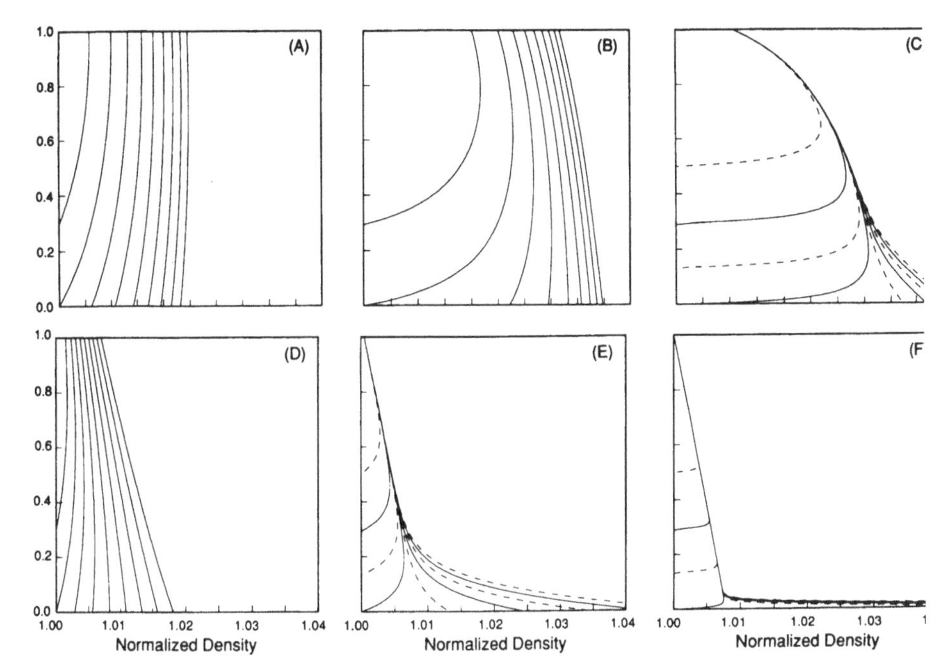

Figure 11.3. Consolidation of an initially uniform layer with 30% porosity following thaw penetration given by (11.15) with $\alpha = 0.264\ \mathrm{mm^2\ s^{-1}}$. Solid curves show density profiles at 0.5 month intervals as functions of height above the active layer base. Dashed curves, where present, show density at intervening 0.25 months. The examples span the range of measured values of permeability and effective soil viscosity. Soil parameters, the time scale τ_0 over which the solution is quantitatively valid and calculated surface settling after 0.15 years, are listed in Table 11.1. (B), (C), and (F) should be regarded as schematic because results are only quantitatively correct for a portion τ_0 of the time shown.

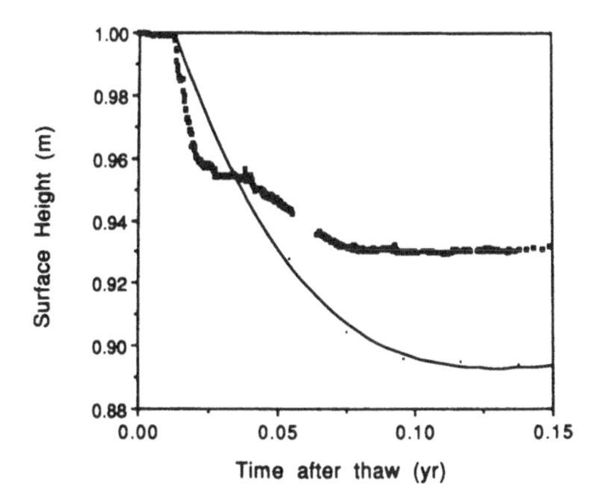

Figure 11.4. Surface settling history during thaw. Solid curve is calculated from model in Figure 11.3C. Locus of points represents measured settling at the surface of a sorted circle in western Spitsbergen

tude of the settling and the rate at which it decreases with time resemble those measured. The very rapid settling observed at the onset of the thaw period reflects the accumulation of ice in the upper few decimeters of soil where ice segregation is so massive that the soil structure is pervasively disrupted, leading to a marked decrease in the soil's effective viscosity. The present model makes no effort to incorporate this effect.

The calculated results in Figure 11.4 were obtained using the coarser grain size, which approaches the average measured grain size of 20 μm. Using a smaller grain size reduces the magnitude and rate of surface settling to unrealistically low values through its marked impact on permeability. Although the use of a smaller grain size and a permeability of 10^{-15} m^2 could be justified on the basis of local permeability measurements, it seems considerably less viable in view of the measured surface settling.

CONSOLIDATION WITH VERTICAL VARIATION IN ICE CONTENT

Observations of ice content in the active layer suggest that in some cases the ice is concentrated near the surface (e.g., Anderson, 1988). Figure 11.5A illustrates the compaction of a 1-m layer of silty soil corresponding to the layer shown in Figure 11.3B, with a uniform particle size $a = 13.4\,\mu$m and effective soil viscosity of 10^{10} Pa s. However, in this case ice content decreases with depth according to

$$\phi_0(z) = 0.3\left[1 + \exp\left(\frac{z - h}{\gamma}\right)\right] \qquad (11.20)$$

where γ is an e-folding depth for the initial ice content. Based on results of Anderson (1988), we use $\gamma = 0.2$ m. Comparison of Figure 11.5A with Figure 11.3B shows that this initial porosity pattern stabilizes the active layer.

In areas of permafrost with very cold temperatures below the active layer, water percolates and freezes into the underlying frozen ground and the active layer freezes upward from the base late in the thaw season (Mackay, 1983). This could lead to a concentration of segregation ice near the base of the active layer (e.g., Harris, 1988), which would enhance the magnitude and duration of buoyant conditions. Figure 11.5B shows the density evolution for the same uniform soil layer as Figure 11.3B, except that the initial porosity upon thaw is given by

$$\phi_0(z) = 0.1\left[3 + \exp\left(\frac{z}{\gamma}\right)\right] \qquad (11.21)$$

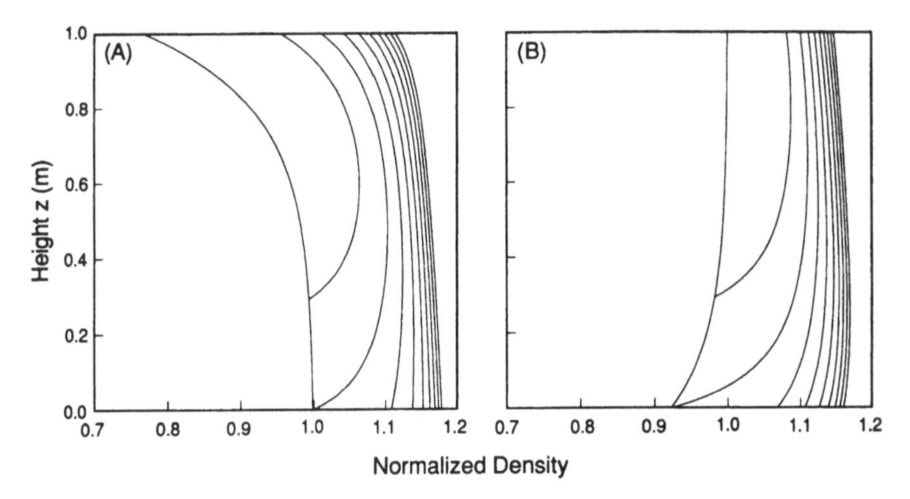

Figure 11.5. Compaction of active layers with initial vertical variation in ice content. Soil parameters same as Figure 11.3B—that is, particle radius $a = 13.4\,\mu\text{m}$, $[\xi + (4\eta/3)] = 10^{10}$ Pa s, and initial compaction length $\delta = 0.707$ m. (A) Porosity (ice content) concentrated near the surface (11.20), preventing or reducing the subsequent development of an unstable density profile. (B) Porosity concentrated near the bed due, for example, to water migration to an upfreezing front (11.21). High porosity at depth enhances and prolongs the density inversion

with $\gamma = 0.2$ m as before, for convenience. Although the upper active layer develops a stable profile, the low density zone at the base persists throughout the thaw period.

This case more clearly violates the assumption of zero density gradients, and so the time over which the McKenzie (1984) solution remains valid (11.14) is much restricted. However, the trends toward a totally stable active layer in Figure 11.5A and a persistent buoyant layer in Figure 11.5B are evident. The physical factors left out of the model will not change this.

CONSOLIDATION WITH VARIATION IN SOIL PERMEABILITY

Noting that the consolidation rate (11.18) is proportional to permeability, we now explore the evolution of a layer of constant initial bulk density that has a zone of low permeability at the base. This model would represent, for example, a relatively coarse-grained soil overlying a fine-grained layer. We can anticipate that slower consolidation in the basal material might result in a persistent gravitational instability.

We model this structure by subdividing a 1-m-thick active layer into three layers. The grain size and thickness are respectively $1.3\,\mu\text{m}$ and 0.25 m for the basal layer and 13.4 μm and 0.60 m for the upper layer. These grain sizes correspond to initial permeabilities of 10^{-15} m^2 and 10^{-13} m^2, respectively (11.9). In the 0.15-m-thick transitional zone, grain size varies continuously following a simple cosine function. The effective soil viscosity is 10^{10} Pa s at all depths. Due to the permeability contrast, the compaction length δ varies from 0.71 m in the upper layer to 0.071 m in the lower layer.

Figure 11.6A shows density profiles at 0.25-month intervals. Figure 11.6B shows the same model result as a time-depth plot of density contours. Due to its relatively low permeability, the basal soil can neither compact as quickly as the overlying layer nor provide significant upwardly migrating expelled water. Thus the clayey-silt acts very much like a rigid impermeable bed for the upper layer. However, any water expelled by the clayey-silt layer can migrate away more freely through the more permeable overlying silty layer, which thus approximates a free upper surface for the lower layer. As a result, each layer compacts independently, following the patterns seen in the uniform layer examples in Figure 11.3, except that the load of the upper layer increases the potential to drive water out of the lower layer.

In Figure 11.6 three separate zones of instability arise with distinct length and time scales. Each layer develops an internal transient zone of buoyant soil associated with the passage of the thaw front. Figure 11.6A shows that the magnitude of this intra-layer density inversion is greater in the upper layer. These intra-layer buoyant zones labeled 1 and 2 in Figure 11.6B, persist for 0.12 years, or 1.5 months, in the surface silty layer and about 0.5 months following thaw penetration in the basal clayey silt. During these periods, buoyant forces would tend to cause mixing within the respective layers. Figure 11.6B shows that the intra-layer instability in the upper silty soil, initially extending nearly the full layer depth, becomes progressively confined toward the base of the layer. The length scale for potential soil mixing decreases, and we might expect the greatest potential for mixing near the base of upper the layer. The instability within the lower layer, however, extends over the whole basal layer depth for much of its short existence.

Now considering the entire active layer, the whole of the fine-grained basal layer compacts much more slowly, so it rapidly becomes, and remains, strongly buoyant relative to the overlying soil throughout the thaw season. The result is the strong and persistent density inversion shown by the marked increase in density above 0.2 m in Figure 11.6A, and labeled 3 in Figure 11.6B. This inter-layer instability constitutes a strong driving force for mixing the lower clayey silt layer with the overlying silty soil.

The existence of multiple length scales and time scales for the various zones of gravitational instability suggests that a comparably wide range of scales may arise for soil diapirs or involutions if these density inversions can be shown to be adequate to trigger differential vertical soil motions.

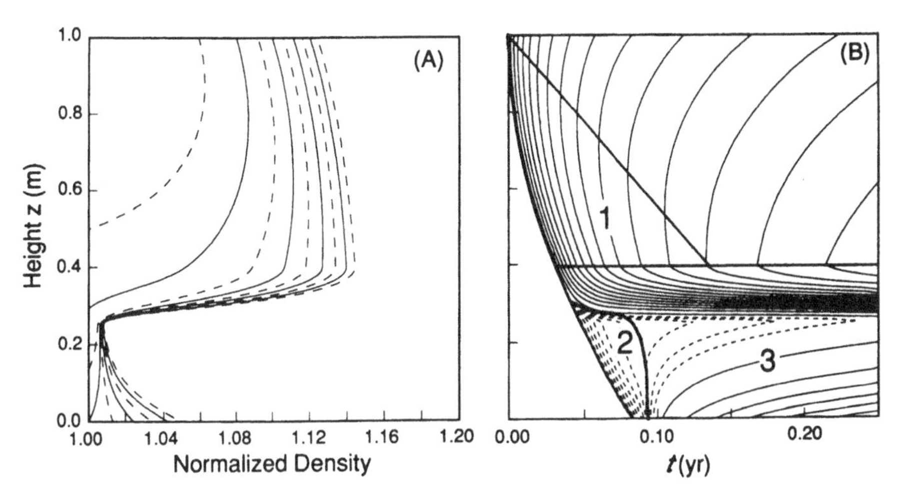

Figure 11.6. Compaction of a permeable silty soil layer (upper 60 cm) overlying a layer of reduced permeability (clayey silt). Soil viscosity $[\xi + (4\eta/3)] = 10^{10}$ Pa s at all depths. In the upper layer soil particle radius $a = 13.4\,\mu$m (initial permeability k and compaction length δ are 10^{-13} m^2 and 0.707 m). In the lower layer the particle radius is $1.3\,\mu$m ($k = 10^{-15}$ m^2 and $\delta = 0.0707$ m). (A) Density profiles at 0.25 = month intervals. (B) Density contours in depth and time. Solid curves show 1% density increase relative to initial value of 2190 kg m^{-3} (30% porosity). Dashed curves show 0.1% increases. Numerals 1 and 2 mark distinct zones of intra-layer instability, and numeral 3 marks the entire basal layer, which is buoyant relative to the upper layer. The leftmost contour indicates both the constant initial density and progressive thaw penetration (11.15)

Response to unstable density profile

CIRCULATORY SOIL MOTION

We now wish to examine the plausibility of buoyancy-driven circulation of soil in the active layer in the context of our thaw consolidation model. Whether buoyancy is sufficient to drive soil circulation can be assessed by evaluating the effective Rayleigh number Ra and comparing it to the critical Rayleigh number, Ra_c (of order 10^3) that must be exceeded for circulation to occur.

For an order of magnitude analysis, thawed soil is idealized as a gravitationally unstable layer of Newtonian viscous fluid with thickness δ and density difference ρ' across the layer. Ascending low density soil rises at a velocity U that is controlled by the effective soil viscosity η. U can be approximated from the constitutive relation for a linear viscous fluid, where σ is the stress resisting the buoyant ascent:

$$\eta \frac{\partial U}{\partial z} = \sigma, \text{ or } \eta \frac{U}{\delta} \approx \frac{\rho' \delta^3 g}{\delta^2} \tag{11.22}$$

so that

$$U \approx \frac{\rho' g \delta^2}{\eta} \tag{11.23}$$

In the absence of compaction, the density contrast between the ascending soil and the surrounding soil increases at the rate

$$U \frac{\partial \rho}{\partial z} \approx U \frac{\rho'}{\delta} \tag{11.24}$$

during the ascent into overlying denser soil, causing enhanced buoyancy.

However, during this time the ascending soil increases in density due to compaction. Combining (11.1), (11.12), and (11.13) gives the order of magnitude compaction rate:

$$\frac{\partial \rho}{\partial t} = -\Delta\rho \frac{\partial \phi}{\partial t} \approx \frac{k \Delta\rho^2 g}{\eta \delta} \left[(1 - \phi) \exp\left(- \frac{z}{\delta} \right) \right] \tag{11.25}$$

The bracketed term in (11.25) averaged through the unstable layer depth δ, with $\phi = 0.3$, is about 0.4.

For sustainable buoyancy-driven circulation, the low density soil must ascend into more dense surroundings more quickly than it compacts. The effective Rayleigh number can now be approximated by combining (11.23) through (11.25), using the definition (11.10) of the compaction length δ, and neglecting the bracketed term in (11.25) to get

$$Ra = U \frac{\partial \rho}{\partial z} \bigg/ \frac{\partial \rho}{\partial t} \approx \frac{\mu \delta^2 \rho'^2}{\eta k \Delta\rho^2} \approx \left(\frac{\rho'}{\Delta\rho} \right)^2 \tag{11.26}$$

That is, free convection depends on the ratio of density differences across the layer to the matrix-pore fluid density contrast. *Ra* is always less than unity because ρ' cannot exceed $\Delta\rho$. Taking 1700 kg m^{-3} for $\Delta\rho$, and a typical value of 100 kg m^{-3} for ρ' from Figure 11.3, yields $Ra = 3 \times 10^{-3}$. This result supersedes a preliminary estimate (Hallet, 1987) of 10^2 for *Ra*, which reflected an unrealistically low hydraulic diffusivity value obtained from the literature and the use of the standard elastic thaw consolidation model. The new estimate of *Ra*, nearly six orders of magnitude below the critical value, implies that buoyancy-driven circulation of active layer soil is not sustainable, contrary to prior suggestions (Wasiutynski, 1946; Hallet and Prestrud, 1986; Hallet et al., 1988). We stress that the inadequacy of buoyancy in driving wholesale circulation in no way implies that soil circulation does not occur; it merely leaves the driving mechanism undetermined.

In this context, it is of particular interest to note that buoyant soil regenerated seasonally at the base of the thawing active layer will nevertheless tend to rise into the overlying material. This could generate small-scale involutions by contorting passive markers, such as wisps of organic matter or distinct sedimentary horizons. Moreover, provided this upward motion is organized on a larger scale through the existence of preexisting soil patterning, it may contribute incrementally to systematic long-term soil displacement patterns in established sorted circles (e.g., Gripp, 1952; Hallet et al., 1988).

To estimate how far a domain of buoyant material would rise each season we can evaluate the ascent rate $U(t)$ from (11.23) and the time t_{stop} required for the domain to stop because it is no longer buoyant (i.e., $\rho'(t_{stop}) = 0$). Buoyancy is lost progressively because compaction stabilizes the density profile; compaction rates increase with depth and, hence, are greatest in the most buoyant material. The distance L a particular domain of buoyant material might rise is given by

$$L = \int_0^{t_{stop}} U(t)\, \mathrm{d}t \tag{11.27}$$

Equations (11.12) and (11.13) provide a relation for the compaction rate as a function of height above the frost table. This leads to an expression for t_{stop} in terms of the initial porosity $\chi_0(z)$. Then (11.27) can be solved for L.

$$L = \frac{\delta}{2} \frac{[\phi_0(\delta) - \phi_0(0)]}{\{[1 - \phi_0(\delta)]^2 - [1 - \phi_0(0)]^2 e^{-1}\}} \tag{11.28}$$

L depends only on the layer thickness, approximated here by the compaction length δ, and on the initial porosity structure $\phi_0(z)$. For representative conditions appropriate for Figure 11.3B at $t = 2$ months, for example, the parameters in (11.28) are $\delta = 0.7\,\mathrm{m}$, $\phi_0(\delta) = 0.17$, and $\phi_0(0) = 0.30$, which leads to $L = 25\,\mathrm{mm}$. This suggests that a domain of buoyant material could rise centimeters in a single season. Repeated annually, the cumulative effect of such incremental contributions could produce considerable long-term soil displacements in the active layer with characteristic rates similar to those deduced from measurements in sorted circles (Hallet et al., 1988).

More quantitative estimates of displacement rates can be obtained by determining displacement rates for a layer of buoyant soil at the base of the active layer. We now examine this diapirism problem for a particularly interesting situation in which a fine-grained soil layer occurs at the base of the active layer.

DIAPIRISM

As can be seen in Figure 11.6, fine-grained material at the base of a stratified active layer consolidates slowly and, hence, retains its low bulk density for the entire thaw phase. Thus a domain of the lower layer protruding into the upper would be buoyant. More importantly, buoyancy forces acting on this basal soil domain would increase with time due to compaction because the coarser material compacts more rapidly, thereby rendering the fine-grained soil increasingly buoyant. This situation contrasts with the uniform layer case in which compaction dissipates buoyancy force.

The layered situation with a lower density basal layer is unconditionally unstable: the lower layer invariably tends to rise to the surface or to a level where it encounters material of equal bulk density. The rate of rise varies widely, however, from negligible for vanishing density gradients or high effective soil viscosities to significant where conditions are most favorable. We now determine whether realistic buoyancy forces can produce significant soil deformation. Here we treat the thawed active layer as if it were comprised of two layers of inherently different densities, more precisely as two layers of immiscible fluids with the upper layer being denser. The simplest version of this Rayleigh–Taylor instability problem examines the behavior of a slightly distorted planar interface between two layers of Newtonian viscous fluids having identical viscosities η. For layers of equal thickness b confined between rigid plates on top and bottom, the solution of the problem is an exponential increase in the amplitude of perturbations in the interface with a characteristic growth time τ (Turcotte and Schubert, 1982). That is, the amplitude A of the interface distortion grows with time t from its initial value A_0:

$$A = A_0 \, e^{t/\tau} \tag{11.29}$$

The characteristic growth time τ depends on the wavelength of the interface distortion. The disturbance with the shortest time constant grows preferentially and dominates the instability. This smallest τ is

$$\tau = \frac{13.04 \, \eta}{\rho' \, g \, b} \tag{11.30}$$

The density difference between top and bottom layers is ρ'. As expected, the instability takes longer to grow the more viscous the fluids, the smaller the density difference, and the thinner the layers.

To evaluate the characteristic growth time for a perturbation in a low density basal layer in the active layer, we use the following representative range of physical properties: 10^8 to 10^{10} Pa s for the effective viscosity of thawed soil; 20 to 200 kg m^{-3} for ρ', which is 1 to 10% of the mean soil density in accord with results summarized in Figure 11.6; and 0.5 m for the layer thickness b to represent a basal layer nearly half the thickness of the active layer. Taking various combinations of these parameters yields values of τ ranging from 10^6 to 10^9 s. As soil motion is insignificant while frozen, diapiric development is limited to a short portion of each year, say about one tenth of a year. This τ range translates to characteristic times of formation of diapirs ranging from 0.3 to 300 years, which corresponds favorably with observations by L. Washburn (personal communication, 1991) suggesting that diapirs can form within 20 years where the ground has been artificially disturbed, and with more general observations suggesting a time scale of centuries (Mackay, 1980).

It is also of interest to compare rates of soil displacements implied in this analysis with the long-term rate of 1 to 10 mm yr^{-1} typically measured in sorted circles (Hallet et al., 1988). Vertical rates scale with the initial rate at which the interface between the two layers distorts, which is $dA/dt = A_0/\tau$ from (11.29). We obtain soil displacement rates equal to those measured provided we select reasonable parameter values: $A_0 = 0.01$ m, and $\tau = 1$–10 years.

Another useful result of the stability analysis is the size and spacing of diapirs. Although the analysis applies only to the development of small disturbances in the planar interface, the spacing between fully developed diapirs is expected to approximate closely the wavelength of the most rapidly growing small disturbance. In a two-layer system with a free surface, the diapir spacing S is a large multiple of the basal layer thickness. It increases with the thickness ratio b_b/b_u and the viscosity ratio μ_u/μ_b, where subscripts u and b refer to upper and basal layers, respectively. According to Fletcher (1972), the normalized spacing S/b_b would be 5, for example, for two layers of equal thickness and viscosity—that is, for $b_b/b_u = \mu_u/\mu_b = 1$. A more appropriate choice of parameters might be $\mu_u/\mu_b = 100$ and $b_b/b_u = 0.5$ for a situation where the lower portion of the active layer, in this case the lower third, is much more mobile because compaction or dessication has hardened the upper layer. In this case, the normalized spacing would be 12.7. Correspondingly, diapirs 1 to 2 m apart would be expected to form from a buoyant basal layer about 0.2 m thick. Such diapir spacings are similar to those of incipient sorted circles observed by the authors in western Spitsbergen and to "ring-bordered sorted circles" and the larger "plugs" observed by Washburn (1991) in the Canadian Arctic. Closer accord is not expected because of the simplistic rheological assumptions and the paucity of data on soil textures in the lower part of the active layer.

Discussion

More definitive results on the time-dependent density structure of thawing soils require more exact calculations and better data on soil permeabilities and effective matrix viscosity as functions of porosity. The analysis has to treat explicitly the finite decrease in porosity with the concomitant reduction in permeability and marked increase in effective soil viscosity (and compaction depth). A logical next step would be to analyze thaw consolidation by means of a numerical model of the governing equations with initial conditions reflecting known thermal conditions and initial porosity profiles in thawed soils. A promising tactic would be to parallel closely the development by Dawson and McTigue (1985) of a model for natural convection in fluid-saturated creeping porous media.

As presently formulated, our initial modeling effort follows much of the previous work on thaw consolidation by including only gravitationally induced consolidation of soil and buoyant ascent of water in a saturated active layer resting on an impermeable substrate. Considerable field data indicate, however, that some water moves downward from unfrozen to frozen ground in the summer (McGaw, Outcalt, and Ng, 1978; Mackay, 1983). This water movement is presumably induced by a chemical potential gradient in water from thawed regions into the underlying frozen layer. Moreover, the upper part of the active layer is commonly unsaturated after the initial phase of the thaw period (Mackay, 1983). Thus, the model could be modified to permit water flow into the frozen ground below and extended to the unsaturated case using as a guideline previous work by Corapcioglu (1983). Alternatively, a general upward flux of water could be included in the model to simulate situations where a large-scale ground water flow leads to an upward force on the soil matrix due to seepage pressures. Including this ground water flow in the model is readily done by altering the boundary condition at the active layer base to a fixed upward flux. The effect is expected to be to slow consolidation and prolong the duration of effective buoyancy forces, which may help explain why involutions and diverse forms of patterned ground are often localized where groundwater tends to converge because of a down slope decrease in slope inclination (e.g., Washburn, 1991) or in permeability.

The analysis of thaw settling of a uniform layer suggests that although the entire active layer can be gravitationally unstable for much of the thaw season, buoyancy is unable to drive wholesale soil circulation. Buoyancy can, however, contribute incrementally to soil circulation on the scale of the active layer thickness by driving soil displacements in a thin compacting boundary layer, provided the displacements are spatially organized on the active layer scale by preexisting soil structures. The conclusion about wholesale circulation is robust and insensitive to the choice of effective viscosity and permeability of the soil. Its surprising insensitivity to soil parameters arises from competing effects. For example, whereas relatively large effective soil viscosities are required to destabilize the entire active layer and to produce buoyancy forces that are both large and persistent, soil motion, including circulation, is strongly impeded by large viscosities.

Where segregation ice is concentrated near the surface, a relatively stable density profile develops in the thawed soil because soil at depth starts off and remains more dense than the overlying soil. In contrast, a fine-grained soil layer occurring near the base of the active layer remains less dense than the overlying soil throughout the thaw season and, hence, will persistently be forced upward by buoyancy. Manifestations of this instability are expected to be ubiquitous because the upward motion of the low permeability layer is not subject to a critical threshold level equivalent to the Rayleigh number for free convection; a light layer invariably ascends diapirically through denser material. The ascent, however, may be slow if buoyancy forces are small or effective viscosities are large. Also, the role of buoyancy is likely to be minimal for coarse-grained soils because their high permeability favors rapid consolidation and, hence, short-lived buoyancy effects.

The gravitational instabilities discussed here arise from buoyancy forces that are caused by thawing ice-inflated soil and decay with time through the thaw season. Detailed measurements of bulk density profiles may reveal unstable conditions in areas currently subjected to freeze-thaw activity (e.g., Hallet and Prestrud, 1986), but measurements in formerly active areas are expected to show the normal increase in soil bulk density with depth due to the increase in effective stress with depth. Therefore documentation of a normal density profile in areas with fossil forms of cryoturbation does not negate the presence of significant destabilizing buoyancy forces in previously active forms (Vandenberghe and Van Den Broeck, 1982).

Finally, we draw attention to an intriguing possibility that emerges from consolidation analysis: that solitary waves, or solitons (Scott and Stevenson, 1984), may form. These would constitute shape-preserving waves of soil with high water content that buoyantly ascend through the soil matrix. This possibility is of special interest in permafrost areas where soliton behavior could perhaps account for enigmatic mud boils or mud eruptions that sporadically spill over the tundra (e.g., Elton, 1927, p. 171; Jahn, 1948, p. 52; Shilts, 1978).

Conclusions

The model provides a quantitative framework for studying buoyancy forces whose role in driving a variety of soil deformation features in the thawed active layer has long been hypothesized. A novel analysis of thaw consolidation of ice-rich soil, in which both soil and water are treated as viscous fluids, permits definition of the length scale for the consolidation process, duration and magnitude of transient density inversions, and effects of specific soil properties. According to the analysis, a zone of decreasing density with depth tends to develop above the thaw front, and its thickness increases with the permeability and effective viscosity of the soil. For soil properties characteristic of a sorted circle area in western Spitsbergen, the maximum potential thickness of the unstable layer exceeds the active layer thickness, suggesting that soil deformation due to buoyancy could occur throughout the active layer. Buoyancy effects arise from bulk density differences of several percent. Such effects are two orders of magnitude larger than those associated with the 0° to 4° C difference in water density. Thus buoyancy forces associated with porosity gradients in saturated thawed soils overwhelm those due to temperature gradients, as recognized long ago by Sørensen (1935).

The model suggests that buoyancy is most likely to be manifested in the diapiric ascent of low permeability material from the lower part of the active layer and, hence, may be responsible for the widely recognized tendency of fine-grained material to be deformed upward by freeze-thaw activity. Moreover, the model provides a plausible rationale for many reported characteristics of periglacial involutions and related structures. These include (1) the multiplicity of sizes of deformation structures generally characteristic of involutions (e.g., Jahn, 1975, Figures 91 and 93, Photographs 97 and 98); (2) the occurrence of deformation structures within individual layers as well as across layers; (3) the rates of growth and spacings of diapirs in simply stratified active layers; (4) the deduction that for buoyancy to drive vertical soil motion thawing must produce water in excess of the pore volume of thawed soil (Sørensen, 1935); and (5) the tendency for diapirism and frost-induced deformation to be best developed where soils have low permeability and low effective viscosity, where downward percolation is inhibited by permafrost (e.g., Vandenberghe, 1988), and where upward seepage forces can be inferred from a downslope decrease in slope (e.g., Washburn, 1991) or permeability, both of which favor groundwater convergence.

Contrary to previous suggestions, buoyancy seems incapable of driving wholesale soil circulation in a laterally uniform active layer. Rather than invalidating the soil circulation concept, which is strongly supported by diverse field evidence, it provides considerable incentive to identify and characterize alternative mechanisms, which may originate from differential frost heaving (Washburn, 1980, 1991; Pissart, 1982, 1990; Van Vliet-Lanoe, 1988) or other very different processes. We stress that buoyancy and differential frost heaving are not mutually exclusive; in fact they may act in concert. For example, buoyancy may play a critical role in the differential heaving process by assuring that frost heave is not fully reversible upon thawing. As we proposed earlier, buoyancy can provide a significant and persistent upward force on frost-susceptible soil at depth through most of the thaw season, which would help transform the seasonal up and down soil motion resulting from freeze/thaw into the net long-term ascent of high porosity soil through the active layer where frost heaving is most active.

Moreover, the role of buoyancy in long-term soil circulation in active sorted soil patterns should not be discounted. Notably, circulation may arise incrementally from the yearly regeneration of buoyant soil above the thaw front, which tends to induce a seasonal ascent of basal material and a corresponding descent of overlying material even in the absence of stratification. Provided such displacements are spatially organized on the scale of the active layer depth by preexisting lateral variations in texture, as would be expected in well-developed sorted circles, buoyancy could account for the observed pattern and rates of long-term soil circulation.

Acknowledgments

This work was supported by the U.S. Army Research Office (Grant No. DAAL03-87-K-0058). We are grateful to R. S. Anderson, S. P. Anderson, E. C. Gregory, and C. S. Stubbs for their key role in our study of sorted circles in Spitsbergen and for numerous discussions. We also thank the Norsk Polarinstitutt and Kings Bay Coal Company for the extensive logistical support they provided for our field work. This paper also benefited from discussions with A. J. Heyneman, B. Murray, A. Pissart, R. S. Sletten, J. Sollid, B. Van Vliet-Lanoe, and A. L. Washburn; and from critical reading and helpful suggestions from R. S. Anderson, S. P. Anderson, A. J. Heyneman, and A. L. Washburn.

References

Anderson, S. P., Upfreezing in sorted circles, western Spitsbergen, *Proceedings of the Fifth International Conference on Permafrost*, **1**, 666–671, 1988.

Butrym, J., J. Cegla, S. Dzulynski, and S. Nakonieczny, New interpretation of "periglacial structures", *Folia Quaternaria*, **17**, 1–34, 1964.

Chamberlain, E. J., and A. J. Gow, Effect of freezing and thawing on the permeability and structure of soils, *International Symposium on Ground Freezing*, 31–44, 1978.

Christensen, R. W., and T. H. Wu, Analysis of clay deformation as a rate process, *Proceedings of the American Society of Civil Engineers, Journal Soil Mechanics and Foundation Division, 90 (SM6)*, 125–157, 1964.

Corapcioglu, M. Y. A mathematical model for the permafrost thaw consolidation, *Proceedings of the Fourth International Conference on Permafrost*, 180–185, 1983.

Dawson, P. R., and D. F. A. McTigue, A numerical model for natural convection in fluid-saturated creeping porous media, *Numerical Heat Transfer*, **8**, 45–63, 1985.

Dzulynski, S., Polygonal structures in experiments and their bearing upon some periglacial phenomena, *Bulletin de l'Académie Polonaise des Sciences*, **11**, 145–150, 1963.

Dzulynski, S., Sedimentary structures resulting from convection-like patterns of motion, *Rocznik Polskiego Towarzystwa Geologicznego*, **36**, 3–21, 1966.

Elton, C., The nature and origin of soil-polygons in Spitsbergen, *Quaternary Journal of Geological Society London*, **83**, 163–194, 1927.

Fletcher, R. C., Application of a mathematical model to the emplacement of mantled gneiss domes, *American Journal of Science*, **272**, 197–216, 1972.

Freeze, R. A., and J. C. Cherry, *Groundwater*, 604 pp., Prentice-Hall, Englewood Cliffs, N.J., 1979.

Gleason, K. J., W. B. Krantz, and N. Caine, Parametric effects in the filtration free convection model for patterned ground, *Proceedings of the Fifth International Conference on Permafrost*, **1**, 349–354, 1988.

Gripp, K., Zwei Beiträge zur Frage der periglacialen Vorgange, *Geological Institute, University of Kiel, Meyniana*, **1**, 112–118, 1952.

Hallet, B., On geomorphic patterns with a focus on stone circles viewed as a free convection phenomenon, in *Irreversible Phenomena and Dynamical System Analysis in Geosciences*, edited by C. Nicolis and G. Nicolis, pp. 533–553, D. Riedel, Hingham, Mass., 1987.

Hallet, B., S. P. Anderson, C. W. Stubbs, and E. C. Gregory, Surface soil displacements in sorted circles, western Spitsbergen, *Proceedings of the Fifth International Conference on Permafrost*, **1**, 770–775, 1988.

Hallet, B., and S. Prestrud, Dynamics of periglacial sorted circles in western Spitsbergen, *Quaternary Research*, **26**, 81–99, 1986.

Harris, S. A., Observations on the redistribution of moisture in the active layer and permafrost, *Proceedings of the Fifth International Conference on Permafrost*, **1**, 364–369, 1988.

Jahn, A., Research on the structure and temperature of soils in western Greenland, *Académie Polonaise des Sciences et Lettres, Bulletin, Series A*, 50–59, 1948.

Jahn, A., Origin and development of patterned ground in Spitsbergen, *Proceedings of the First International Conference on Permafrost*, 140–145, 1963.

Jahn, A., *Problems of the Periglacial Zone*, 223 pp., Polish Scientific Publishers, Warsaw, 1975.

Krantz, W. B., Self-organization manifest as patterned ground in recurrently frozen soils, *Earth-Science Reviews*, **29**, 117–130, 1990.

Lun, P. T., and A. K. Parkin, Consolidation behavior determined by the velocity method, *Canadian Geotechnical Journal*, **22**, 158–165, 1985.

Mackay, J. R., The origin of hummocks, western Arctic coast, Canada, *Canadian Journal of Earth Sciences*, **17**, 996–1006, 1980.

Mackay, J. R., Downward water movement into frozen ground, western Arctic coast, Canada, *Canadian Journal Earth Sciences*, **20**, 120–134, 1983.

McGaw, R. W., S. I. Outcalt, and E. Ng, Thermal properties of wet tundra soils at Barrow, Alaska, *Proceedings of the Third International Conference on Permafrost*, 47–53, 1978.

McKenzie, D., The generation and compaction of partially molten rock, *Journal of Petrology*, **25**, 713–765, 1984.

McRoberts, E. C., Field observations of thawing in soils, *Canadian Geotechnical Journal*, **12**, 126–129, 1975.

Mitchell, J. K., *Fundamentals of Soil Behavior*, 422 pp., John Wiley, New York, 1976.

Morgenstern, N. R., and J. F. Nixon, One-dimensional consolidation of thawing soils, *Canadian Geotechnical Journal*, **8**, 558–569, 1971.

Morgenstern, N. R., and L. B. Smith, Thaw-consolidation tests on remoulded clays, *Canadian Geotechnical Journal*, **10**, 25–40, 1973.

Mortensen, H., Uber die physikalische Möglichkeit der "Brodel"—Hypothese, *Centralblatt fur Mineralogie, Geologie und Palaontologie, Abt. B*, 417—422, 1932.

Nicholson, F. H., Pattern ground formation and description as suggested by Low Arctic and Subarctic examples, *Arctic and Alpine Research*, **8**, 329–342, 1976.

Nixon, J. F., and N. R. Morgenstern, Practical extensions to a theory of consolidation for thawing soils, *Proceedings of the Second International Conference on Permafrost*, 369–377, 1973.

Nordenskjold, O. L., Uber die Natur des Polarländer, *Geographische Zeitschrift*, **13**, 563–566, 1907.
Pissart, A., Déformations de cylindres de limon entourés de graviers sous l'action d'alternances gel/dégel, *Biuletyn Peryglacjalny*, **29**, 119–229, 1982.
Pissart, A. Advances in periglacial geomorphology, *Zeitschrift für Geomorphologie*, **79**, 119–131, 1990.
Richter, F. M., and D. McKenzie, Dynamical models for melt segregation from a deformable matrix, *Journal of Geology*, **92**, 729–740, 1984.
Scott, D. R., and D. J. Stevenson, Magma solitons, *Geophysical Research Letters*, **11**, 1161–1164, 1984.
Scott, D. R., and D. J. Stevenson, Magma ascent by porous flow, *Journal of Geophysical Research*, **91**, 9283–9296, 1986.
Shilts, W. W., Nature and genesis of mudboils, central Keewatin, Canada, *Canadian Journal of Earth Sciences*, **15**, 1053–1068, 1978.
Sørensen, T., Bodenformen und Pflanzendecke in Nordostgrönland, *Meddelelser om Grønland*, **93**, 1–69, 1935.
Terzaghi, K., *Theoretical Soil Mechanics*, 510 pp., John Wiley, New York, 1943.
Turcotte, D. L., and G. Schubert, *Geodynamics*, 450 pp., John Wiley, New York, 1982.
Vandenberghe, J., Cryoturbations, in *Advances in Periglacial Geomorphology*, edited by M. J. Clark, pp. 179–198, John Wiley, Chichester, England, 1988.
Vandenberghe, J., and P. Van Den Broeck, Weichselian convolution phenomena and processes in fine sediments, *Boreas*, **11**, 299–315, 1982.
Van Vliet-Lanoe, B., The origin of patterned grounds in N. W. Svalbard, *Proceedings of the Fifth International Conference on Permafrost*, **2**, 1008–1013, 1988.
Washburn, A. L., *Geocryology: A Survey of Periglacial Processes and Environments*, 406 pp., John Wiley, New York, 1980.
Washburn, A. L., Near-surface soil displacement in sorted circles, Resolute area, Cornwallis Island, Canadian High Arctic, *Canadian Journal of Earth Sciences*, **26**, 941–955, 1989.
Washburn, A. L., Plugs: Origin and transitions to some associated forms of patterned ground, Cornwallis Island, Canadian High Arctic, paper presented at Twenty-second Annual Geomorphology Symposium on Periglacial Geomorphology, Sept. 21–22, 1991.
Wasiutynski, J., Studies in hydrodynamics and structure of stars and planets, *Astrophysica Norvegica*, **4**, 1–497, 1946.

12 Formation of Seasonal Ice Bodies

Wayne H. Pollard
Department of Geography, McGill University

Robert O. van Everdingen
Arctic Institute of North America, University of Calgary

Abstract

This review focuses upon the distribution, formation, and significance of two types of periglacial features associated with the seasonal formation of ice: (1) icings, and (2) seasonal frost mounds. The term icing refers to sheet-like masses of layered surface ice formed by freezing of successive flows of water seeping from the ground, discharging from springs, or emerging from below river or lake ice through fractures. The term seasonal frost mound refers to small to medium-sized mounds of ice or frozen ground or both that develop during a single winter season in response to increased pressures in ground water trapped within residual unfrozen zones in the active layer during freeze-back. Based on the analysis of structure and genetic process, seasonal frost mounds are subdivided into icing mounds, icing blisters, and frost blisters. Seasonal frost mounds often form in close association with icings. Implicit in these definitions is the seasonal nature of the ice formation and degradation processes. Under extreme environmental conditions, however, individual icings and seasonal frost mounds may persist for more than one year. Permafrost is not a prerequisite for either icings or seasonal frost mounds, but its presence facilitates their formation.

Introduction

The seasonal formation of ice in periglacial environments can produce a variety of morphological features. This review focuses upon two groups of features associated with seasonal freezing: (1) icings, and (2) seasonal frost

Periglacial Geomorphology. Edited by J. C. Dixon and A. D. Abrahams

mounds. Even though icings and some types of seasonal frost mounds may occur in association with the formation of seasonal ice covers on lakes and rivers, the latter will not be discussed in this paper. The literature abounds with icing and frost-mound synonyms, many of which are summarized in Carey (1973), Muller (1945), Troll (1958), and Williams (1965). Their existence and frequently inconsistent usage generate significant terminological problems. In addition, the random use of Soviet, German, Scandinavian, and Inuktitut terms in the North American literature (e.g., bugor, bulgannyakh, kotchi, naled, aufeis, karnig, pingo) is also confusing, particularly as some have English language equivalents, while others have taken on new meanings to fill terminological gaps.

As used in this review, the term icing refers to sheet-like masses of layered ice formed on the ground surface or on river or lake ice by freezing of successive flows of water that seep from the ground (ground icings), flow from a spring (spring icings), or emerge from below river ice through fractures (river icings) (Carey, 1973; Permafrost Subcommittee, 1988, p. 52).

The term frost mound is the family name for mound-shaped landforms produced by freezing combined with groundwater movement or the migration of soil moisture (Permafrost Subcommittee, 1988). This basic definition has been widely adopted, either in whole or in part (e.g., Frederking, 1979; Hennion, 1955; Maarleveld, 1965; Pollard and French, 1984, 1985; Tóth, 1971, 1972; van Everdingen, 1978, 1982a; Pollard, 1991a). On the basis of genetic processes, one can distinguish two general categories of frost mounds: (1) seasonal frost mounds, which develop during a single winter season, and (2) perennial frost mounds, which are formed by growth processes acting over periods of decades or centuries (Mackay, 1990; Muller, 1945; Pollard, 1988). Although these terms provide a relative indication of mound duration, their genetic connotation is more consistent with the process approach to landform analysis.

Icings

In winter, much of the groundwater discharged through seeps and springs and in streambeds in northern regions will freeze and form icings. This term, in common North American usage, is synonymous to aufeis (widely used German term) or naled (Russian term) (Carey, 1973; Sokolov, 1973).

The distance between the point of discharge and the point where icing formation starts is a function of the relatively constant discharge rate, temperature, and dissolved-solids content of the water; the geometry and gradient of the discharge channel(s); and the variable meteorological conditions of air temperature, humidity, and wind direction and speed.

Formation of icings associated with discharge of groundwater is not restricted to permafrost regions. Icings will form wherever groundwater discharge occurs under the appropriate conditions of water temperature, mineralization, discharge rate, air temperature, and discharge-channel geometry. They occur most widely in mountainous regions at higher latitudes.

All icings are to some extent related to the discharge of groundwater (van Everdingen, 1982b). In permafrost regions the source of the discharge has a significant influence on the potential severity of natural icing problems (Sokolov, 1973; Tolstikhin and Tolstikhin, 1974). Icings fed by discharge of suprapermafrost water commonly stop growing long before the end of the winter, when reserves are exhausted or the discharge conduits closed off by freezing. Because water temperatures are low they generally form relatively close to the points of discharge. Icings fed by discharge of subpermafrost water and some of those so fed by intrapermafrost water will keep growing so long as mean daily air temperatures remain below 0°C. They often form much farther from the points of discharge because water temperatures and discharge rates are commonly higher.

STUDY METHODS

Field studies of icings can include measurements of spatial parameters, temperature, and melt rate, and sampling for ice-fabric, chemical, and isotopic analyses (Pollard and French, 1984, 1985; van Everdingen, 1982a,b). In addition, three remote sensing techniques, using satellite imagery, aerial photography, and time-lapse photography, have proven useful in icing studies.

Satellite imagery of discharge phenomena

The resolution of MSS imagery obtained by LANDSAT 1 and 2 was adequate and that of more recent satellites is more than adequate to reveal the discharge of large springs during the winter and associated large icings during late spring, summer, and early fall (Dean, 1984; Hall, 1976; van Everdingen, 1975, 1976).

Water-filled discharge channels show black against white or grey on MSS band 7 winter imagery; icings show white on MSS band 4 summer imagery. Caution should be exercised not to mistake small clouds (white) for small icings. The presence of black shadows, all offset in one direction from the white spots representing the clouds, helps to distinguish clouds and icings (van Everdingen, 1976). Summer imagery has been used to document the progressive melting of large icings (Figure 12.1).

Large active icings can also be detected during winter. They show as dark grey to black against the white to light grey of the surrounding terrain, as a result of the water running over their surface (van Everdingen, 1976).

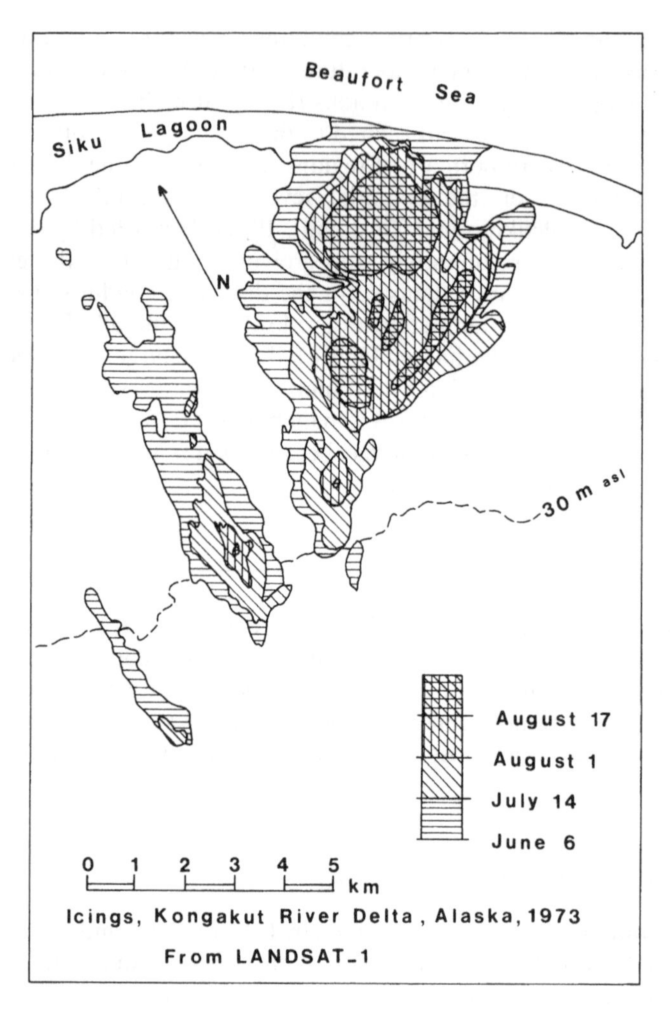

Figure 12.1. Example of the progressive melting of a large icing extracted from satellite imagery, Kongakut River Delta, northern Alaska, June 6 to August 17, 1973

Terrestrial time-lapse photography

Additional site-specific information on growth and decay of icings and seasonal frost mounds can be obtained through the use of automatic time-lapse photography (van Everdingen, 1982a,b). Daily photographs of staff gauges installed in an icing area will provide a record of the start and end of the icing development and of the daily icing activity and growth. The maximum extent of the icing should be determined in late winter, before noticeable melting occurs. The simplest method requires that ice levels be marked on trees or special rods installed throughout the icing area. After the icing has melted, a topographic survey of the ground surface and of the position and height of the markers will enable calculation of the surface area and volume of the icing. This information and the data from the time-lapse photography will allow determination of the average rate of groundwater discharge during the winter (see Discharge Rates below).

CHARACTERISTICS

Distribution and dimensions

Distribution of large icings can be mapped either in the field or by using satellite imagery; smaller icings and icing areas can often be identified on airphotos taken in late spring. Systematic mapping has been carried out in Alaska (Dean, 1984; Sloan, Zenone, and Mayo, 1976), in northern Yukon (van Everdingen, 1975), on Spitsbergen (Akerman, 1982a), and in Siberia (e.g., Alekseyey et al., 1969).

A summary of published values for mean and maximum thicknesses and maximum volumes of icings in Alaska, Canada, and the U.S.S.R. was presented by Grey and MacKay (1979), who listed maximum thicknesses ranging up to 13 m and maximum volumes up to 400×10^6 m^3

Ice fabric

Icing ice is generally characterized by small equigranular subhedral crystals that do not show discernible structure dimensional or *c*-axis orientation patterns. This reflects the growth process of icings by the addition of thin layers of ice to the icing surface. Each layer results from moderately rapid freezing of localized overflows; inclusion of snow imparts a distinct texture to affected layers.

ICINGS FROM SALINE SPRING DISCHARGE ON AXEL HEIBERG ISLAND

Location

Perennial springs with high NaCl content near the base of Gypsum Hill along Expedition River on Axel Heiberg Island, N.W.T. (79°24′N, 90°43′W), were described by Beschel (1963) and Pollard (1991b). The springs occur in two groups. The outlets of at least 15 springs of the first group are located in bedrock and colluvial deposits along the lower slopes of Gypsum Hill. The second group of springs discharges at several points within the main channel and floodplain of the Expedition River. They were observed within the icing area during May and June 1989 and are inferred from the presence of small icings and icing mounds, which are isolated by active channel flow, up to 100 m from the edge of the floodplain. The discharge of gas bubbles in pools and river channels is another indication of the presence of springs. Gas discharge is also associated with the first group of springs.

Discharge rates

Spring discharge rates range from 0.2 to 1.2 $l\,s^{-1}$, and the average discharge rate for 15 springs in group 1 in July 1988 was 0.45 $l\,s^{-1}$. Discharge temperatures ranged from −1.5 to +6.5°C, averaging 2.4°C. These values are similar to observations made by Beschel (1963). The springs with the highest temperatures also had the highest discharge rates. These springs are clustered in two distinct zones roughly 70 m apart; both temperature and discharge rate decrease with increasing distance away from these zones.

Dimensions

Spring discharge continues year round and produces a large groundwater icing that merges with a river icing caused by runoff from the White and Thompson Glaciers. When fieldwork began on June 28, 1988, the icing covered roughly 30,000 m^2 and had been dissected into three parts, the largest associated with the main spring group. In 1989 the groundwater icing extended nearly 300 m into the floodplain and more than 700 m downstream, covering approximately 200,000 m^2. The thickness was generally between 1.0 and 1.5 m with a maximum of 2.1 m. A yellowish mineral precipitate covered the icing and the ground surface around the springs.

SIGNIFICANCE OF ICINGS

Hydrologic significance

Icings represent temporary above-ground storage of groundwater discharged during the winter. The stored water is released by melting of the icing during the following spring and summer. Large icings may thus cause significant interseasonal redistribution of water resources. For instance, the icings developed over a period of about 6 months in the valleys of upper Babbage River, upper Firth River, and Joe Creek, northern Yukon, take about 3 months to melt, increasing streamflow by 1.0 to 2.8 $m^3\,s^{-1}$ during late spring and summer (van Everdingen, 1987). The estimated $123 \times 10^6\,m^3$ of icings forming in the Sagavanirktok River basin in Alaska during an 8-month freezing period, represent an average groundwater discharge rate of about 6 $m^3\,s^{-1}$. Melting of the icings adds approximately 24 $m^3\,s^{-1}$ to streamflow in the basin during the 2-month summer season (Williams and van Everdingen, 1973). In general terms, the rate of runoff from melting icings appears to be 1.5 to about 4 times as high as the associated rate of groundwater discharge (Williams and van Everdingen, 1973). The actual ratio depends on the shape and exposure of individual icings and on weather conditions.

Icing volumes can provide quantitative information on the groundwater resources of an area, as demonstrated by long-term studies in the U.S.S.R. (Sokolov, 1973). Those studies showed that the measured areal extent of individual icings (e.g., from air photos) can be converted into approximate volumes by the equation

$$V_i = 0.96\, A_i^{1.09} \qquad (12.1)$$

where V_i is volume in 1000 m^3 and A_i is area in 1000 m^2. Icing volumes can also be determined in the field at the end of the winter. If the duration of the freezing period or icing season t_i is known (e.g., from meteorological records), then the average groundwater discharge rate Q_w can be determined from

$$Q_w = V_i / t_i \qquad (12.1)$$

This method can also be applied to icings that stop developing before the end of the winter, if information is available on the actual duration of their growth period.

In the summer, after icings have melted, it is often still possible to determine their earlier presence and areal extent. The presence of an icing in a river valley during spring runoff may lead to accelerated lateral erosion by water channeling around the icing and to braided channels. These characteristics assist in identifying icing areas during the summer. Mineral precipitates (calcite, dolomite, gypsum), formed during freezing of the discharged groundwater and left behind after melting of the ice, generally persist for some time as a powdery coating on vegetation and rocks throughout the icing area (van Everdingen, 1974; Hall, 1980). The approximate extent of the icing can often be determined from a careful survey of the distribution of the mineral precipitate.

Hydrogeochemical significance

The chemical and isotopic composition of icing ice and meltwater is not a reliable indicator for the quality of the associated groundwater discharge because of uneven distribution of mineral precipitates and liquid inclusions in the icing (van Everdingen, 1974, 1982b). The uneven distribution commonly results from several causes. During gradual freezing of water flowing over and around an icing, dissolved-solid concentrations gradually increase in the remaining water until saturation is reached with respect to one or more minerals. Precipitate formation starts at that point and will continue until the last dissolved material is either precipitated or incorporated as brine inclusions at the point where the last water freezes. At any particular stage in the process the mineral composition of the precipitate will depend on the chemical composition of the water and the solubility of potential precipitate minerals.

The colder the weather, the closer to the discharge source the above sequence of events takes place. In addition, the continuously changing position of runoff channels on an icing tends to distribute the water unevenly. Incorporation of snow also affects the chemical and isotopic composition of both ice and meltwater (van Everdingen, 1982b).

During melting of the ice the more soluble constituents (e.g., NaCl) go back into solution immediately, whereas calcite, dolomite, and gypsum redissolve much more slowly, forming a mineral slush on the surface and around the edges of a melting icing. The meltwater therefore generally has a lower mineral content than the associated groundwater and a higher ratio of chloride to bicarbonate and sulfate (van Everdingen, 1974, 1982b). These properties will affect the water quality in streams receiving meltwater from large icings.

Morphological significance

The presence of icings reduces the carrying capacity of affected streams during snowmelt runoff. Where icing levels are below the river bank, diversion of streamflow against erodible banks leads to intensive lateral erosion; the resulting widening of the main channel into braided reaches increases the icing-susceptibility (Froehlich and Słupik, 1982; Grey and MacKay, 1979). Where icing levels are sufficiently high, fluvial activity on the floodplain during the snowmelt will create secondary channels that may eventually take over from the main channel (Froehlich and Słupik, 1982). Bed scour beneath icings may also affect the channel morphology (Carlson, 1979).

Geotechnical significance

Icings may present serious problems for a variety of construction projects; areas of natural icing occurrence should therefore be avoided, if at all possible (Carey, 1973). However, construction of highway, railroad, and pipeline embankments may cause compaction of underlying water-bearing materials and upward growth of permafrost beneath such embankments, or both. The ensuing reduction in transmissivity causes build-up of pressure in the water-bearing materials on the uphill side of the embankment, which often leads to discharge of groundwater. During the winter this may cause formation of icings (Carey, 1973; Eager and Pryor, 1945; Thomson, 1966; van Everdingen,

1982b) that may plug culverts, which in turn may lead to serious flooding and washouts during the snowmelt. Installation of subdrains may alleviate such problems once they have been identified (Carey, 1973; van Everdingen, 1982b).

Seasonal frost mounds

Seasonal frost mounds are small to medium-sized mounds or upwarps of ice or frozen sediments or both. They develop in a single winter in response to increased hydraulic or hydrostatic pressures of groundwater trapped within residual unfrozen zones of the active layer during freeze-back (Pollard and French, 1983, 1984; van Everdingen 1978, 1982a).

The term seasonal frost mound, first used by Muller (1945), has been extended to include all icing-related phenomena resulting directly from the presence of groundwater under pressure and characterized by a seasonal growth cycle. Three types of seasonal frost mounds can be distinguished: icing mounds, icing blisters, and frost blisters. They commonly occur in close association, and they are frequently difficult to differentiate on morphological grounds.

ICING MOUNDS AND ICING BLISTERS

The term icing mound was defined by Muller (1945, p. 218) as a "localized icing of substantial thickness but more or less limited in areal extent." In the past, the term has also been used to describe other types of seasonal or perennial frost mounds; for example, frost blisters (e.g., Gell, 1978).

Muller (1945, p. 218) included mounds formed "entirely or in part by the upwarp of a layer of ice by hydrostatic pressures" in his definition. The term icing blister was proposed by van Everdingen (1978) for mounds formed through the hydraulic lifting of icing layers by water under high hydraulic potential. He thus distinguished between mounds formed by localized icing accumulation and mounds formed by injection of groundwater between icing layers and the simultaneous upward deformation of overlying layers of the icing. In both cases hydraulic potential or hydrostatic pressure is the primary growth mechanism, either forcing water to the surface to cause icing mounds or heaving overlying icing layers to form icing blisters. In some cases a single mound may be a combination of both types.

Distinguishing icing mounds from icing blisters in the field can be difficult and can often be done only on the basis of internal structure. Indeed, when the water chamber inside an icing blister is completely frozen, only detailed examination of ice texture and fabric may provide an indication of the mound type (Pollard and French, 1983, 1984, 1985).

FROST BLISTERS

Frost blisters are "produced by localized hydrostatic pressure of groundwater" (Muller, 1945, p. 216). van Everdingen (1978) recommended that Muller's definition be modified to include the term high hydraulic potential in place of hydrostatic pressure to stress the hydrodynamic character of frost-blister genesis.

Frost blisters are sometimes confused with morphologically similar perennial frost mounds, especially palsas and small pingos (e.g., Academia Sinica, 1975; Åkerman, 1982b; Browne, Nelson, Brocket, Outcalt, and Everett, 1983; Hughes, Rampton, and Rutter, 1972; Lewis, 1962; Maarleveld, 1965; Sharp, 1942; Washburn, 1983). Unlike pingos, which tend to occur as solitary features, frost blisters often occur in groups and/or as compound features. A consequence of their seasonal nature is their tendency to shift location and to vary in size and shape from year to year.

Icing mounds may be described as groundwater-eruption features, growing progressively higher as water issuing from a single orifice or fracture freezes. Similarly, icing blisters and frost blisters may be described as groundwater-injection phenomena.

OCCURRENCE AND DISTRIBUTION

The conditions favoring formation of seasonal frost mounds include: (1) perennial discharge of groundwater with a low temperature, (2) the presence of a low-permeability layer (e.g., permafrost) close to the ground surface, and (3) a long cold winter with daily mean temperatures below freezing.

Seasonal frost mounds thus occur most frequently in areas of high relief, where a thick active layer is underlain by permafrost, and where perennial spring discharge provides suitable hydrologic conditions. They may also form outside permafrost regions, in areas where other geological materials (e.g., clay) act as aquitards while deep penetration of seasonal frost constricts groundwater circulation (Tóth, 1971, 1972).

Icing mounds and icing blisters have been reported from the Mackenzie Valley (van Everdingen, 1978), interior Yukon (Hughes and van Everdingen, 1978; Pollard and French, 1983), South Victoria Land, Antarctica (Autenboer, 1962; J. R. Keys, personal communication, 1978), Baffin Island (Frederking, 1979), and Axel Heiberg Island (Pollard, 1991b).

Frost blisters have been reported from many areas in arctic North America. Leffingwell (1919) found them in the Canning River region of northern Alaska, and Porsild (1938) in Greenland and the Mackenzie Delta, Northwest Territories. Mackay (1977, 1979) described "frost blister" features occurring on the flanks of pingos on the Tuktoyaktuk Peninsula. Frost blisters have also been reported from the Fairbanks area, Alaska (Linell, 1973), the

Mackenzie Valley (van Everdingen, 1978, 1982a), and interior Yukon (Hughes and van Everdingen, 1978; Pollard and French 1983, 1984; van Everdingen, 1982b).

Frost blisters have recently been identified in the area of saline spring discharge at Gypsum Hill on Axel Heiberg Island (Pollard, 1991b). There are only a few reported occurrences of seasonal frost mounds at such high latitude (79°24′). Åkerman (1980) described a variety of frost mounds in the Kapp Linné area, west Spitsbergen (78°04′N, 13°38′E), including forms that closely resemble frost blisters. Although the Kapp Linné area has a high Arctic location, its climate displays a strong maritime influence resulting in a mean annual air temperature of −4.6°C and mild winters, with a mean monthly temperature for January of −11.2°C.

Outside North America, frost blisters have also been reported from the Khangay Mountains, Mongolia (Froehlich and Slupik, 1978), Yakutia and other areas in the permafrost region of the USSR (Gokeov, 1939; Shumskii, 1964; Sumgin, 1941), and the Tangha and Fenhuoshan areas in China (e.g., Academia Sinica, 1975).

DIMENSIONS

Icing mounds at North Fork Pass (Pollard and French, 1983) tended to be circular in shape ranging from 1.0 to 3.1 m in height and up to 40 m in diameter. Icing blisters were typically oval shaped, with their long axis oriented parallel to the direction of water movement. They ranged in length from 2 to 23 m, while their widths were characteristically 0.3 to 0.5 times their lengths. Their heights ranged from less than 1 to 2.3 m, and their long-axis profiles were smooth and slightly asymmetrical in shape.

At Gypsum Hill on Axel Heiberg Island (Pollard, 1991b), icing blisters (1 in 1988, 5 in 1989) near the spring outlets were up to 2.3 m high. Icing mounds (1 in 1988, 1 in 1989) were up to 2.5 m high.

Frost blisters seldom exceed 8 m in height (Muller, 1945) and most often measure between 0.5 m and 5 m (Froehlich and Slupik, 1978; Linell, 1973; Muller, 1945; van Everdingen, 1978; Zoltai and Johnson, 1978). Their horizontal dimensions can range from less than 10 m to more than 100 m.

Frost blisters at Gypsum Hill range from 0.9 to 3.5 m high and 6.0 to 41.0 m long with an average height of 1.9 m and average length of 18 m. The majority of the frost blisters are oval and elongated parallel to the local slope. They have an average length to width ratio of about 2.0. In places two or three mounds coalesce, generating irregular or complex forms. In general, frost blisters have simple shapes and display smooth to slightly asymmetrical longitudinal profiles with very steep side slopes (25–48°) and flat tops. Compound and complex forms have undulating longitudinal profiles. Dilation cracks dissect mound surfaces but quickly infill with loose sand once thawing starts.

INTERNAL STRUCTURE

Stratigraphic and ice-fabric analyses may be needed to infer the origin of any seasonal frost mound.

Icing mounds and icing blisters

Icing mounds are composed entirely of thin, horizontal to gently dipping layers of pale-white and clear icing ice. In contrast, icing blisters consist of 40 to 130 cm of icing ice arched over an oval or disk-shaped ice core or cavity 35 to 97 cm high. In spring the cavity is usually drained.

Frost blisters

Most descriptions of frost-blister structure are based on shallow drilling and observation of manually excavated cross-sections or carefully cleaned natural exposures of partially collapsed mounds (e.g., Pollard and French, 1983, 1984; van Everdingen, 1978, 1982a). The typical structure of a frost blister consists of a dome of seasonally frozen ground, overlying a layer of clear ice vaulted over a water-filled cavity (Figure 12.2). Many frost blisters also carry a thin cover of icing ice. A vegetation and peat mat helps to maintain the strength of surficial materials during mound formation and collapse.

Two aspects that distinguish the frost blisters at Gypsum Hill from others described in the literature are (1) their occurrence in horizontally bedded, medium to coarse fluvial sand with thin layers of fine gravel situated in an active river channel, and (2) the absence of a peat or vegetation cover. The bedded sand was arched over a core of pure ice between 40 and 200 cm thick. The ice core was underlain by similar bedded sand. Both the upper and lower contacts between the ice and sand were abrupt and unconformable, suggesting an intrusive origin for the ice.

The base of the core was observed in only three mounds; in other cases it lay below the water table. The massive ice core had a faintly foliated milky-white appearance and was composed of either solid ice or layered ice, with an internal cavity up to 60 cm high. High concentrations of small spherical to vertically elongated tubular gas inclusions caused the milky-white, layered appearance. The central cavity indicated that water had drained from the residual reservoir.

Although some frost blisters may resemble small palsas, they can be easily distinguished because their structures differ significantly. In contrast to the massive ice core or water-filled cavity of a frost blister, the main constituents

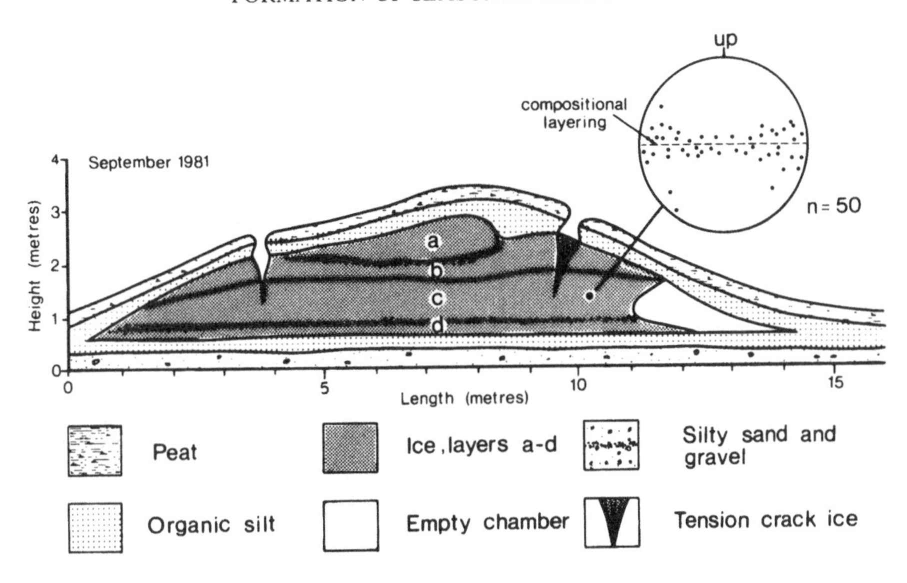

Figure 12.2 Cross-section showing the internal structure and ice fabric of a frost blister in the North Fork Pass area, northern Yukon. This frost blister experienced growth in two consecutive winters, resulting in the incorporation of the ice core of the original blister (layer a) into a larger structure. The ice-fabric diagram is for a vertical thin section and presents a lower-hemisphere projection of *c*-axis distributions plotted on a Schmidt equal-area net (modified from Pollard, 1988)

of a palsa include a thick insulating layer of peat overlying a core of ice-rich mineral soil characterized by segregation-ice lenses (Åhman, 1976, Seppälä, 1980). Ice lenses rarely exceed 2 to 3 cm in thickness (Lundquist, 1969), although thicker lenses have been reported (e.g., Forsgren, 1968). Where trees are present, tree-ring analysis may assist in revealing recent frost-blister activity (van Everdingen and Allen, 1983).

ICE FABRIC

Icing mounds and icing blisters

The ice fabric in icing mounds and in the surficial layers of icing blisters is similar to that of icing ice (Figures 12.3A and 12.3B). The ice core, formed by freezing of water in the icing-blister cavity, is characterized by large, vertically oriented, columnar anhedral crystals elongated normal to the overlying layering (parallel to the inferred growth direction). Crystal *c*-axes showed preferred orientations normal to the growth direction. According to Glen (1974), this ice fabric is indicative of rapid growth of ice in bulk water, with the crystals growing along their basal plane in the *a*-axis direction. An initial

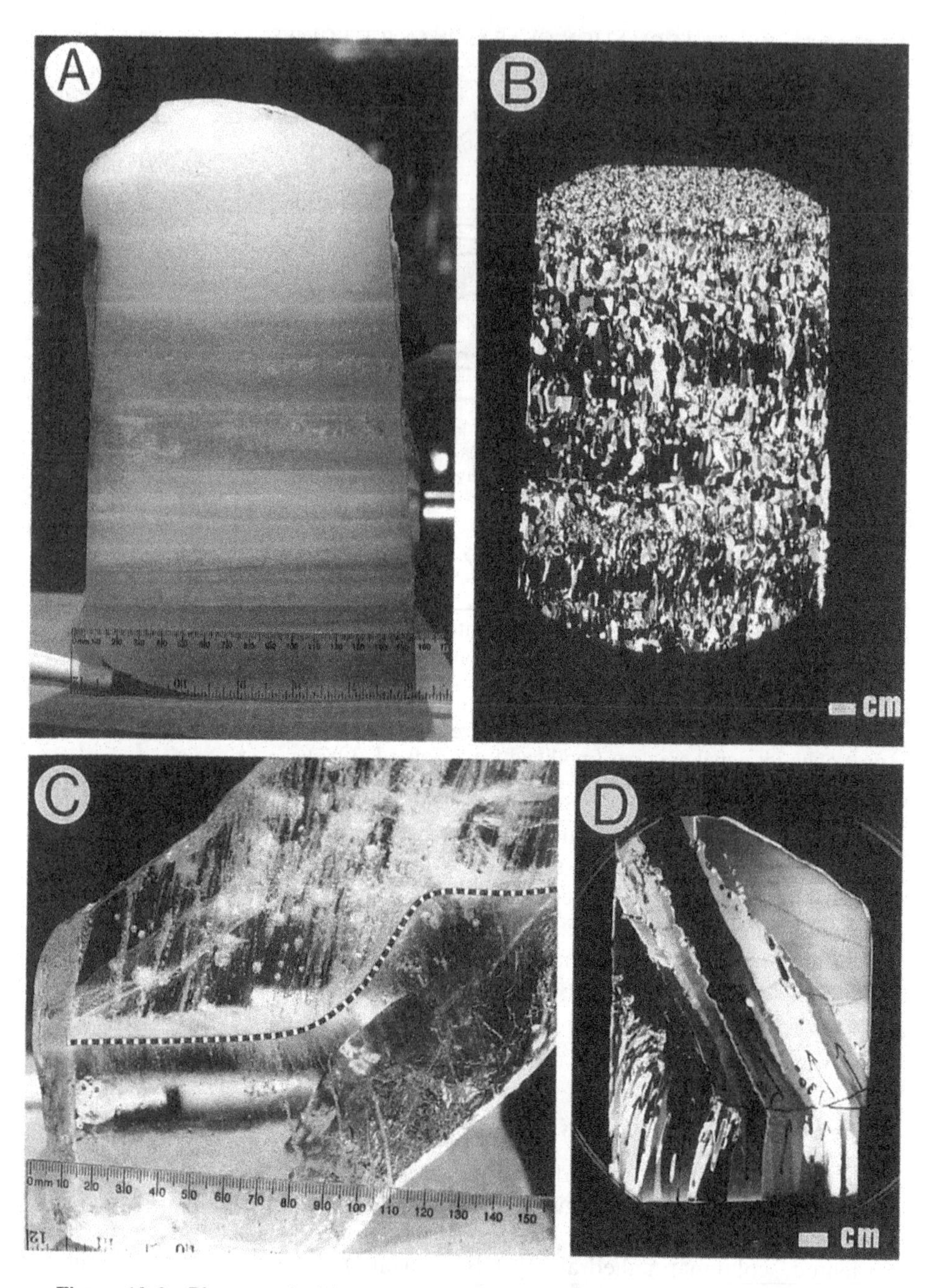

Figure 12.3 Photographs illustrating petrographic characteristics of icing and frost-blister ice. (A) Oriented block sample of icing ice. (B) Vertical thin section of icing ice shown in (A) between crossed polarizers. (C) Oriented block sample of frost-blister ice, including contact zone between ice layers a and b from Figure 12.2. (D) Vertical thin section of frost-blister ice between crossed polarizers, illustrating tabular crystal structure

chill zone at the contact between the icing ice and the injection ice of the core is difficult to detect.

Frost blisters

Two fabric and texture patterns can be distinguished in frost-blister ice below the frozen sediments (Pollard and French, 1985; Pollard, 1991b). First, a chill zone of small subhedral ice crystals with *c*-axis orientations ranging from random to a loose girdle parallel to the internal layering and perpendicular to the freezing direction. The second pattern was best developed deep within the ice core where textures were characterized by large, vertically oriented, columnar anhedral crystals elongated normal to the compositional layering and parallel to their growth direction (Figures 12.3C and 12.3D). Crystal *c*-axes displayed a strong preferred pattern normal to the inferred growth direction. These fabrics are consistent with an injection origin for water forming the ice core.

PRESSURE STUDIES

Although the role of groundwater under pressure has been suggested by several researchers as the dynamic component of seasonal frost mound growth (e.g., Maarleveld, 1965; Muller, 1945; van Everdingen, 1978), the magnitude of these pressures is still not fully understood. An approximation of minimum hydraulic potential was calculated by van Everdingen (1978), using the thicknesses and densities of the layers of various materials displaced but ignoring overburden strength. It was calculated that frost blisters 2 to 5 m high require a minimum hydraulic potential of 2.6 to 4.9 m (25 to 50 kPa). A Soviet study (Petrov, 1934; abstracted by Williams, 1965) determined the groundwater pressure of an 'icing mound' (frost blister) by measuring the depression of the freezing point due to pressure. It was concluded that the pressure was 52 atmospheres, or 5.2676×10^3 kPa; the possible depression of the freezing point by dissolved solids was apparently not taken into account.

In the North Fork Pass area, frost blister pressures were measured using closed-system antifreeze-filled piezometers (Pollard and French 1983, 1984). They reported pressures ranging between 25 and 80 kPa over a seven-day period for mounds between 1 and 2 m high. A number of other studies have investigated active-layer pressures during freeze-back (Mackay, 1980; Mackay and MacKay, 1976; Pissart, 1970; Slusarchuk, Watson, and Speer, 1937) but have failed to record similar high values.

FORMATION AND DECAY—PROCESSES AND TIMING

Observations of the growth and decay of frost blisters over a period of several seasons have been carried out at three locations in Canada: the Bear Rock area of the Mackenzie Valley (by van Everdingen), the North Fork pass area in northern Yukon (by Pollard), and Gypsum Hill on Axel Heiberg Island (by Pollard). Winter observations were undertaken at both Bear Rock and North Fork Pass. At Bear Rock, growth rates were recorded during the winter of 1977 to 1978, using time-lapse photography (van Everdingen and Banner, 1979).

Growth of icing mounds and icing blisters

Development of icing mounds can start when mean daily temperatures drop below 0°C. Increasing hydraulic potential does not have to lift either soil materials or ice; it only has to force water to the surface of the ground or icing either through tension cracks or along some plane of weakness. Icing blisters can start forming only some time after the onset of winter, when some minimum amount of icing has developed. They may rupture, sometimes explosively, drain, and subsequently reseal one or more times during a single winter.

Growth of frost blisters

The main process forming frost blisters is injection of water from perennial, spring-fed supra-permafrost groundwater systems. As the active layer refreezes in early winter, groundwater circulation gradually becomes constricted. The penetration of the freezing front into the water-bearing active layer reduces hydraulic transmissivity and increases pressure in the groundwater system. As hydraulic potential increases, it may exceed the weight and deformation resistance of the frozen overburden, deforming these materials into a mound enclosing a reservoir of water. The mound will increase in size as the pressure and reservoir volume increase.

The uplift process is assumed to be slow enough to allow the overlying frozen material to deform plastically. Nevertheless, growth rates of more than 0.5 m d^{-1} have been recorded using time-lapse photography (van Everdingen, 1982a). If internal stresses exceed the strength of the confining materials, frost blisters may rupture, sometimes explosively (Academia Sinica, 1975; Bogomolov and Sklyarovskaya, 1969; Shumskii, 1964; Sloan, Zenone, and Mayo, 1976; van Everdingen, 1982a), drain, reseal, and resume growth several times during a single winter. Thermal contraction cracking during periods of extreme cold may help trigger some of the ruptures.

Depending on several factors, such as the volume of the water reservoir, the thermal diffusivity of the overlying materials, and the temperature regime, the water reservoir may freeze completely, forming a core of solid ice (Hughes, Rampton, and Rutter, 1972; Muller, 1945; van Everdingen, 1978),

Figure 12.4. Photograph of the drained cavity inside a frost blister that formed on the side of a pingo in the Mackenzie Delta area (photograph provided by J. R. Mackay)

or the mound may stabilize with a water-filled chamber inside the epigenetic ice core. If the supply of water is cut off when a frost blister ruptures or if the mound fails to reseal itself, then the cavity inside the frost-blister will remain empty (Figure 12.4) (e.g., Mackay, 1977; van Everdingen, 1978).

Decay of icing mounds and icing blisters

The decay of icing mounds and icing blisters start when air temperatures rise above freezing; it is often completed by early summer. Degradation of icing blisters may be accelerated by collapse owing to the presence of dilation cracks and the draining of any remaining water from their cavities.

Decay of frost blisters

Collapse of structurally intact frost blisters begins when thaw intersects the ice core. As the ice core is often 20 to 40 cm below the ground surface, frost blister collapse often begins only in midsummer. The presence of dilation cracks exposing the ice core and early drainage of unfrozen water from inside the mound will accelerate the degradation process.

Thawing of the dark sandy materials of the frost blisters at Gypsum Hill proceeded rapidly once the snow and icing cover melted sufficiently to expose the frost-blister surface. The coarsely granular cover had little cohesive strength and exhibited rapid drainage of melt water, which enhanced thawing. By mid to late June, thawing on most frost blisters had penetrated to their ice cores. At this stage, the sand on the steep sides began to slough-off as blocks, or as loose sediment, gradually assuming its natural angle of repose. Continued melting of the ice core maintained the instability of the cover sediment. Occasionally, the ice core became exposed, accelerating degradation. In some cases sloughing began after only 10 to 15 cm of thaw. As many of the mounds were located on the active floodplain and river bed, fluvial erosion was responsible for collapse of several of the mounds. As a result, the collapse of all frost blisters was complete by late July or early August.

CLIMATIC SIGNIFICANCE

As climate is an important factor controlling frost mound occurrence, frost blisters indicate cold winter temperatures and a climate potentially suitable for the maintenance of permafrost. The winter temperature regime and the timing and thickness of snow cover determine the timing and rate of active-layer freeze-back and, therefore, the timing and intensity of frost blister occurrence (van Everdingen, 1982a). Despite this, a comparison of several climatic parameters (e.g., period with air temperature < 0°C, freezing index, daily temperature regime, timing and thickness of snow cover) with frost blister activity at Bear Rock, Northwest Territories, by van Everdingen (1982a) failed to reveal any obvious connection between weather factors and the number, position or dimension of resulting frost blisters. However, based on a freezing index of 1115 degree days, the time of mound initiation varied by only 15 days from one year to the next. In a year with heavy snowfall at North Fork Pass, frost blister initiation was delayed due to retarded freeze-back.

GEOTECHNICAL SIGNIFICANCE

The examples described above deal primarily with natural occurrences of seasonal frost mounds. However, the potential geotechnical significance of their association with groundwater discharge must also be recognized. On the one hand, their occurrence may provide engineers with an indication of areas of groundwater activity sensitive to freezing-related problems. On the other hand, human-induced modifications of the groundwater and thermal regimes may inadvertently create a combination of conditions suitable for the formation of seasonal frost mounds at locations where they do not occur naturally.

The most common cause of the induced formation of seasonal frost mounds appears to be the obstruction of local groundwater circulation described in the section on icings. Examples of such occurrences have been described from the Dempster Highway 6 km east of Fort McPherson, Northwest Territories (van Everdingen, 1982a), and from the Alaska Highway during the winter of 1944 to 1945 (Eager and Pryor, 1945). "Seasonal pingos" or frost blisters described in Academia Sinica (1975) occurred apparently in response to the obstruction of groundwater circulation during construction of a railroad on the northeast Tibetan Plateau. Linell (1973) described the formation of a frost blister resulting from uncontrolled artesian flow from a water-supply well drilled near Fairbanks, Alaska.

In another case, a number of "ice-cored mounds" and "palsas" were observed at various locations along the Dalton Highway–Trans-Alaska Pipeline corridor (Brown et al., 1983; Nelson and Outcalt, 1982; Nelson, Outcalt, Goodwin, and Hinkel, 1985). Based on descriptions of these mounds (J. Brown, personal communication, 1984), it is likely that at least some of them were frost blisters. However, it is not known whether or not they were formed as a result of the highway and pipeline construction.

POTENTIAL AS LOCAL WINTER WATER SOURCES

Water flows from ruptured icing blisters and frost blisters have been observed in a number of cases (e.g., van Everdingen, 1978; Pollard, 1991b). Coring of an icing blister on Axel Heiberg Island in 1988 produced a water fountain some 25 cm high; the flow decreased rapidly and stopped completely after 10 minutes. This indicated either that the blister was cut off from the springwater supply or that the height of the mound exceeded the local piezometric level (Pollard, 1991b).

Chacho, Collins, Delaney, and Arcone (1990) surveyed some 100 icing mounds and icing blisters on each of two rivers on the eastern North Slope, Alaska, using an airborne short-pulse radar system. They were able to distinguish water-bearing mounds (presumably icing blisters) from dry mounds (icing mounds or completely drained icing blisters), and they obtained estimates of ice thicknesses and mound heights. Ice thickness of 1.3 to 1.5 m appeared to separate water-bearing and dry mounds, whereas a mound height of 1.25 m or more commonly identified water-bearing mounds.

Suggestions for future research

In view of their geomorphological, hydrological, and potential geotechnical significance, future research on icings should be directed towards quantifying their effect on channel morphology and measuring their effects on spring and summer stream flow and water quality. Future studies of proposed transportation corridors should include mapping of the distribution of icings.

Long-term studies in a variety of settings will be needed to establish to what extent the locations of individual frost blisters and the timing and rate of their growth are controlled by the air-temperature regime and snow-cover thickness, and whether maximum dimensions of frost blisters are limited by such overburden characteristics as mineral composition, grain size distribution, and type of vegetation cover, if any.

Acknowledgments

Much of the information presented in this paper is based on the authors' field research on icings and seasonal frost mounds. Funding for Pollard's recent field work was provided by a grant from the National Sciences and Engineering Research Council of Canada; logistical support was provided by the Polar Continental Shelf Project, Department of Energy, Mines and Resources. J. Ross Mackay, University of British Columbia, kindly provided the photograph showing the drained cavity of a frost blister.

References

Academia Sinica, Permafrost, Research Institute of Glaciology, Permafrost and Desert Research, Lanzhou, China (*Technical Translation 2006*, 146 pp., National Research Council of Canada, Ottawa, 1981), 1975.

Åhman, R., The structure and morphology of minerogenic palsas in northern Norway, *Biuletyn Peryglacjalny*, **26**, 25–31, 1976.

Åkerman, J. H., Studies on periglacial geomorphology in West Spitsbergen, *Meddelanden fra Lunds Universitets Geografiska Institution Avhandlingar*, **LXXXIX**, 297 pp., 1980.

Åkerman, J. H., Studies on naledi (icings) in West Spitsbergen, in *The Roger J. E. Brown Memorial Volume*, edited by H. M. French, pp. 189–202, National Research Council of Canada, Ottawa, 1982a.

Åkerman, J. H., Observations of palsas within the continuous permafrost zone in eastern Siberia and in Svalbard, *Geografisk Tidsskrift, 82*, 42–51, 1982b.

Alekseyev, V. R. et al. (eds.), Naledi Sibirii, USSR Academy of Sciences, Siberian Branch, Izdatel'stvo Nauka, Moscow, *Siberian Naleds*, 206 pp., U.S. Army CRREL Draft Translation 399, 1969.

Autenboer, T., Ice mounds and melt phenomena in the SorRondance, Antarctica, *Journal of Glaciology*, **4**, 349–354, 1962.

Beschel, R. E., Sulphur springs at Gypsum Hill, Preliminary Report, in *1961–1962 Axel Heiberg Island Research Reports*, pp. 183–187, McGill University, Montreal, 1963.

Bogomolov, N. S., and A. N. Sklyarevskaya, On explosion of hydrolaccoliths in the southernpart of Chitinskaya Oblast, in *Siberian Naleds* (V. R. Alekseyev et al., eds.), U.S. Army CRREL Draft Translation 399, 187–192, 1969.

Brown, J., F. Nelson, B. Brooket, S. I. Outcalt, and K. R. Everett, Observations on ice-cored mounds at Sukakpak Mountain, South Central Brooks Range, Alaska, *Proceedings of the Fourth International Permafrost Conference*, **1**, 91–96, 1983.

Carey, K., Icings developed from surface and ground water, *CRREL Monograph III-D3*, 67 pp., U.S. Army Cold Regions Research an Engineering Laboratory, Hanover, N.H., 1973.

Carlson, R. F., A theory of aufeis and streambed erosion, *Proceedings of the Canadian Hydrology Symposium: 79—Cold Climate Hydrology*, 197–205, 1979.

Chacho, E. F., Jr., C. M. Collins, A. J. Delaney, and S. A. Arcone, River icing mounds. A winter water source on the eastern North Slope of Alaska, *Proceedings of the Northern Hydrology Symposium*, in press, 1990.

Dean, K. G., Stream-icing zones in Alaska, *Report of Investigations 84–16*, 20 pp., Division of Geological and Geophysical Surveys, Alaska Department of Natural Resources, 1984.

Eager, W. L., and W. T. Pryor, Ice formation on the Alaska Highway, *Public Roads*, 24(3), 55–82, 1945.

Forsgren, B., Studies of palsas in Finland, Norway and Sweden 1964–1966, *Biuletyn Peryglacjalny*, **17**, 177–223, 1968.

Frederking, R. M., Rupture of an ice mound near Cape Dorset, N.W.T., *Canadian Geotechnical Journal*, **16**, 604–609, 1979.

Froehlich, W., and J. Slupik, Frost mounds as indicators of water transmission zones in the active layer of permafrost during the winter season (Khangai, Mts., Mongolia), *Proceedings of the Third International Conference on Permafrost*, **1**, 188–193, 1978.

Froehlich, W., and J. Slupik, River icings and fluvial activity in extreme continental climate: Khangai Mountains, Mongolia, in *The Roger J. E. Brown Memorial Volume*, edited by H. M. French, pp. 203–211, National Research Council of Canada, Ottawa, 1982.

Gell, A. W., Fabrics of icing-mound and pingo ice in permafrost, *Journal of Glaciology*, **20**, 563–569, 1978.

Glen, J. W., The physics of ice, *CRREL Monograph II-C2a*, 83 pp., U.S. Army Cold Regions Research and Engineering Laboratory, Hanover, N.H., 1974.

Gokeov, A. G., Frost mounds and hydrolaccoliths in the Kazakh Steppe, *Vses. geog. obshch. Izv*, **71**, 541–6 (in Russian; abstracted in Williams, 1965, pp. 86–87, 1939.

Grey, B. J., and D. K. MacKay, Aufeis (overflow ice) in rivers, *Proceedings of the Canadian Hydrology Symposium: 79—Cold Climate Hydrology*, 139–163, 1979.

Hall, D. K., Analysis of some hydrological variables on the North Slope of Alaska using passive microwave, visible and near-infrared imagery, *Proceedings of 1977 Conference, Falls Church, Virginia*, 344–361, 1976.

Hall, D. K., Mineral precipitation in North Slope river icings, *Arctic*, **33**, 343–348, 1980.

Hennion, F., Frost and permafrost definitions, *Highway Research Board Bulletin 111*, 107–110, 1955.

Hughes, O. L., V. N. Rampton, and N. W. Rutter, Quaternary geology and geomorphology, southern and central Yukon, *Guidebook Field Excursion A-11, XXIV International Geological Congress*, 59 pp., J. D. McAra, Calgary, Alta, 1972.

Hughes, O. L., and R. O. van Everdingen, Central Yukon-Alaska, *Guidebook Field Trip No. 1. Third International Conference on Permafrost*, 32 pp., National Research Council of Canada, Ottawa, 1978.

Leffingwell, E. de K., The Canning River region, northern Alaska, *U.S. Geological Survey Professional Paper 109*, 251 pp., 1919.

Lewis, C. R., Icing mound on the Sadlerochit River, Alaska, *Arctic, 15*, 145–150, 1962.

Linell, K. A., Risk of uncontrolled flow from wells through permafrost, *North American Contribution to the Second International Conference on Permafrost*, 462–469, 1973.

Lundquist, J., Earth and ice mounds: A terminological discussion, in *The Periglacial Environment*, edited by T. L. Péwé, pp. 203–215, McGill-Queens University Press, Montreal, 1969.

Maarleveld, G. D., Frost mounds—a summary of literature of the last decade, *Mededelingen van de Geologische Stichting, Nieuwe Serie, 17*, 16 pp., 1965.

Mackay, J. R., Pulsating pingos, Tuktoyaktuk Peninsula, N.W.T., *Canadian Journal of Earth Sciences*, **14**, 209–222, 1977.

Mackay, J. R., Pingos of the Tuktoyaktuk Peninsula area, Northwest Territories, *Géographie Physique et Quaternaire*, **33**, 3–61, 1979.

Mackay, J. R., The origin of hummocks, western Arctic coast, Canada, *Canadian Journal of Earth Sciences*, **17**, 996–1006, 1980.

Mackay, J. R., Seasonal growth bands in pingo ice, *Canadian Journal of Earth Sciences*, **27**, 1115–1125, 1990.

Mackay, J. R., and D. K. MacKay, Cryostatic pressures in non-sorted circles (mud hummocks), Inuvik, North West Territories, *Canadian Journal of Earth Sciences*, **13**, 889–898, 1976.

Muller, S. W., Permafrost or permanently frozen ground and related engineering problems, *U.S. Geological Survey Special Report, Strategic Engineering Study No. 62*, 231 pp., 1945.

Nelson, F., and S. I. Outcalt, Anthropogenic geomorphology in northern Alaska, *Physical Geography*, **1**, 17–48, 1982.

Nelson, F. E., S. I. Outcalt, C. W. Goodwin, and K. M. Hinkel, Diurnal thermal regime in a peat-covered palsa, ToolikLake, Alaska, *Arctic*, **38**, 310–315, 1985.

Permafrost Subcommittee, Glossary of permafrost and related ground-ice terms, *Technical Memorandum No. 142*, 156 pp., National Research Council of Canada, Ottawa, 1988.

Petrov, V. G., An attempt at ascertaining the pressure of groundwater in icing mounds, *Akademiia Nauk S.S.S.R.,* Kom izuch vechnoimerzloty trudy V.G. 1930 (in Russian; abstracted in Williams, 1965, pp. 160–161), 1934.

Pissart, A., The pingos of Prince Patrick Island (76°N, 120°W), *Technical Translation 1401*. 46 pp., National Research Council of Canada, Ottawa, 1970.

Pollard, W. H., Seasonal frost mounds, in *Advances in Periglacial Geomorphology*, edited by M. J. Clark, pp. 201–229, John Wiley, Chichester, England, 1988.

Pollard, W. H., Canadian Landforms Examples 21, Seasonal frost mounds, *Canadian Geographer*, **35**, 214–218, 1991a.

Pollard, W. H., A high Arctic occurrence of seasonal frost mounds, *Proceedings of the Northern Hydrology Symposium*, in press, 1991b.

Pollard, W.H., and H. M. French, Seasonal frost mound occurrence, North Fork Pass, Ogilvie Mountains, Northern Yukon, *Proceedings of the Fourth International Conference on Permafrost*, 1000–1004, 1983.

Pollard, W. H., and H. M. French, The groundwater hydraulics of seasonal frost mounds, North Fork Pass, Yukon Territory, *Canadian Journal of Earth Sciences*, **21**, 1073–1081, 1984.

Pollard, W. H., and H. M. French, The internal structure and ice crystallography of seasonal frost mounds, *Journal of Glaciology*, **31**, 157–162, 1985.

Porsild, A. E., Earth mounds in unglaciated arctic northwestern America, *Geographical Review*, **28**, 46–58, 1983.
Seppälä, M., Stratigraphy of a silt-cored palsa, Atlin region, British Columbia, Canada, *Arctic*, **33**, 357–365, 1980.
Sharp, R. P., Ground ice mounds in tundra, *Geographical Review*, **32**, 417–423, 1942.
Shumskii, P. A., *Principles of Structural Glaciology: The Petrography of Fresh Water Ice as a Method of Glaciological Investigation*, translated from Russian by D. Kraus, 497 pp., Dover Publications, New York, 1964.
Sloan, C. E., C. Zenone, and L. R. Mayo, Icings along the Trans-Alaska pipeline route. *U.S Geological Survey Professional Paper No. 979*, 31 pp., 1976.
Slusarchuk, W. A., G. H. Watson, and T. L. Speer, Instrumentation around a warm oil pipeline buried in permafrost, *Canadian Geotechnical Journal*, **10**, 227–245, 1973.
Sokolov, B. L. Regime of naleds, *U.S.S.R. Contribution to the Second International Conference on Permafrost*, 99–108, 1973.
Sumgin, M. I., Naledi i nalednye bugry (Icings and icing mounds), *Priroda*, **30**, 26–33 (abstracted in Williams, 1965, pp. 212–213), 1941.
Thomson, S., Icings on the Alaska Highway, *Proceedings of the First International Conference on Permafrost*, 526–529, 1966.
Tolstikhin, N. I., and O. N. Tolstikhin, Groundwater and surface water in the Permafrost region, in *General Permafrost Studies*, edited by P. I. Melnikov and O. N. Tolstikhin, USSR Academy of Sciences, Siberian Branch, Novosibirsk (translation in Technical Bulletin 97, 25 pp., Inland Waters Directorate, Fisheries and Environment Canada, Ottawa, 1976), 1974.
Tóth, J., Groundwater discharge: A common generator of diverse geologic and morphologic phenomena, *Bulletin of the International Association of Scientific Hydrology*, **16**(1), 7–24, 1971.
Tóth, J., Properties and manifestations of regional groundwater movement, *Proceedings of the 24th International Geological Congress, Section 12*, 153–163, 1972.
Troll, K., Structure soils, solifluction and frost climates of the earth, *U.S. Army, Snow, Ice and Permafrost Research Establishment, Translation 43*, 121 pp., 1958.
van Everdingen, R. O., Groundwater in permafrost regions of Canada, *Proceedings of the Workshop Seminar on Permafrost Hydrology*, 83–93, 1974.
van Everdingen, R. O., Use of ERTS-1 imagery for monitoring of icings, N. Yukon and N. E. Alaska, *Hydrology Research Division Report Series No. 42*, pp. 75–87, Environment Canada, Ottawa, 1975.
van Everdingen, R. O., Use of LANDSAT imagery in studies of spring icings and seasonally flooded karst in permafrost areas, *Proceedings of the Workshop on Remote Sensing of Soil Moisture and Groundwater*, 213–235, 1976.
van Everdingen, R. O., Frost blisters of the Bear Rock near Fort Norman, Northwest Territories, 1975–1976, *Canadian Journal of Earth Sciences*, **15**, 263–276, 1978.
van Everdingen, R. O., Frost blisters of the Bear Rock spring area near Fort Norman, N.W.T., *Arctic*, **35**, 243–265, 1982a.
van Everdingen, R. O., Management of groundwater discharge for the solution of icing problems in the Yukon, in *The Roger J. E. Brown Memorial Volume*, edited by H. M. French, pp. 212-226, National Research Council of Canada, Ottawa, 1982b.
van Everdingen, R. O., The importance of permafrost in the hydrological regime, in *Canadian Aquatic Resources*, edited by M. C. Heayley and R. R. Wallace, pp. 243–275, Canadian Bulletin of Aquatic Sciences 215, Fisheries and Oceans Canada, Ottawa, 1987.
van Everdingen, R. O., and H. D. Allen, Ground movements and dendrogeomor-

phology in a small icing area on the Alaska Highway, Yukon, Canada, *Proceedings of the Fourth International Conference on Permafrost*, **2**, 1292–1297, 1983.

van Everdingen, R. O., and J. A. Banner, Use of long-term automatic time-lapse photography to measure the growth of frost blisters, *Canadian Journal of Earth Sciences*, **16**, 1632–1635, 1979.

Washburn, A. L., What is a palsa? *Abhandlungen der Akademie der Wissenschaften in Göttingen, Mathematischen-Physikalische Klasse, Dritte Folge, 35*, 34–47, 1983.

Williams, J. R., Ground water in permafrost, an annotated bibliography, *U.S. Geological Survey Water Supply Paper 1792*, 294 pp., 1965.

Williams, J. R., and R. O. van Everdingen, Groundwater investigations in permafrost regions of North America, *North American Contribution to the Second International Conference on Permafrost*, 435–446, 1973.

Zoltai, S. C., and J. D. Johnson, Vegetation-soil relationships in the Keewatin District, *Arctic Islands Pipeline Program Report*, 95 pp., Canadian Forestry Service, Fisheries and Environment Canada, Ottawa, 1978.

13 Palsa-scale Frost Mounds

Frederick E. Nelson
Department of Geography, Rutgers University

Kenneth M. Hinkel
Department of Geography, University of Cincinnati

Samuel I. Outcalt
Department of Geological Sciences, University of Michigan

Abstract

Palsa-scale frost mounds have great potential in paleoecological studies and for monitoring the effects of global change on permafrost in peatlands of the northern hemisphere. Incorporation of such applications into the larger scientific enterprise is jeopardized, however, by the overly complex and exotic terminology employed by periglacial geomorphologists. The term palsa should be used only in a morphologic context but can be modified adjectivally to communicate genetic information. Palsa-scale frost mounds can be divided into four genetic categories: ice-segregation, hydrostatic, hydraulic, and buoyant. Criteria for recognition of the genetic varieties include stratigraphy, ice fabrics, and patterns of ionic concentration and bubble inclusions. Palsas may be of aggradational or degradational origin, although these distinctions may be blurred in some cases.

Introduction

During the past decade, several authors have reviewed the literature of "palsas" from a geomorphic perspective (Washburn, 1983b; Worsley, 1986; Pissart, 1985; Pollard, 1986; Seppälä, 1986, 1988a). Although interest in these permafrost mounds has until now been quartered primarily within periglacial geomorphology (and, to a lesser extent, in ecology), the significance and utility of the features transcends disciplinary boundaries. As geoscience

Periglacial Geomorphology. Edited by J. C. Dixon and A. D. Abrahams

adopts an increasingly integrative outlook, it is appropriate to explore the potential of periglacial landforms for such larger issues as paleoclimatology and global change. In this paper we address nomenclature from an interdisciplinary perspective and examine the mechanisms responsible for what might be termed palsa-scale frost mounds.

Terminological considerations

THE TERM PALSA

Use of the term palsa was rather controversial throughout much of the 1980s; we believe this issue to be larger than questions posed in a narrow genetic context by some periglacial geomorphologists. Indeed, the logic and accessibility of geomorphology's terminology will to a very large extent influence how well the larger scientific community is able to make use of our discipline's results. This paper details our objections to terminology advocated by some other writers and suggests a more objective and accessible lexicon.

The terminology used with reference to frost-mound phenomena by many periglacial geomorphologists is both arcane and inadequate. Indeed, even geomorphologists from other branches of the science dismiss the terminology applied to frost mounds by periglacial geomorphologists as "bewildering" (Chorley, Schumm, and Sugden, 1984, p. 496). Much of the problem stems from an unfortunate tendency to mix morphologic and genetic inferences (Thorn, 1983) together neologistically or in colloquial analogies. The result is a baroque, multifarious, overly differentiative argot well suited for challenging the tenacity and random-access memorization capabilities of graduate students, but which provides a poor basis for communicating with scientists outside our discipline, or for creating a rational, hierarchical classification system. Thorn (1988, Chapter 4) provides examples of similar problems from other branches of geomorphology.

Two viewpoints on the terminology of palsa-scale frost mounds have been advocated during the 1980s. The first, a less restrictive position, is represented by Washburn (1983b, p. 44), who defined the term palsa as referring to

> peaty permafrost mounds, ranging from about 0.5 to about 10 m in height and exceeding about 2 m in average diameter, comprising (1) aggradation forms due to permafrost aggradation at an active-layer/permafrost contact zone, and (2) similar-appearing degradation forms due to disintegration of an extensive peaty deposit.

This morphologically based definition contrasts with the beliefs of several other writers, who restrict the term to permafrost mounds such as those shown in Figure 13.1, whose relief is believed attributable solely or predominantly to the accretion of segregation ice. The generally excellent Canadian

Figure 13.1. Dome-shaped palsa, MacMillan Pass area, Yukon-Northwest Territories border, Canada. See Kershaw and Gill (1979) for detailed discussion of palsas in this locality. Photo taken in July 1975

Glossary of Permafrost and Related Ground-Ice Terms (Permafrost Subcommittee, 1988, p. 60), for example, suggests that palsa should be used only with reference to

> those features where the internal structure shows the presence of *segregated ice* and where the environment lacks high hydraulic potentials, provided that other parameters (size, shape, location in wetlands) are also satisfied. The term "*frost mound*" should be used as a non-genetic term to describe the range of morphologically similar, but genetically different, features that occur in permafrost terrain.

We find the second of these definitions untenable on at least five grounds:

1. *Idiosyncrasy.* Examination of various other entries in the *Glossary* reveals a very uneven treatment of the possible combination of morphologic/genetic characteristics associated with frost mounds. Frost mound applies to the entire range of morphologic, dimensional, genetic, and duration-of-freezing possibilities, but within this continuum only palsa, pingo, frost blister, and icing blister were given separate definitions. Moreover, of these four categories, only palsas and pingos were assigned dimensional ranges. Thus, a 1-m-high, peaty permafrost mound of hydrostatic origin could only be termed a frost mound by the *Glossary's* definition, while a feature with similar

dimensions but of ice-segregation origin is referenced by a separate term, palsa. To assign each genetic category a separate term, replete with its own genetic inference, is an untenable solution because it would add complexity to an already plethoric lexicon. Undue terminological rigidity also makes transitional forms difficult to classify (Akerman and Malmström, 1986; Lagerbäck and Rodhe, 1986).

2. *Tradition.* Some periglacial geomorphologists have expressed concern that applying the term palsa to mounds with origins other than ice segregation is contrary to tradition or may result in confusion between investigators (e.g., Seppälä, 1988a). According to Seppälä (1988b, p. 249), "the term palsa was originally used by the Lapps and Finns, and in their languages it means a peat hummock with a frozen core rising above the surface of a mire." He also noted that this "initial definition was morphologically descriptive" and that "people using the term did not imply any specific genesis or consequence of development." Original usage therefore supports using palsa in an entirely morphologic sense.

According to Thorarinsson (1951, p. 149), the first scientific investigation of palsas ("rústs") in Scandinavia was by the appropriately named Icelandic scientist Pálsson in 1792. Thorarinsson noted that Pálsson's diary suggests that palsas are likely to result from frost action. Although subsequent visitors to the northern parts of the Scandinavian countries published descriptions of palsa mires (e.g., Keilhau, 1831), detailed investigations and genetic interpretations were not made until the end of the nineteenth century. Kihlman (1890) believed palsas to be erosional remnants from a former mire surface of higher elevation, thereby raising the question of their utility as indicators of climatic amelioration (Friedman, Johansson, Oskersson, and Svensson, 1971). In two papers published soon thereafter, Fries and Bergström (1910) and Fries (1913) introduced the term palse into the Swedish literature and advocated the importance of local factors, particularly variations in snow-cover depth, for palsa formation. Nearly 40 years elapsed, however, before Lundqvist (1951), drawing on Beskow's (1935) work, advanced the hypothesis that ice segregation may be responsible for the formation of what Seppälä (1988b, p. 249) has called "classic Fennoscandian palsas." At approximately the same time in North America, Muller (1947) suggested that peat mounds (a term that he equated with palsen; cf. Seppälä, 1988a) may be the result of localized ice segregation "in swamp or tundra country." In the intervening years, such distinguished workers as Lozinski, the generally acknowledged "father of periglacial studies" (Washburn, 1980a), applied the term to features that do not meet the more restrictive definition (Lozinski, 1933).

In the Soviet Union, early investigations of palsas ("bugristye torfyaniki" or hummocky peat bogs) were conducted by Gorodkov (1928, 1932), Govorukin (1947), and P'iavchenko (1949). According to Kachurin (1964, p. 16), the most common hypothesis of palsa origin among these writers was an erosional remnant hypothesis. Again, only in the 1950s was an ice-segregation explanation for the formation of the mounds advanced (Baranov, 1951; Ruoff, 1951; Popov, 1953).

It is clear, therefore, that the term palsa, for most of the time it has been in existence, has carried no genetic inference about ice segregation, a hypothesis that has been applied to frost mounds only relatively recently. Washburn's (1983a) plea for use of the term in an exclusively morphologic sense is therefore supported by tradition, as well as by original usage, sound classificatory procedure, and terminological coherence.

3. *Inconsistency*. Pingos, according to most authorities (e.g., Mackay, 1979; Washburn, 1980a; Permafrost Subcommittee, 1988), can be formed by a variety of mechanisms, often working in concert. The term pingo has only a morphological connotation, however, and can be modified adjectivally to convey genetic information (e.g., hydrostatic (closed-system) or hydraulic (open-system) pingo). These terms are taken directly from the Canadian *Glossary* (Permafrost Subcommittee, 1988, pp. 71–72); it seems both logical and parsimonious to employ the same terms for communicating genetic information about frost mounds of other dimensional and positional characteristics.

4. *Operational considerations*. If the restrictive definition is adopted, inventory of palsas, sensu stricto, is inhibited by logistical, financial, and temporal constraints owing to the necessity for examining the core of each mound included in the survey. Since inventory is considered desirable in many quarters (e.g., Seppälä, 1988b, pp. 274–275), a morphologically based definition is preferable. One of the most interesting facets of these landforms is that, through their close association with peatlands in subarctic regions, they represent the southernmost extent of permafrost in many regions of the northern hemisphere.

5. *Communicability*. If other scientists are to make use of the very significant results obtained by periglacial geomorphologists about palsa-scale frost mounds and their climatic affinities, terminology must be made more rational and accessible. Extending the logic used in the Canadian *Glossary's* treatment of frost mounds would result in a polyglot collection of terms, each representing a separate growth mechanism. Such a situation is a serious impediment to dissemination of periglacial geomorphology's results to a wider audience. Time is also likely to exacerbate problems stemming from an overly multipartite vocabulary; given the increasing importance and imminent dominance of online bibliographic search and retrieval, the use of multiple terms to refer to similar features will eventually lead to unnecessary and undesirable fragmentation of the literature.

PRESENCE OF PEAT

Many investigators have reported permafrost mounds of dimensions compatible with those assigned to palsas, often of ice-segregation origin, but lacking peat as a constituent material. Although such mounds have been referred to as minerogenic palsas (e.g., Åhman, 1976) or mineral palsas (e.g., Pissart and Gangloff, 1984; Pissart, 1985), other authors (e.g., Seppälä, 1988b, pp. 249–251) have objected strenuously to such usage, advocating terms such as mineral frost mound, which could apply equally well to mounds of any dimensions or position relative to the active layer and permafrost.

We can find little reason for assigning yet more names to features genetically and morphologically identical to many mounds designated palsas, simply on the basis of the absence of peat. Although Seppälä (1988b, pp. 249–251) warns that "confusion arises if we speak of 'mineral palsas' (i.e. palsas without peat), since this would imply a necessity to recognize also the term 'mineral mire' (i.e. a peat bog without peat)," we concur with Pissart (1985) and Jahn (1986) that to stipulate that a feature must contain peat in order to qualify as a palsa is a rather artificial distinction and more likely to obfuscate than to clarify. Peat plays an important role in palsa genesis and maintenance in localities that are climatically marginal for permafrost, but its hydrophysical qualities are not particularly conducive to ice segregation (Williams and Smith, 1989, p. 153), and frost-susceptible soil is a far more critical and universal requirement for the development of ice-segregation mounds than is the presence of peat. Washburn's (1983b) definition is based primarily on morphology, dimensions, and position with respect to the permafrost/active-layer interface; individual mounds, with and without peat, that meet these criteria exist, in some regions in close proximity (e.g., Allard and Seguin, 1987; Pissart and Gangloff, 1984, Figure 10), and may be identical apart from erosion of a peat cover on some individuals (Lagarec, 1982). To refer to these highly similar forms by different names invites confusion. Moreover, precedent argues for such inclusions; G. Lundqvist, who originally posited the ice-segregation hypothesis of palsa formation, suggested using the terms peat-palses, loam-palses, and boulder-palses to differentiate between mounds of varying constituent materials (Lundqvist, 1953). We conclude that the presence of peat is not a relevant consideration in the designation of particular features as palsas.

EPHEMERAL VERSUS PERENNIAL MOUNDS

Several authors (e.g., Pollard and French, 1983) have referred to transitory ice-cored mounds as seasonal frost mounds. Because such mounds often meet the strict definition of permafrost (see Hinkel, Nelson, and Outcalt, 1987 for extended discussion), this term is often inappropriate and could be replaced by ephemeral frost mound, a more suitable counterpoint to perennial frost mounds (i.e., permafrost mounds). A distinction can be drawn between mounds that survive for one or two years and those of longer duration. The observations of Jonasson and Sköld (1983), Luken and Billings (1983), Jonasson (1986), and Svensson (1986) suggest that ephemeral and perennial frost mounds could in many cases be distinguished by phytosociological criteria, even in the field. Indeed, the fact that the composition of plant communities changes following mound inception provides the basis for the most successful method of dating palsa incipience (Vorren, 1972; Vorren and Vorren, 1975).

PERMAFROST ZONE

Despite the publication of Washburn's (1983a) well-reasoned contribution, some individuals insist, as did an anonymous reviewer of one of our manuscripts, that "there are no palsas in continuous permafrost." Such views apparently derive from the idea that little moisture is available for mound growth in polar regions, owing to the immobility of water in surrounding frozen terrain. It is well to remember, however, that no definition of the term continuous permafrost requires that all points below the ground surface be perennially frozen; small discontinuities and variations in active-layer thickness are the norm, even in the high arctic. The notion that palsas of ice-segregation origin occur only in the discontinuous zone (e.g., Seppälä, 1988a, p. 108) may well be a result of unintentional and arbitrary mixing of spatial scales (Nelson, 1989).

FORM

The morphology of palsa-scale frost mounds is varied. The classic form is a conical or dome-shaped mound (Lundqvist, 1969), but many other plan shapes have been reported. Following Washburn (1983b), five morphological categories circumscribe the various shapes assumed by palsa-scale frost mounds: dome-shaped (conical), sinuate, string (rectilinear), plateau, and palsa complexes (cf. Åhman, 1977). The various forms can grade into one another and can evolve from one category to another.

Origin and criteria for recognition

Most palsa-scale frost mounds form as a result of preferential migration of water toward a localized area in a near-surface soil layer. Subsequent freezing results in volumetric expansion due to phase change from water to ice; topographic uplift results as additional water arrives and is frozen. Water migrates in response to two types of pressures: it can be "pushed", that is, subjected to positive pore-water pressures, or "pulled" by a negative pressure potential (suction). In general, positive pore water pressures are associated with a hydrologically closed system, an elevation-induced head, or lithostatic pressure. In contrast, suction of water toward a region of the soil occurs in a hydrologically open system in response to a thermal gradient. Subsequent freezing of the water produces petrofabrics, ionic concentration patterns, and in some cases a stratigraphic record that is diagnostic of particular water-migration processes.

Other periglacial features result from preferential migration of water, including earth hummocks and pingos. In earlier literature (e.g., Lundqvist, 1969) these terms have been used as distinct categories of landforms developing through differing water-migration processes and/or freezing environments. More recent theoretical and field research supports a view of frost mounds as lying on a scale continuum, with hummocks at the smaller extreme, pingos at the larger end, and palsas occupying an intermediate position. Although formative conditions may vary within and between types, they may all to some degree be the product of ice-segregation processes. The size of the feature along this continuum reflects both the extent to which site-specific factors influence water migration processes and, in a crude fashion, the time duration over which these processes operate.

In the absence of tectonic forces, an aggradation landform can achieve topographic expression by two general processes: local accumulation of material or uplift produced by density differences. In the study of aggradational frost mounds, attention is focused on preferential migration of water to a localized area, where subsequent freezing and uplift occur. The processes by which water movement and crystallization occur are referred to as growth mechanisms.

Because mound morphology and geologic setting often are not diagnostic of a particular growth mechanism, it is necessary to examine the mound core. Typically, a cursory examination of the internal stratigraphy is sufficient, particularly when pore ice or segregation-ice lenses clearly dominate. However, when massive ice exists, it is useful to note crystal size, texture and fabric, crystallographic *c*-axis orientation, and patterns of both particulate and gas bubble inclusions (e.g., Pollard and Dallimore, 1988). *c*-axis orientation can be measured by standard thin-section techniques (e.g., Gell, 1978; Pollard and French, 1985). In the field, exposing polycrystalline ice to solar radiation causes candling and promotes development of Tyndall figures along the basal (0001) plane. This provides a quick and reasonably accurate method for establishing crystal size, ice fabric, and *c*-axis orientation in relatively pure, bubble-free ice (Shumskii, 1964).

The next section describes the internal petrographic and chemical characteristics associated with four mechanisms by which aggradation palsas can form. Keeping in mind that a mound's growth may be a response to several mechanisms over the course of its history, these characteristics may vary gradationally or abruptly within a mound's core.

AGGRADATION FORMS

Mackay (1979) identified several possible growth mechanisms to explain pingo formation that were expanded and adapted for palsa-scale frost mounds by Outcalt and Nelson (1984b). These represent a refinement of the simple "open/closed" classification scheme often applied to pingos and allow various water-migration processes to be identified through use of an adjectival modifier. Four growth mechanisms were identified by Outcalt and Nelson (1984b) to explain the genesis of the aggradational varieties of palsa-scale frost mounds. The features can be initiated and maintained by ice segregation, hydrostatic pressure, elevation-induced hydraulic pressure, or buoyancy. Mound morphology is generally not diagnostic of a particular growth mechanism. The growth history of a mound is, however, often preserved in the frozen core; geological and glaciological techniques can be applied to ice-rich cores to ascertain the ice-crystal fabric and texture, ionic concentration and bubble inclusion patterns with depth, and core stratigraphy. Because each growth mechanism produces a characteristic suite of fabrics, inclusion patterns, and stratigraphy, field data can be used to infer the process(es) responsible for mound formation (Figure 13.2).

The inclusive, flexible vocabulary we advocate uses a variety of adjectives to communicate geometric and genetic information about individual occurrences. Washburn's (1983b) definition of palsa forms the basic unit, the requirement for the presence of peat excepted. Using this vocabulary, the description of a so-called classic palsa, such as that in Figure 13.1, would be "a peat-covered, dome-shaped palsa of ice-segregation origin."

Ice segregation

Ice segregation has in recent years been the most widely invoked mechanism to explain the formation of palsa-scale frost mounds (Lundqvist, 1951; Brown, 1968; French, 1976; Seppäla, 1982, 1986). Ice crystallization occurs at, or just behind, the advancing frost and is accompanied by transport of heat and moisture from warmer regions at depth (Kay and Perfect, 1988). A layer of ice, often lens-shaped in profile, grows behind the "frozen fringe," a partially frozen layer where pore ice coexists with an unfrozen water film that surrounds soil particles (Miller, 1972). The lens will thicken if thermodynamic conditions favor continued segregation-ice growth, and the ground surface will be displaced upward by primary and secondary heaving.

Ice segregation and frost heaving processes have been the subject of numerous recent investigations that utilize different conceptual approaches to

Mechanism and References	Simplified Graphic Representation	Fabric and Texture	Ion Inclusion Pattern	Miscellaneous, Structure, Stratigraphy
Ice Segregation Lundqvist, G., 1951 French, 1976 Seppala, 1982, 1986	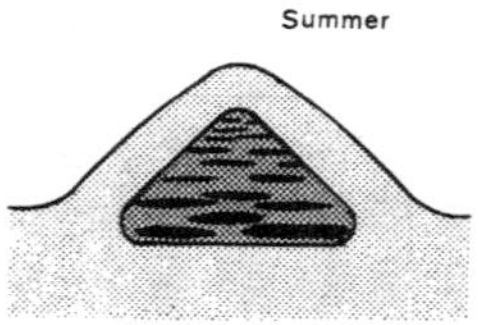 Summer	small crystals; c-axes random to preferred normal often with mineral inclusions. Texture diverse, usually granular, subhedral to anhedral crystals	very complex, due to concentration of solutes and unidirectional flow of water	thin ice lenses parallel to ground surface, within peat or mineral soil; lenses tend to thicken with depth
Buoyancy Zoltai, 1972; Kershaw & Gill, 1979 Outcalt & Nelson, 1984a	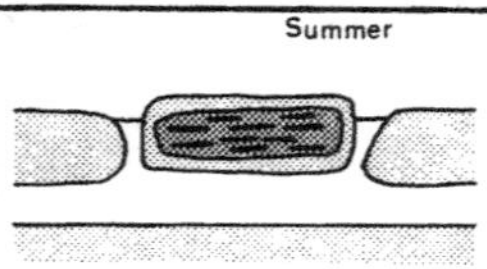 Summer	pore ice and/or segregation ice lenses	as in ice segregation mechanism	low elevation floating peat bodies; water in bore holes
Hydraulic van Everdingen, 1978; 1982 Pollard and French 1983; 1984; 1985	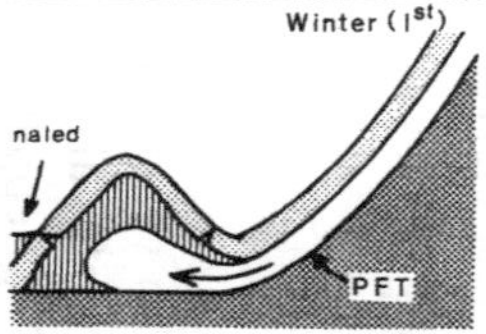 Winter (1st) naled PFT	bulk freezing characteristics with upper chill zone; bubbles along crystal boundaries; c-axes preferred normal, crystals elongated parallel to freezing direction; subhedral to anhedral; complicated by periodic pressure relief	relatively high in chill zone; increasing coreward until rupture and flushing introduce new source of water; varied across unconformaties	varied ice stratigraphy; probable structural deformation of ice; restricted spatially to areas of high relief, associated with perennial springs
Hydrostatic Akerman, 1982 Brown et al., 1983 Outcalt et al., 1986 Hinkel et al., 1987 Hinkel, 1988	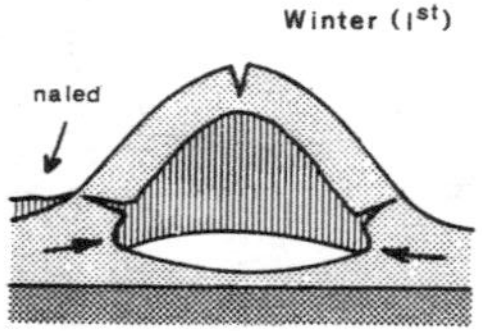 Winter (1st) naled	radially inward toward unfrozen core; possible chill zone and/or unfrozen chambers; bubble density increases inward. Vertical prismatic texture; anhedral crystals, c-axes preferred normal	initially decreasing, then increasing toward center as melt progressively enriched	condordant injection of water, often evidence of overburden deformation; unconformities common

Water — Seasonally Frozen — Permafrost — Ice

Figure 13.2. Summary of growth mechanisms and characteristics of palsa-scale frost mounds

describe and model the thermodynamic conditions and rheological properties of freezing soils. Although a discussion of the relative merits of the hydrodynamic model of Harlan (1973) and the rigid ice model proposed by Miller (1978) is beyond the scope of this paper, an excellent summary is available in Williams and Smith (1989, pp. 217–232).

Repetition of the driving process(es) results in a characteristic pattern of alternating lenses, which generally increase in thickness downward and are separated by frozen material (soil or peat) containing pore ice (Taber, 1930; Konrad and Morgenstern, 1980). This pattern reflects a general reduction in the thermal gradient with depth and a concurrent increase in lithostatic pressure. The mound achieves topographic expression by upward displacement of the surface, which is approximately equal to the sum of the thicknesses of the segregation-ice lenses.

Ice lenses in palsa-scale frost mounds are often thin (less than 1 cm) and rarely exceed 2 to 3 cm in thickness (Lundqvist, 1969). The ice is clear to opaque in appearance and often grades into surrounding ice-rich soil. Although crystal size and *c*-axis orientation vary in response to a number of factors (Shumskii, 1964), segregation ice can be identified by the following characteristics: (1) columnar crystals elongated in a direction nearly perpendicular to the surface and increasing in diameter coreward, whose form is subhedral to anhedral (Shumskii, 1964); (2) the presence of foreign inclusions, especially soil particles, within the lens, which often imparts a dirty appearance to the ice; (3) stratigraphic banding defined by differing amounts of particulate matter or gas bubbles, which reflects the degree of segregation efficiency; and (4) the absence of gross structural deformation, dike-like intrusions, discordant contacts, chilled margins, or chambers of unfrozen water, all of which are indicative of freezing in a closed system.

Because it is relatively easy to identify segregation lenses and pore ice, it is usually unnecessary to evoke chemical analyses of the melted ice to identify this growth mechanism in mounds of this size. Chemical techniques are discussed later.

Buoyancy

Buoyancy can initiate low "floating" mounds and may be an important factor in plateau palsas. Pore water in the highly porous peat is frozen in situ or with some enrichment from the ice-segregation process. This results in a layer of relatively low density that floats if permafrost aggradation does not "anchor" the palsa. Such features have been investigated in the field by Zoltai (1972), Allard, Seguin, and Levesque (1987), and Kershaw and Gill (1979); their formation and maintenance was modeled by Outcalt and Nelson (1984a). Palsas formed or maintained by buoyancy may contain both pore and segregation ice. In some cases, they can be recognized in the field by a combination of their low topographic expression, a frozen core composed predominantly of pore ice, and the presence of free water at the base of bore holes. Conversely, mound cores formed predominantly by the ice-segregation mechanism are often connected to, and may represent upward extensions of, the permafrost table.

Hydraulic pressure

A third mechanism responsible for palsa-scale frost mounds is hydraulic pressure exerted by groundwater flowing in a conduit formed by frozen layers above and below. Forced by an elevation-induced head, water can be injected concordantly between weakly bonded stratigraphic layers to produce a tabular ice-cored feature. Discordant injections often result in localized uplift, dome-shaped mounds, and icings, with which mounds formed by this mechanism are often associated.

These mounds are often found near ascending springs. As winter progresses, spring discharge may cease as the active layer eventually refreezes, and closed-system freezing commences. In such cases hydraulic pressures cease, and further mound formation or growth results from hydrostatic pressures (see next section). The hydraulic mechanism was invoked to explain ephemeral frost mounds (frost blisters) by van Everdingen (1978, 1982) and Pollard and French (1983, 1984, 1985). If the resulting mounds retain a frozen core and topographic expression over several years, they are defined as permafrost and qualify as palsas under the criteria adopted here (Figure 13.3).

As described, mounds formed by the hydraulic pressure mechanism are restricted to regions of high relief. These features are expected to display internal characteristics similar to those created by the hydrostatic pressure mechanism, as they both involve water injection and subsequent freezing. However, the mounds vary significantly in their geologic setting, as those formed by the hydraulic mechanism are located near the base of slopes (e.g., Pollard and French, 1984. Additionally, the water is not derived locally but has been transported into the region by the flow of subsurface springs. Pollard and French (1984) reported hydraulic pressures in March ranging from 30 to 81 kPa for mounds approximately 2 m high. They noted water discharge from a dilation crack that lasted for 2 hours and reasoned that:

> The constant discharge over such a period of time indicates an integrated hydrologic system linked to a perennial spring source, in contrast to the limited flow that would be expected if the water chamber had been part of a closed system. Late in the winter, the integrated nature of the hydrologic system may give way to a series of closed systems with a corresponding change from hydraulic to hydrostatic conditions. (p. 1077)

Figure 13.3. Palsa at the base of Sukakpak Mountain, Alaska. This feature appears to have been formed by the hydrostatic growth mechanism. Although some writers would call this mound a frost blister or seasonal frost mound, the J-shaped tree suggests that it has been a positive relief element over an extended period. The feature therefore qualifies as a true permafrost mound. See Brown, Nelson, Brockett, Outcalt, and Everett (1983) and Outcalt and Nelson (1984b). Photo taken in October 1982.

Michel (1986), working in the same area, reported that ^3H analysis indicated the spring water had fallen as precipitation some 10 to 15 years prior to reemergence and its incorporation in the ice core. As these are ephemeral mounds, the water must have been in transit during the intervening period.

Hydrostatic pressure

Mounds often form in response to hydrostatic pressures generated within the active layer in a hydrologically closed system. The mechanism is known to operate at a variety of scales and can lead to the formation of closed-system pingos (e.g., Mackay, 1979), ice-cored plateaus (Hinkel, 1988) or palsas (Hinkel, Nelson, and Outcalt, 1987). In all cases freezing occurs along fronts advancing toward a central location, and the water reservoir is isolated from surrounding groundwater. Mounds tend to develop in stream valleys and mires, where the active layer is thicker and water more plentiful. They differ from features formed by the hydraulic mechanism in that they are not necessarily associated with regions of high relief. In addition, the massive ice core is derived from surface streams and local soil water in the active layer. These features are referred to in the Russian literature as hydrolaccoliths. This term is apt, since it recognizes the similarity of the formative processes; in both cases internal pressure forces the fluid laterally between weakly bonded strata, thereby doming the overburden.

Ephemeral or perennial mounds of plateau (tabular), string or sinuate (anticlinical), or domal form can develop when water is forced between stratigraphic layers of relatively weak cohesion. Positive pore water pressures, which are developed in response to the volume increase associated with freezing, are sufficient to deform the frozen overburden. Rupture often results in formation of icings near the base or summit of the mound, where the tensile stresses are maximized.

Water injection, overburden deformation, and subsequent freezing result in a mound with a core of polycrystalline ice. Typically, ice crystals are oriented radially inward toward the center of the mound, reflecting the influence of topography on the heat-flow vectors (Pollard and French, 1985; Outcalt, Nelson, Hinkel, and Martin, 1986; Hinkel, Nelson, and Outcalt, 1987). A chill zone can sometimes be recognized at the frozen slab/ice-body interface. Crystals here tend to be small and equigranular with randomly oriented *c*-axes, and the ice tends to be bubble free; these characteristics are typical of rapid freezing (Gell, 1978). Plant roots or other debris can often be observed "dangling" into the ice from above (Shumskii, 1964; Hinkel, 1988), which is diagnostic of water injection. As a rule, the ice is very clean and contains little particulate matter (Figure 13.4). Toward the center of the mound, those crystals with *c*-axes oriented perpendicular to the direction of heat flow usually predominate (Ketchum and Hobbs, 1967), and crystal diameters increase. Candling reveals long, columnar anhedral crystals. This pattern is, however, often masked by an increase in the frequency of gas bubbles coreward.

Gas inclusions are common in ice bodies associated with closed-system freezing and form when the gas is excluded from the selective ice-crystal lattice. The resulting bubbles can occur individually or as bubble trains along the crystal boundaries and are thus useful for identifying the direction of heat flow and for interpreting a mound's deformation history (Outcalt et al., 1986; Hinkel, Nelson, and Outcalt, 1987). Variations in the freezing rate can be

Figure 13.4. Ephemeral mound of hydrostatic origin formed at Slope Mountain, Alaska. Note clean ice and dangling plant roots. See text and Hinkel (1988). Photo taken in July 1983

inferred by changes in bubble density, which forms indistinct layers that parallel surface topography. Occasionally, chambers are found at depth that contain ion-enriched water (van Everdingen, 1978) or air, resulting from mound rupture and reservoir drainage. Outcalt et al. (1986) and Hinkel (1988) reported finding chambers that were drained and subsequently refilled with water. The fabrics, texture, and ionic concentration of the chamber ice were distinctly different from those across the discordant chamber boundary.

In a closed system the concentration of ions in the residual water reservoir will increase as these impurities are excluded from the ice. At depth, where growth tends to proceed at a slower rate, the fractionation process is more efficient. Eventually, however, the enriched melt results in enhanced retention of ions in the ice crystals coreward, and this fact has been used to help identify mounds formed by the hydrostatic mechanism (Outcalt et al., 1986; Hinkel, Nelson, and Outcalt, 1987; Hinkel, 1988). When ice segregation occurs in soil, even in an open system, the unidirectional flow of water to the freezing front inhibits ion flushing (Anisimova and Kritsuk, 1983; Kay and Groenevelt, 1983). The concentration of ions in segregation ice may therefore increase with depth and for this reason cannot serve as a stand-alone diagnostic technique.

In a similar manner the relative abundance of H and O heavy isotopes in ice has been used to infer hydraulic conditions during freezing in ephemeral frost mounds (e.g., Pollard and French, 1984; Michel, 1986). In an open system, with continuous replenishment of water of constant composition, δ values remain constant with depth. Conversely, closed-system freezing results in the depletion of heavy isotopes in the water reservoir.

DEGRADATION FORMS

Washburn (1983b, p. 42) stated that degradation palsas develop "due to disintegration of an extensive peaty body." In accordance with the process-oriented terminology presented here, it should be recognized that degradation forms can also result from the erosion of a larger feature that may itself have developed by any of the four growth mechanisms described previously. For example, Hinkel (1988) described an extensive (2400-m^2) ice-cored plateau that was formed annually by the hydrostatic-pressure mechanism in a stream-valley mire. Concordant injection of water at the base of the organic mat-mineral soil interface, uplift, and subsequent freezing resulted in a layer of massive ice more than a meter thick. Mechanical and thermal erosion of the plateau in summer by stream flow caused locally intensive ablation. In addition, thermal contraction cracks, which developed in the overlying organic mat, provided channels for snow meltwater and precipitation and were enlarged and deepened over several days. After several weeks, only remnants of the ice-cored plateau remained as stable mounds, which were protected from further ablation by the draping of the organic mat over the flanks. After several years, a palsa complex developed. The horizontal stratigraphic patterns in the original ice core were truncated along the flanks, allowing easy interpretation of mound growth history.

Degradation palsas can also develop in low-relief peatlands with preexisting permafrost, where disintegration may be concentrated in thermal or desiccation cracks (e.g., Lagarec, 1976). More difficult from a classificatory standpoint are the palsas described by Allard, Seguin, and Levesque (1987) and Allard and Seguin (1987); these dome-shaped palsas in Québec appeared to be remnants of degrading palsa (peat) plateaus.

Conclusions

Palsa-scale frost mounds have unusual and largely unrecognized potential for relating permafrost studies to problems in Quaternary paleoecology, climatic geomorphology, and global change (e.g., Vorren, 1972; Thie, 1974; Allard and Seguin, 1987). Development of interest in these landforms and integration of their information potential in the larger scientific enterprise will, however, require periglacial geomorphologists to dispense with counterproductive terminological conventions and adopt a more flexible, inclusive, and accessible lexicon. The nomenclature suggested here meets these requirements and decouples morphologic and genetic inferences.

Many large-scale investigations cannot be accomplished without extensive international cooperation and relatively large amounts of financial support. In times of intense competition for scientific funding, it is critical to make the importance of periglacial geomorphology's results readily apparent to those responsible for disbursing pecuniary resources. An overly complex, arcane vocabulary does little to assist in establishing clear lines of communication with those operating in the realm of "big science" (Price, 1963).

Acknowledgments

We gratefully acknowledge support from the U.S. National Science Foundation, Grants DPP-8117124, DPP-8721922, and SES-8722676. Useful comments were provided by two anonymous reviewers. The Department of Geography in the University of Wisconsin-Madison kindly provided space, facilities, and a stimulating intellectual atmosphere to the first author during the 1990–1991 academic year.

References

Åhman, R., The structure and morphology of minerogenic palsas in northern Norway, *Biuletyn Peryglacjalny*, **26**, 25–31, 1976.

Åhman, R., Palsar i Nordnorge. En studie av palsars morfologi, utbredning och klimatiska förutsattningar i Finnmarks och Troms fylke, *Meddelanden Fran Lunds Universitets Geografiska Institution Avhandlingar*, **78**, 1–165, 1977.

Akerman, H. J., Observations of palsas within the continuous permafrost zone in eastern Siberia and in Svalbard, *Geografisk Tidsskrift*, **82**, 45–51, 1982.

Akerman, H. J., and B. Malmström, Permafrost mounds in the Abisko area, northern Sweden, *Geografiska Annaler*, **68A**, 155–165, 1986.

Allard, M., and M. K. Seguin, The Holocene evolution of permafrost near the tree line, on the eastern coast of Hudson Bay (northern Quebec), *Canadian Journal of Earth Sciences*, **24**, 2206–2222, 1987.

Allard, M., M. K. Seguin, and R. Levesque, Palsas and mineral mounds in northern Quebec, in *International Geomorphology*, vol. II, edited by V. Gardiner, pp. 285–309, John Wiley, Chichester, England, 1987.

Anisimova, N. P. and L. N. Kritsuk, Use of cryochemical data in the study of the origin of underground ice deposits (in Russian), in *Problems in Geocryology*, edited by P. Melnikov, pp. 230–239, Nauka, Moscow, 1983.

Baranov, I. Y., Geothermic features of the Kola Peninsula (in Russian), *Izvestiya Vostochnykh Filialov Akademii Nauk SSSR*, Moscow, 1951.

Beskow, G. Tjälbildningen och tjällyftningen, med särskild hänsyn till vägar och järnvägar, *Sveriges Geologiska Undersökning*, **375C**, 1–242, 1935.

Brown, J., F. Nelson, B. Brockett, S. I. Outcalt, and K. R. Everett, Observations on ice-cored mounds at Sukakpak Mountain, South-Central Brooks Range, Alaska, *Proceedings of the Fourth International Conference on Permafrost*, **1**, 91–96, 1983.

Brown, R. J. E., Occurrence of permafrost in Canadian peatlands, *Proceedings of the Third International Peat Congress*, 174–181, 1968.

Chorley, R. J., S. A. Schumm, and D. E. Sugden, *Geomorphology*, 605 pp., Methuen, New York, 1984.

French, H. M., *The Periglacial Environment*, 309 pp., Longman, New York, 1976.

Friedman, J. D., C. E. Johansson, N. Oskersson, and H. Svensson, Observations on Icelandic polygon surfaces and palsa areas. Photo interpretation and field studies, *Geografiska Annaler*, **53A**, 115–145, 1971.

Fries, T., *Botanische Untersuchungen im nördlichsten Schweden*, 361 pp., Akademische Abhandlungen, Stockholm, 1913.

Fries, T., and E. Bergström, Några iakttagelser öfver palsar och deras förekomst i nordligaste Sverige, *Geologiska Föreningen Förhandlingar*, **32**, 195–205, 1910.

Fujino, K., S. Sato, K. Matsuda, O. Shimizu, and K. Kato, Characteristics of the massive ground ice body in the western Canadian Arctic, *Proceedings of the Fifth International Conference on Permafrost*, **1**, 143–147, 1988.

Gell, W. A., Fabrics of icing-mound and pingo ice in permafrost, *Journal of Glaciology*, **20**, 563–569, 1978.

Gorodkov, B. N., Large peat mounds and their geographical distribution (in Russian), *Priroda*, **17**, 599–601, 1928.

Gorodkov, B. N., Permafrost in northern regions (in Russian), *Trudy SOPS AN SSSR*, **1**, 1–109, 1932.

Govorukin, D. S., Mound-studded swamps of northern Asia and warming of the arctic (in Russian), *Uchenye Zapiski Moskovskogo Oblastnogo Pedagogicheskogo Instituta*, **9**, 1947.

Harlan, R. L., Analysis of coupled heat-fluid transport in partially frozen soil, *Water Resources Research*, **9**, 1314–1323, 1973.

Hinkel, K. M., Frost mounds formed by degradation at Slope Mountain, Alaska, USA, *Arctic and Alpine Research*, **20**, 76–85, 1988.

Hinkel, K. M., F. E. Nelson, and S. I. Outcalt, Frost mounds at Toolik Lake, Alaska, *Physical Geography*, **8**, 148–159, 1987.

Jahn, A., Remarks on the origin of palsa frost mounds, *Biuletyn Peryglacjalny*, **31**, 123–130, 1986.

Jonasson, S., Influence of frost heaving on soil chemistry and on the distribution of plant growth forms. *Geografiska Annaler*, **68A**, 185–195, 1986.

Jonasson, S., and S. E. Sköld, Influences of frost-heaving on vegetation and nutrient regime of polygon-patterned ground, *Vegetatio*, **53**, 97–112, 1983.

Kachurin, S. P., Cryogenic physico-geological phenomena in permafrost regions, in *Principles of Geocryology*, pp. 1–91, National Research Council of Canada, Technical Translation 1157, Ottawa, 1964.

Kay, B. D., and P. H. Groenevelt, The redistribution of solutes in freezing soil: Exclusion of solutes, *Proceedings of the Fourth International Conference on Permafrost*, 584–588, 1983.

Kay, B. D., and E. Perfect, State of the art: Heat and mass transfer in freezing soils, in *Ground Freezing*, **1**, edited by R. H. Jones and J. T. Holden, pp. 3–21, Balkema, Rotterdam, 1988.

Keilhau, B. M., *Riese i Öst-og Vest-Finmarken samt til Beeren-Eiland og Spitsbergen i Aarene 1827 og 1828*, Christiana, 1831.

Kershaw, G. P., and D. Gill, Growth and decay of palsas and peat plateaus in the MacMillan Pass-Tsichu River area, Northwest Territories, Canada, *Canadian Journal of Earth Sciences*, **16**, 1362–1374, 1979.

Ketchum, W. M., and P. V. Hobbs, The preferred orientation in the growth of ice from the melt, *Journal of Crystal Growth*, **1**, 263–270, 1967.

Kihlman, O. Pflanzenbiologische studien aus Russich Lappland, *Acta Societatis pro Fauna et Flora Fennica*, **6**, 1890.
Konrad, J. M., and N. R. Morgenstern, A mechanistic theory of ice lens formation in fine-grained soils, *Canadian Geotechnical Journal*, **17**, 473–486, 1980.
Lagarec, D., Étude géomorphologique de palses dans la région de Chimo, Nouveau-Québec, Canada, *Cahiers Géologiques*, **92**, 153–162, 1976.
Lagarec, D., Cryogenetic mounds as indicators of permafrost conditions, northern Quebec, *Proceedings of the Fourth Canadian Permafrost Conference*, 43–48, 1982.
Lagerbäck, R., and L. Rodhe, Pingos and palsas in northernmost Sweden: Preliminary notes on recent investigations, *Geografiska Annaler*, **68A**, 149–154, 1986.
Lozinski, W., Palsenfelder und periglaziale Bodenbildung, *Nues Jahrbuch für Mineralogie, Geologie, und Paleontologie*, **71B**, 18–47, 1933.
Luken, J. O., and W. D. Billings, Changes in bryophyte production associated with a thermokarst erosion cycle in a subarctic bog, *Lindbergia*, **9**, 163–168, 1983.
Lundqvist, G., En palsmyr sydost om Kebnekaise, *Geologiska Föreningen Förhandlingar*, **73**, 209–225, 1951.
Lundqvist, G., Tillägg till palsfrågan, *Geologiska Föreningen Förhandlingar*, **75**, 149–154, 1953.
Lundqvist, J., Earth and ice mounds: A terminological discussion, in *The Periglacial Environment Past and Present*, edited by T. L. Péwé, pp. 203–215, McGill-Queen's University Press, Montreal, 1969.
Mackay, J. R., Pingos of the Tuktoyaktuk Peninsula Area, Northwest Territories, *Géographie physique et Quaternaire*, **33**, 3–61, 1979.
Michel, F., Isotope geochemistry of frost-blister ice, North Fork Pass, Yukon, Canada, *Canadian Journal of Earth Sciences*, **23**, 543–549, 1986.
Miller, R. D., Freezing and thawing of saturated and unsaturated soils, *Highway Research Record*, **393**, 1–11, 1972.
Miller, R. D., Frost heaving in non-colloidal soils, *Proceedings of the Third International Conference on Permafrost*, **1**, 708–713, 1978.
Muller, S. W., *Permafrost or Permanently Frozen Ground and Related Engineering Problems*, 231 pp., J. W. Edwards, Ann Arbor, Mich., 1947.
Nelson, F. E., Permafrost in eastern Canada: A review of published maps, *Physical Geography*, **10**, 233–248, 1989.
Outcalt, S. I., and F. Nelson, Computer simulation of buoyancy and snow-cover effects in palsa dynamics, *Arctic and Alpine Research*, **16**, 259–263, 1984a.
Outcalt, S. I., and F. Nelson, Growth mechanisms in aggradation palsas, *Zeitschrift für Gletscherkunde und Glazialgeologie*, **20**, 65–78, 1984b.
Outcalt, S. I., F. E. Nelson, K. M. Hinkel, and G. D. Martin, Hydrostatic-system palsas at Toolik Lake, Alaska: Field observations and simulation, *Earth Surface Processes and Landforms*, **11**, 79–94, 1986.
Permafrost Subcommittee, *Glossary of Permafrost and Related Ground-Ice Terms*, 156 pp., National Research Council of Canada Technical Memorandum 142, Ottawa, 1988.
P'iavchenko, N. I., O genezise Bugristogo rel'efa Torfiaikov v Severo-Vostochnoi chasta Evropelskoi Rossii, *Pochvovedenie*, **5**, 276–284, 1949.
Pissart, A., Pingos and palsas: A review of the current state of knowledge, *Polar Geography and Geology*, **9**, 171–195, 1985.
Pissart, A., and P. Gangloff, Les palses minerales et organiques de la vallee de l'Aveneau, pres de Kuujjuaq, Québec subarctique, *Géographie physique et Quaternaire*, **38**, 217–228, 1984.
Pollard, W. H., A morphogenetic classification of frost mound phenomena, in *Geological Association of Canada/Mineralogical Association of Canada/Canadian*

Geophysical Union, Joint Annual Meeting, Program with Abstracts, **11**, 115, Energy, Mines, and Resources, Earth Physics Branch, Ottawa, 1986.

Pollard, W. H., and S. R. Dallimore, Petrographic characteristics of massive ground ice, Yukon Coastal Plain, Canada, *Proceedings of the Fifth International Conference on Permafrost*, **1**, 224–229, 1988.

Pollard, W. H., and H. M. French, Seasonal frost mound occurrence, North Fork Pass, Ogilvie Mountains, northern Yukon, Canada, *Proceedings of the Fourth Interna-tional Conference on Permafrost*, **1**, 1000–1004, 1983.

Pollard, W. H., and H. M. French, The groundwater hydraulics of seasonal frost mounds, North Fork Pass, Yukon Territory, *Canadian Journal of Earth Sciences*, **21**, 1073–1081, 1984.

Pollard, W. H., and H. M. French, The internal structure and ice crystallography of seasonal frost mounds, *Journal of Glaciology*, **31**, 157–162, 1985.

Popov, A. I., Permafrost in western Siberia, *Izvestiya Vostochnykh Filialov Akademii Nauk SSSR*, Moscow, 1953.

Price, D. J. S., *Little Science, Big Science*, 119 pp., Columbia University Press, New York, 1963.

Ruoff, Z. F., On the question of age and origin of peat mounds in the northern pre-Ural region, *Problemy Fiziki Geografiya*, **16**, 1951.

Seppälä, M., An experimental study of the formation of palsas, *Proceedings of the Fourth Canadian Permafrost Conference*, 36–42, 1982.

Seppälä, M., The origin of palsas, *Geografiska Annaler*, **68A**, 141–147, 1986.

Seppälä, M., Frozen peat mounds in continuous permafrost, northern Ungava, Québec, Canada, *Zeitschrift für Geomorphologie, Supplement Band*, **71**, 107–116, 1988a.

Seppälä, M., Palsas and related forms, in *Advances in Periglacial Geomorphology*, edited by M. J. Clark, pp. 247–278, John Wiley, Chichester, England, 1988b.

Shumskii, P. A., *Principles of Structural Glaciology*, 497 pp., Dover, New York, 1964.

Svensson, G., Recognition of peat-forming plant communities from their peat deposits in two south Swedish bog complexes, *Vegetatio*, **66**, 95–108, 1986.

Taber, S., The mechanics of frost heaving, *Journal of Geology*, **38**, 303–317, 1930.

Thie, J., Distribution and thawing of permafrost in the southern part of the discontinous zone in Manitoba, *Arctic*, **27**, 189–200, 1974.

Thorarinsson, S., Notes on patterned ground in Iceland, with particular reference to the Icelandic "flas", *Geografiska Annaler*, **33**, 144–156, 1951.

Thorn, C. E., Seasonal snowpack variability and alpine periglacial geomorphology, *Polarforschung*, **53**, 31–35, 1983.

Thorn, C. E., *Introduction to Theoretical Geomorphology*, 247 pp., Unwin Hyman, Boston, 1988.

van Everdingen, R. O., Frost mounds at Bear Rock, near Fort Norman, Northwest Territories 1975–1976, *Canadian Journal of Earth Sciences*, **15**, 263–276, 1978.

van Everdingen, R. O., Frost blisters of the Bear Rock Spring area near Fort Norman, N.W.T., *Arctic*, **35**, 243–265, 1982.

Vorren, K.-D., Stratigraphical investigations of a palsa bog in northern Norway, *Astarte*, **5**, 39–71, 1972.

Vorren, K.-D. and B. Vorren, The problem of dating a palsa: two attempts involving pollen diagrams, determination of moss subfossils, and C-14 datings, *Astarte*, **8**, 73–81, 1975.

Washburn, A. L., *Geocryology: A Survey of Periglacial Processes and Environments*, 406 pp., Halsted Press, New York, 1980a.

Washburn, A. L., Permafrost features as evidence of climatic change, *Earth-Science Reviews*, **15**, 327–402, 1980b.

Washburn, A. L., Palsas and continuous permafrost, *Proceedings of the Fourth International Conference on Permafrost*, **1**, 1372–1377, 1983a.
Washburn, A. L., What is a palsa? *Abhandlungen der Akademie der Wissenschaften in Göttingen, Mathematisch-Physikalische Klasse*, **35**, 34–47, 1983b.
Williams, P. J., and M. W. Smith, *The Frozen Earth: Fundamentals of Geocryology*, 306 pp., Cambridge University Press, Cambridge, England, 1989.
Worsley, P., Periglacial environment, *Progress in Physical Geography*, **10**, 265–274, 1986.
Zoltai, S. C., Palsas and peat plateaus in central Manitoba and Saskatchewan, *Canadian Journal of Forest Research*, **2**, 291–302, 1972.

14 Recent Ground Warming Inferred from the Temperature in Permafrost near Mayo, Yukon Territory

C.R. Burn
Department of Geography, University of British Columbia

Abstract

Mayo, Yukon Territory, lies in the widespread discontinuous permafrost zone. Nearby, permafrost thicknesses of up to 40 m have been measured in valleys, and of up to 135 m at higher elevations. Ice-rich glaciolacustrine sediments near Mayo, in which permafrost is 34 m thick, show evidence of recent warming. The mean annual ground temperature at the depth of zero annual amplitude (10 m) is −1.3°C. At present temperatures are slightly warmer, between 0 and 5 m, almost constant between 5 and 15 m, and then increase with depth. Upward projection of the temperature gradient at depth suggests a former mean temperature at the surface of permafrost of at most −2.0°C. The ground is close to the temperature at which the unfrozen water content characteristic begins to change during warming, indicating that some ice is thawing in the soil. Geothermal modeling suggests that if ground temperatures were previously in equilibrium with a near-surface temperature of approximately −3.0°C, then it has taken about 20 years for permafrost to reach present conditions. Observed changes in mean winter temperature and snowfall have probably caused the ground warming.

Introduction

Permafrost in the discontinuous zone is comparatively warm and thin and, hence, may be eradicated by climatic warming (e.g., Osterkamp and Lachen-

---

Periglacial Geomorphology. Edited by J. C. Dixon and A. D. Abrahams

bruch, 1990; Smith, 1988, 1990). The surface layers of thick permafrost on the North Slope of Alaska have warmed as a result of climatic change during the twentieth century (Lachenbruch and Marshall, 1986; Lachenbruch, Sass, Marshall, and Moses, 1982; Osterkamp, 1988). The warming is characterized by curvature in the ground temperature profile, such that the gradient changes at depths well below the zone of annual temperature fluctuation. Similar inflections can be observed in ground temperature profiles from arctic Canada (Judge, Taylor, Burgess, and Allen, 1981). Such inflections are more difficult to detect in "warm" permafrost—that is, where ground temperatures are above −2.0°C—because (1) frozen ground is considerably thinner, and (2) the unfrozen water content is more sensitive to changes in temperature than at cooler temperatures. A small change in temperature of the warm permafrost may involve a considerable exchange of latent heat. The response of such ground to warming or cooling at the surface is slower than the response to a similar absolute change in surface temperature of "cold" permafrost, where only conductive processes may be involved (Riseborough, 1990). However, the impact on surface conditions of changes in active layer thickness may be as great as in the continuous permafrost zone, as the uppermost layers of permafrost are frequently ice-rich (e.g., Burn, Michel, and Smith, 1986; Hughes, Veillette, Pilon, Hanley, and van Everdingen, 1973; Outcalt, 1982).

Research on permafrost by the author has been conducted since 1982 at a field site in the Stewart River valley, 3 km south of Mayo, Yukon Territory. Reports have concentrated on near-surface permafrost conditions accessible by portable CRREL drill (to 5 m depth) or visible in the headwalls of two retrogressive thaw slumps (e.g., Burn, Michel, and Smith, 1986; Burn, 1988). During 1988 several thermistor cables were installed to determine the thermal regime throughout permafrost. The purposes of this paper are (1) to describe the thermal regime of permafrost in Pleistocene glaciolacustrine sediments; and (2) to examine the association of the temperature profile in permafrost with recent changes in air temperature and snowfall.

Regional conditions

Central Yukon lies in the boreal forest and the widespread discontinuous permafrost zone (Brown, 1978) (Figure 14.1). The terrain comprises tablelands separated by broad, interconnected, glaciated valleys (Bostock, 1948). The plateau elements stand about 600 m above the valley floors. The principal settlement is Mayo (63°35′N, 135°35′W), elevation 480 m, at the confluence of the Stewart and Mayo Rivers (Figure 14.2). Stewart River flows westwards for 200 km to join the Yukon River 80 km upstream of Dawson.

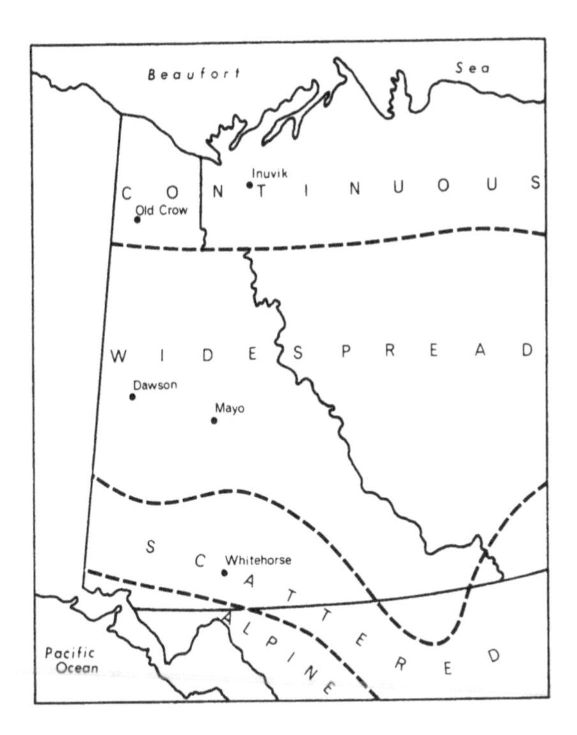

Figure 14.1. Permafrost map of the Yukon Territory and adjacent Northwest Territories, Canada (after Brown, 1978)

CLIMATE

Mean annual and monthly air temperatures and annual precipitation at Keno Hill, Elsa, and Mayo Airport, 2 km north of the village, are presented in Table 14.1 (see Figure 14.2 for locations of stations). Air temperatures during winter may be quite variable. If stable conditions are disturbed by incursions of Pacific air, rapid warming often occurs. At Mayo temperatures in February have ranged from −62.2 to 12.2°C. During the period of record (1926–1990) mean monthly temperatures in winter have been an order of magnitude more variable than in summer (Burn, 1990, Figure 2). As a result, variations in mean annual temperature at Mayo are primarily associated with fluctuations in winter temperatures (Figure 14.3). The smoothed record of air temperatures recorded at Mayo Airport (Figure 14.3) indicates warming beginning in the late 1960s and continuing through the 1970s to a peak in the late 1980s. Mean annual air temperatures observed during the late 1980s at Mayo Airport and at other stations in Yukon are amongst the highest that have been recorded in the region.

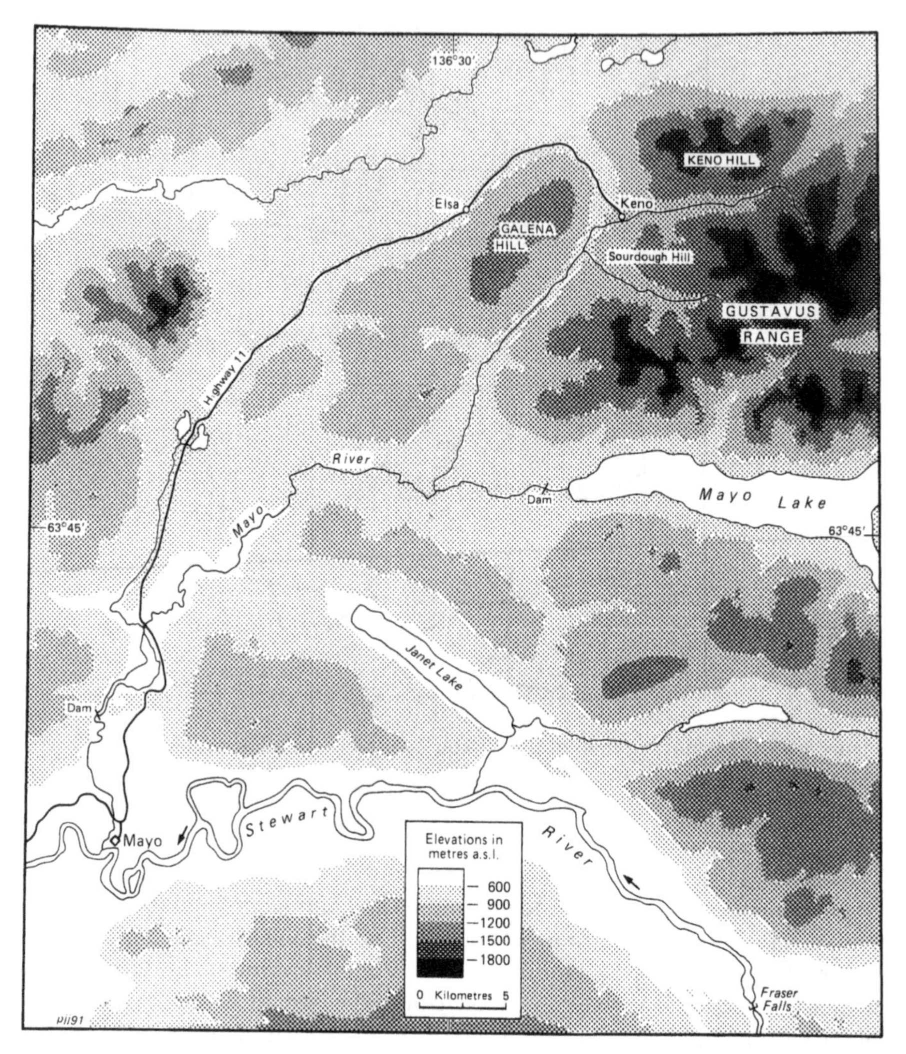

Figure 14.2. Topography and drainage of the Mayo region, central Yukon Territory (part of NTS sheet 105 M)

Precipitation is derived from maritime air masses, and stable conditions, with frigid temperatures in valley bottoms, are dry. Over periods during winter on the order of weeks, temperatures and snowfall may be positively correlated. However, over periods of months, higher frequency variability obscures such relationships (Figure 14.4). During the 1980s, when winter air temperatures were milder than the previous decade, snowfall also increased, particularly after 1984 (Figure 14.4).

Table 14.1. Climatic Statistics from Stations in Central Yukon, 1951–1980[a]

Station	Elevation (m)	Mean Air Temperature (°C) Jan.	Feb.	March	April	May	June	July	Aug.	Sep.	Oct.	Nov.	Dec.	Annual
Mayo	504	–29.0	–19.9	–11.7	–0.4	7.5	13.4	15.2	12.6	6.5	–2.3	–15.2	–24.2	–4.0
Elsa	814	–25.1	–18.1	–12.3	–2.9	5.7	12.5	14.1	11.0	4.6	–4.9	–15.3	–21.5	–4.4
Keno Hill	1472	–19.6	–13.9	–13.7	–6.6	1.4	7.8	10.0	7.9	2.0	–6.3	–12.3	–16.9	–5.0

Station	Mean Annual Rainfall (mm)	Mean Annual Snowfall (cm)	Mean Annual Precipitation (mm)
Mayo	185.1	130.5	306.3
Elsa	219.5	202.9	413.0
Keno Hill	241.3	365.7	590.2

[a]Source: Environment Canada (1982).

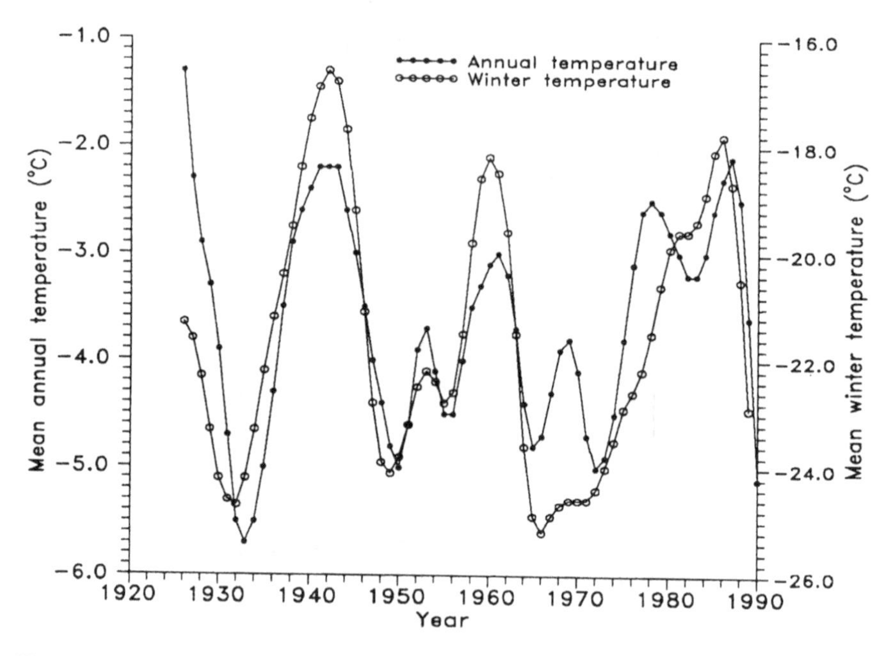

Figure 14.3. Mean annual and mean winter (November to February) air temperature at Mayo, 1926 to 1990, smoothed with running medians of 3 and Hanning filtering (Tukey, 1977). Four passes of the filter were required before the series were unaltered by further smoothing. The smoothing filter does not handle end values well; little significance should be attached to the two points at each end of each series

QUATERNARY GEOLOGY

The Stewart Valley near Mayo is filled with thick sequences of glacial deposits. North, south, and east of Mayo and at the village itself outwash deposits are mixed with glaciolacustrine sediments deposited toward the end of the last glaciation (13,600 yr B.P.) (Hughes, 1983, 1987). Surficial glacial deposits have been eroded by Stewart River over much of the valley upstream (east) of Mayo, which is now covered by sandy, alluvial material.

The Holocene environmental history of central Yukon comprises three principal phases: (1) deglaciation and subsequent drainage of glacial lakes under climatic conditions sufficiently severe for the establishment of permafrost and the growth of large bodies of segregated ice in glaciolacustrine sediments (Burn, Michel, and Smith, 1986); (2) an early Holocene climatic optimum, ca 8500 yr B.P., during which active layer thicknesses were approximately twice those of today and thermokarst lakes developed (Burn, Michel, and Smith, 1986; Burn and Smith, 1990); and (3) a period since 5000 yr B.P. when climatic conditions have been broadly similar to the present (Cwynar and Spear, 1991).

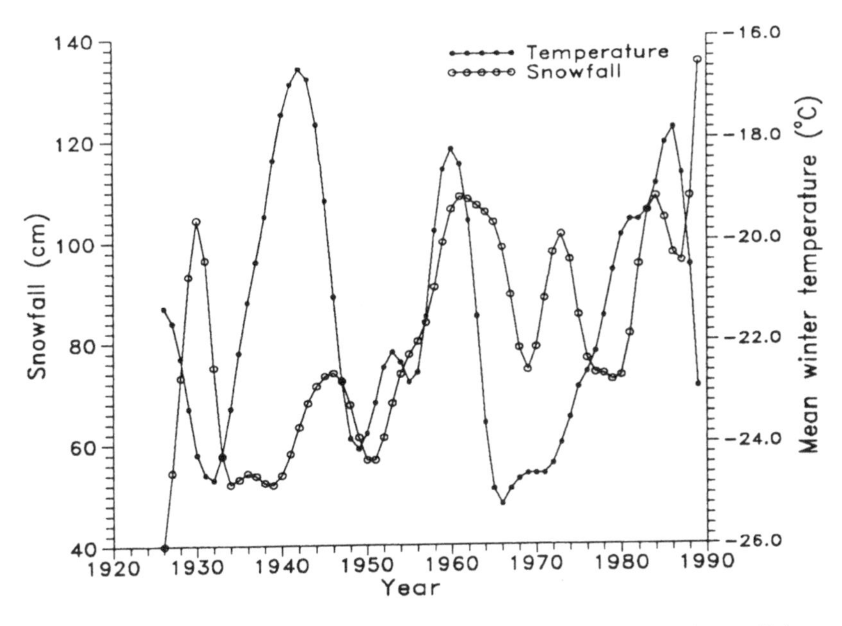

Figure 14.4. Total snowfall and mean temperature at Mayo, November to February, 1926/27 to 1989/90. Both variables were smoothed as in Figure 14.3, and both required four passes of the median filter before further passes left the series unaltered

PERMAFROST

Permafrost is widespread but discontinuous in central Yukon. Reconnaisance reports by Brown (1967) and Hughes and van Everdingen (1978) suggest that mesic valley bottoms and north-facing slopes are underlain by perennially frozen ground, but the surficial layers of dry substrates and the lower elevations of south-facing slopes are only frozen seasonally. Ground ice was not reported in gravel pits quarried from glacial outwash materials during initial road construction in the broad, main valleys of the area (O.L. Hughes, personal communication, 1987). However, about one quarter of the shallow (5-m) geotechnical holes drilled during autumn 1987 for assessment of gravel resources in terraces along Yukon Highway 11 between Mayo and Keno (Figure 14.2) encountered ground ice below the depth of seasonal thaw (Paine, 1988). Ground ice is almost always encountered by placer miners in the surficial gravels and organic mucks of small creeks at higher elevations. Near Mayo ground ice has been encountered at depths of up to 40 m (Midnight Sun Drilling Company, personal communication, 1990).

Permafrost thicknesses determined by temperature measurements are not reported as frequently as observations of ground ice. However, Pike (1966) reported a thickness of 135 m at elevations between 1350 m and 1980 m on

Figure 14.5. Part of aerial photograph A27482-84 of the Mayo area, central Yukon, July 6, 1989. A and B indicate the locations of temperature cables in the permafrost research area

Keno Hill from measurements of ground temperature. Wernecke (1932) determined the geothermal gradient in permafrost on Keno Hill to be 0.025 K m^{-1}. Faults, fractures and solution cavities within the permafrost zone are often filled with ice veins. Some ice veins contain native silver (Wernecke, 1932; Boyle, 1965).

Research by the author on ground ice and permafrost in central Yukon has been concentrated at a site 3 km south of Mayo in glaciolacustrine deposits where there are several thermokarst lakes (Figure 14.5) (Burn and Smith, 1990). The sediments at the site are ice-rich and contain both occasionally active and relict ice wedges (Burn, 1990). Detailed examination of the cryostratigraphy has been provided by Burn, Michel, and Smith (1986), who suggested that most of the ground ice in the sediments formed at the end of the glacial period during permafrost aggradation after the glacial lake drained. More recent ground ice is preserved between the base of the present active layer and the early Holocene thaw unconformity.

Field activities

During July 1988 a series of five holes was drilled by water-jet through permafrost in the ice-rich glaciolacustrine sediments. At two sites, 60 m apart (A1 and B; see Figure 14.5), one-inch plumbing pipe was installed in the holes as casing for thermistor strings. Thermistor (YSI 44030A) cables were assembled, calibrated in an ice bath, and placed in the pipe casing. The precision of the calibrated thermistors is better than ±0.05°C. The sensors were located at intervals of approximately 3 m. The access pipes were filled with dry sand to minimize convection within them. During the period between pipe and cable installation, the lower parts of the pipes were blocked by water seeping in and freezing. As a result, the cables only monitor the thermal regime to 23 m depth. Another hole (A2) was drilled in these sediments 5 m from cable A1 during September 1989, and a calibrated thermistor (YSI 44033) cable was successfully installed through permafrost. Sensors were placed at 5-m intervals in the upper portion of the profile and every 2 m below 25 m. The access tube was again filled with dry sand.

Although jet drilling does not provide core samples from which to determine stratigraphy, a qualitative impression may be obtained from the behaviour of the pipe during drilling and from sediments returned uphole. In particular, the lower limit of ground ice may be inferred from an acceleration of drilling speed associated with weakening of sediment. In sands, the acceleration may occur suddenly.

Over 25 holes drilled with a CRREL barrel near the cables in the upper 5 m of sediment have returned core samples that provide a quantitative assessment of near-surface ground ice contents. The thaw slump exposures, located less than 500 m from cables A1, A2, and B have also indicated the nature of

Figure 14.6. Segregated ground ice exposed in a thaw slump headwall near Mayo, November 14, 1987

ground ice conditions in the upper 10 m of permafrost (Figure 14.6). Core samples collected by drilling to 5 m depth just behind the thaw-slump headwall have yielded ice contents similar to those collected near the cables. Stratigraphic inferences made during jet-drilling indicated that ground ice content was high and relatively constant within the upper portion of permafrost between the base of the thaw unconformity and a depth of 10 m. The cryostratigraphy of the headwall is apparently indicative of ground ice conditions near the thermistor cables.

The near-surface ground thermal regime is variable within the study area owing to microclimatic factors. Temperatures at 15-cm intervals were obtained from the uppermost 2 m almost weekly at seven sites within the study area between June 1984 and August 1985. The measured mean annual surface temperature of permafrost—that is, the mean temperature at the base of the active layer—ranged between −2.8 and −0.2°C for the seven sites: at a site 30 m from Cables A1 and A2 it was −2.2°C (Burn and Smith, 1988). Surface vegetation at this site is similar to the community near cables A1, A2, and B. The mean air temperature at Mayo Airport for the 12 months beginning in June 1984 was −4.4°C, slightly below normal (Table 14.1) and relatively cool for the 1980s (Figure 14.3).

The near-surface ground thermal regime has also been monitored at seven other sites in the area on cables installed in July 1985 and read approximately once a month subsequently (Smith and Burn, 1986). These cables have sensors to 5 m depth, which are spaced at 0.5- or 1-m intervals. One of these 5-m cables, also 30 m from cables A1 and A2, is adjacent to the 2-m cable mentioned above, which was monitored in 1984–1985. The record from the 5-m cable, collected monthly over a year beginning August 30, 1987, was used in the numerical simulation presented later.

Field results

Figure 14.7 indicates ground temperatures measured on thermistor strings installed within and through permafrost. Measurements are from cables A1, A2, and B. The cables installed in 1988 (A1 and B) equilibrated by summer 1989; cable A2, installed in September 1989, equilibrated during the follow-

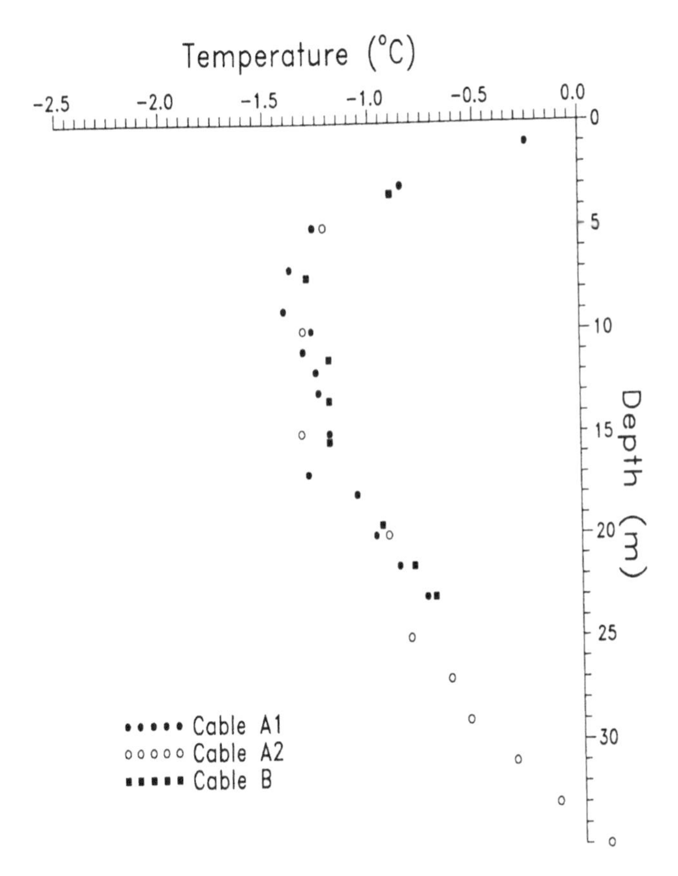

Figure 14.7. Temperature profile through permafrost in glaciolacustrine sediments at sites A and B, December 15, 1990

ing year. Measurements have been taken every three months since installation; the data reported in Figure 14.7 were collected in mid December 1990. The active layer at the sites is approximately 1 m deep.

PERMAFROST STRATIGRAPHY

The base of ice-bonded permafrost in these sediments is at approximately 30 m at a temperature of −0.4°C. During drilling of the six holes (five in 1988, one in 1989), the base of ice-rich permafrost was consistently encountered between 29 and 31 m. The uppermost 12 m of sediments are predominantly ice-rich glaciolacustrine silty clay (Burn and Smith, 1990). Between 12 and 20 m depth the sediments appear to contain fewer ice lenses but are of similar grain size. However, sands appear below 15 m and dominate below 23 m. The lowermost 4 m of ice-bonded permafrost are ice-rich.

The permafrost is underlain by a sandy aquifer. The piezometric surface, determined by blockage of access tubes, is 23 m below the ground surface, about 7 m above the base of ice-bonded permafrost. The presence of water under pressure below the ice-bonded permafrost and the direction of the ground temperature gradient have likely contributed to the accumulation of ice at the base of permafrost (Harlan, 1974).

At the base of permafrost the freezing temperature of ice depends on the overburden pressure, capillary and soil surface area effects in pores, and the solute concentration in subpermafrost groundwater. At 30 m the overburden pressure is on the order of 0.4 mPa, leading to a minute freezing point depression of 0.003°C (Williams and Smith, 1989, p. 185). Capillary and soil surface area effects in sands are also minimal. The temperature depression at the base of ice-bonded permafrost is likely due to solutes in subpermafrost groundwater. There are no measurements of the electrical conductivity of subpermafrost groundwater at the study sites, but four analyses of Mayo well water reported by Brandon (1965, Table 13) returned electrical conductivities between 190 and 2692 μmhos cm^{-2}. Conductivities of this magnitude indicate solute concentrations sufficient to account for the freezing-point depression (0.4°C) of subpermafrost pore water (see Tables in Weast, 1982).

THERMAL PROFILE OF PERMAFROST

The ground temperature profile (Figure 14.7) exhibits a relatively uniform temperature of −1.3°C between 5 and 15 m. Measurements from all three cables are consistent in this interval. The depth of zero annual temperature fluctuation (<0.1°C) at these sites is 10 m; at depths of 5 m or more the annual range in temperature is less than 0.4°C. Monthly observations from other cables in the area, including one 30 m from A1 and A2, indicate that maximum temperature at 5 m occurs in late February. Therefore the temperature profile below 4 m displayed in Figure 14.7 is relatively unaffected by summer warming or autumn freeze-back of the active layer.

Permafrost warms with depth below 15 m. Temperature measurements from all three cables are consistent to 23 m. The gradient between 15 m and 23 m is 0.076 K m^{-1}. Measurements below 23 m are available from cable A2 only. The profile at these depths is offset from measurements taken above. The gradient between 25 m and the base of permafrost at 34 m is approximately 0.1 K m^{-1}. Projection of the gradient between 15 and 23 m upwards from 15 m indicates a mean annual temperature at 1.5 m of, at most, −2.0°C. The profile suggests that permafrost thickness is not in equilibrium with present (1990) conditions but reflects a previously cooler near-surface thermal regime. Osterkamp (1988) has associated similar near-surface warming of permafrost on the North Slope of Alaska with recent (1977–1985) climatic warming.

Numerical simulation

Following Osterkamp's (1988) observations, it is of interest to note the association of increased snowfall with warming of winter and mean annual air temperatures that has been recorded at Mayo since 1968 (Figure 14.4). Numerical simulations of the permafrost thermal regime were conducted to determine whether the 1990 profile could be attributed to changes in near-surface temperatures driven by warmer conditions with increased snowfall in winter.

Smith and Riseborough (1983) and Lachenbruch, Cladouhous, and Saltus (1988) have pointed out that the relationship between changes in the air temperature regime and surface permafrost temperatures is complex and ill-defined. Microclimatic processes may dampen or enhance the effects of changes in atmospheric temperature T_a on the ground surface temperature T_s. T_s drives the ground thermal regime, not T_a. In addition, the zero-curtain effect within the active layer moderates the impact of fluctuations in T_a on the temperature of near-surface permafrost during spring thaw and autumn freeze-back (Outcalt, Nelson, and Hinkel, 1990). Therefore the simulations did not attempt to derive changes in near-surface permafrost temperature from the atmospheric record. Instead, the simulations investigated the time required to reproduce the present ground temperature profile from near-equilibrium conditions (a linear temperature gradient). The intention was to determine whether two decades, the period of most recent atmospheric warming, would be sufficient to cause the inflection in the ground temperature profile.

MODEL

The model comprised a finite difference implementation of the Fourier heat conduction equation:

$$\frac{dT}{dt} = \varkappa \frac{d^2T}{dz^2} \tag{14.1}$$

where T is temperature (°C), t is time (s), z is depth (m), and $\varkappa$ is thermal diffusivity ($m^2 s^{-1}$). The finite difference algorithm was drawn from Goodrich (1982, equation (5)). A node-spacing of 10 cm and time-steps of 1 hour were used to allow computation with an explicit or forward difference technique (Goodrich, 1982, p. 3).

BOUNDARY CONDITIONS

Initially a linear ground temperature profile was set between two points: (1) the base of permafrost at 34 m (Figure 14.7), and (2) the projection of the gradient between 23 and 15 m up to −2.32°C at 1.5 m. The gradient between these points was 0.071 K m^{-1}. The base of the profile at 35 m, 0.1°C (Figure 14.7), was held constant throughout the simulations. The top of the profile was placed below the active layer at 1.5 m in order to avoid explicit association of ground warming with absolute changes in air temperature and to shorten calculation time by omitting freezing and thawing in the active layer.

The simulations were run in response to an imposed annual temperature regime at 1.5 m. This was derived from field observations at the 5-m cable 30 m from cables A1 and A2, which has yielded five years of ground temperature measurements at approximately monthly intervals. The records from cables A1, A2, and B are neither long enough nor of sufficient frequency to derive a reliable annual regime. Warmest conditions at the site were monitored in the year beginning August 30, 1987, when the mean temperature at 1.5 m was −1.2°C. The mean air temperature for this period was −0.5°C, considerably warmer than the 1951–1980 mean (Table 14.1). As mentioned above, surface conditions at the site are similar to those at cables A1, A2, and B. A cubic spline interpolation of the ground temperature record at 1.5 m for the year beginning August 30, 1987 is presented in Figure 14.8. The record was used to drive the finite difference simulations in order to approximate the annual temperature oscillation below the active layer and to include, in an indirect way, the impact of active-layer development on near-surface permafrost temperatures. The warmest available record was chosen because of its association with a climatically warm year and because it represented a recent ground temperature regime. The simulation was started on day 106 of the spline (December 15), so that results from full years of

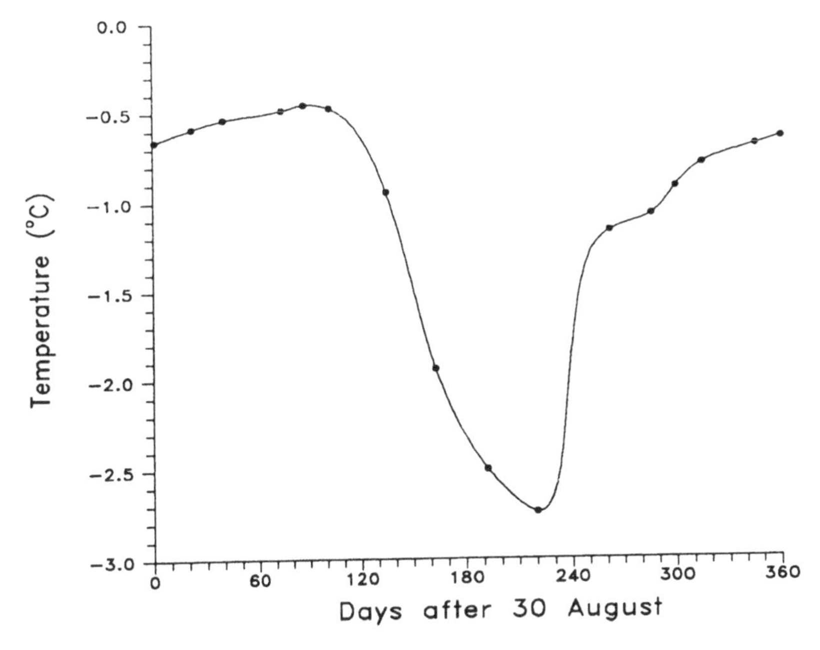

Figure 14.8. Cubic spline interpolation of monthly temperatures at 1.5 m at a site 30 m from cables A1 and A2, August 30, 1987 to August 30, 1988

calculations could be compared with observations taken on December 15, 1990.

THERMAL PROPERTIES OF GROUND MATERIALS

The freezing characteristic curve of the glaciolacustrine silty clay determined by Time Domain Reflectometry (e.g., Patterson and Smith, 1981) is presented in Figure 14.9. The volumetric water content at saturation (no excess ice) is 38% at a dry bulk density of 1.6×10^3 kg m^{-3}. The characteristic can be approximated by a constitutive equation of the form:

$$\theta_u = a(-T)^b \tag{14.2}$$

where θ_u is the unfrozen water content (cm^3 cm^{-3}); $a = 0.195$ and $b = -0.3331$ in this case.

The sediments are mostly chlorite, hydrous mica, and quartz (Burn, Michel, and Smith, 1986). Thermal properties of the sediments are dependent upon mineralogy, temperature, total water content and excess ice content—that is, the volume fraction of ice in the soil above the thawed soil porosity. They may be calculated using (14.2) to estimate the volume proportions of

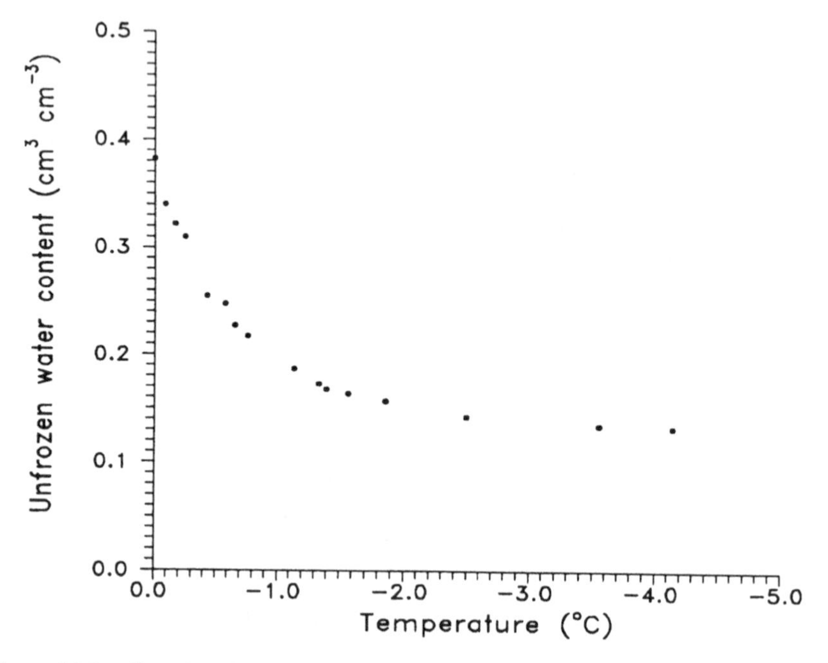

Figure 14.9. Freezing characteristic curve of Mayo glaciolacustrine silty clay determined by Time Domain Reflectometry. Analysis by D.W. Riseborough, November 1984 (Smith and Riseborough, 1985, Figure 1)

ice, water, and mineral material in the sediments. Thermal conductivity at temperature T, K_T, was determined from (Johansen, 1975)

$$K_T = K_m^{(1-e_i)\,(1-\varrho)}\; K_i^{e_i+(1-e_i)\,(\varrho-\theta_u)}\; K_w^{\theta u(1-e_i)} \qquad (14.3)$$

where K_m, K_i and K_w are the thermal conductivities of constituent materials (mineral soil, ice, and water: 5, 2.3 and 0.56 W m^{-1}K^{-1}, Williams, 1982), ϱ is the porosity, and e_i the excess ice content. $(1-e_i)$ is the unit volume of soil matrix and pore contents. Heat capacity at temperature T, C_T, was determined from

$$C_T = C_m(1-e_i)\,(1-\varrho) + C_i[e_i + (1-e_i)\,(\varrho-\theta_u)] + C_w\theta_u(1-e_i) \qquad (14.4)$$

where C_m, C_i, and C_w are the heat capacities of the constituent materials (2.0, 1.83, and 4.2 MJ K^{-1}m^{-3}). The latent heat of fusion of pore ice ($L = 3.33 \times 10^2$ MJ m^{-3}) was incorporated into the heat capacity to provide an apparent heat capacity AC_T (Goodrich, 1982, equation (27)):

$$AC_T = C_T + L\frac{d\theta_u}{dT}\,(1-e_i) \qquad (14.5)$$

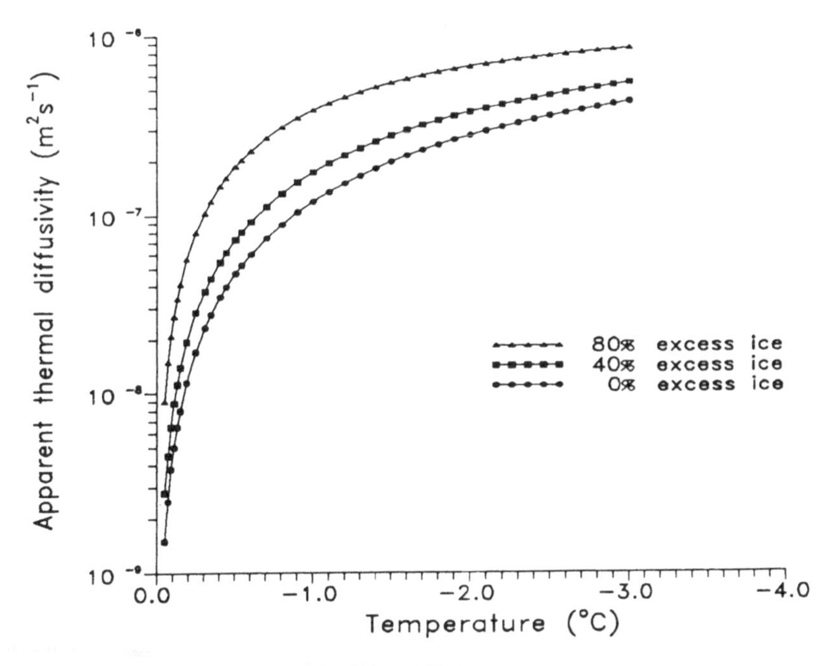

Figure 14.10 Apparent thermal diffusivity of Mayo silty clay for excess ice contents of 0, 40, and 80%. Values calculated from Figure 14.8 using the method described in the text

An apparent thermal diffusivity, $A\varkappa_T$ was calculated from

$$A\varkappa_T = \frac{K_T}{AC_T} \tag{14.6}$$

and used in (14.1).

Estimated apparent thermal diffusivities of Mayo silty clay for excess ice contents of 0, 40, and 80% over the temperature interval 0 to −4°C are presented in Figure 14.10. $\varkappa_T$ of ice is $1.26 \times 10^{-6}\,m^2\,s^{-1}$. Figure 14.10 indicates that $A\varkappa_T$ changes at around -1.0°C. At higher temperatures the thermal response ($A\varkappa_T$) is low because $d\theta_u/dT$ is large; there is considerable exchange of latent heat which moderates conduction and results in small apparent diffusivity. At lower temperatures changes in apparent heat capacity occur in similar proportion to changes in thermal conductivity, leaving the apparent diffusivity relatively constant.

The excess ice is assumed to thaw at 0°C. This assumption is accurate near the surface of the ground but not at depth, where the presence of solutes in subpermafrost groundwater effects a freezing-point depression of about

Table 14.2. Cryostratigraphy used in Numerical Simulation of Permafrost

Depth (m)	Excess Ice Content (%)
1.5– 2.5	35
2.5–10.0	60
10.0–13.0	30
13.0–17.0	15
17.0–20.0	10
20.0–26.0	5
26.0–30.0	30
30.0–35.0	0

0.4°C. However, for the purposes of the modeling the discrepancy is not serious (1) because attention is focused on the upper portion of the profile; and (2) because over the time interval considered here, 20 years, little thawing may occur at the base of permafrost.

The cryostratigraphy used in the simulations is described in Table 14.2. The excess ice contents between 1.5 and 10 m were estimated from analysis of core samples and from the thaw-slump exposures, as described above. The ice contents below 10 m were estimated from logs compiled during jet-drilling. These values are unlikely to be accurate but are precise in relative terms. Other cryostratigraphies were used to investigate the sensitivity of the model to changes in ground ice conditions.

Sediments returned during drilling indicated that sand is the dominant material below 23 m. The implications of this observation for the calculations are discussed in the sensitivity analysis. No quantitative assessment of the sandy sediments is available, so for simulation the profile was assumed to comprise silty clay.

RESULTS

Three simulations were completed: (1) a simple step change in 1.5 m temperature from initial conditions (−2.32°C) to the imposed temperature regime (mean annual temperature −1.20°C); (2) a ramped change in 1.5 m temperature from mean −2.32°C to −1.20°C over 20 years, increasing by 0.071°C yr^{-1}; and (3), as a consequence of results from (1) and (2), a ramped change in 1.5 m temperature from mean −3.0°C to mean −1.2°C over 18 years, increasing by 0.1°C yr^{-1}, with an initial linear ground temperature profile between −3.0°C at 1.5 m and the base of permafrost (0°C) at 34 m. The results of these simulations are summarized in Tables 14.3 and 14.4. The simulations do not purport to represent the actual changes in 1.5 m tempera-

Table 14.3. Temperature Profiles from Numerical Simulation of Permafrost Warming at Mayo

Depth (m)	1990 Field Observations	Initial Profile	Step Change		Ramped Change		
			5 years	15 years	15 years	18 years	20 years
5.0	-1.25	-2.07	-1.20	-1.11	-1.37	-1.25	-1.18
7.5	-1.30	-1.89	-1.30	-1.14	-1.43	-1.31	-1.24
10.0	-1.30	-1.72	-1.30	-1.10	-1.36	-1.27	-1.21
12.5	-1.26	-1.54	-1.27	-1.06	-1.28	-1.21	-1.17
15.0	-1.26	-1.36	-1.21	-1.00	-1.18	-1.13	-1.10
17.5	-1.18	-1.18	-1.10	-0.92	-1.07	-1.03	-1.00
20.0	-0.95	-1.00	-0.96	-0.84	-0.94	-0.91	-0.89
22.5	-0.74	-0.83	-0.81	-0.72	-0.79	-0.77	-0.75
25.0	-0.82[a]	-0.65	-0.64	-0.59	-0.62	-0.61	-0.60

[a]Measurement from cable A2 offset from profile above (see Figure 14.7).

ture that have occurred recently at Mayo; they represent three models of changes in ground temperatures between initial conditions, deduced from the measured profile at depth, and the present regime, observed in the field.

Step change

The results presented in Table 14.3 indicate that the 1990 ground temperature profile is reasonably well represented after 5 years of simulation. Subsequently the profile warms, but the isothermal portion, measured in 1990 between 5 and 15 m, shrinks, so that by 15 years a gradient on the order of 0.02°C m^{-1} is established below 12.5 m. In order to reproduce 1990 temperatures after more than a decade of this simulation, cooler initial conditions must be prescribed.

Ramped change over 20 years

The ground temperature profiles generated by this simulation after 15, 18, and 20 years are also indicated in Table 14.3. The profile is closest to 1990 field observations after 18 years of simulation, but by this stage temperatures below 12.5 m are higher than the field observations, as in the step-change simulation. Since the ramped change is a more likely model of changes in near-surface temperatures than the step change (see Figure 14.4), these results suggest (1) that a period of about two decades may be required to produce a layer isothermal at −1.3°C in the upper portion of the profile; and (2) that ground temperatures below 15 m before warming may have been lower than the profile observed in 1990.

Table 14.4 Temperature Profile from Numerical Simulation of Permafrost Warming over 18 Years

Depth (m)	1990 Field Observations	Initial Profile[a]	Ramped Change 18 years[a]	Initial Profile[b]	Ramped Change 18 years[b]
5.0	−1.25	−2.68	−1.26	−2.59	−1.25
7.5	−1.30	−2.45	−1.38	−2.36	−1.37
10.0	−1.30	−2.22	−1.40	−2.14	−1.37
12.5	−1.26	−1.98	−1.37	−1.92	−1.35
15.0	−1.26	−1.75	−1.32	−1.70	−1.29
17.5	−1.18	−1.52	−1.23	−1.47	−1.20
20.0	−0.95	−1.29	−1.10	−1.25	−1.10
22.5	−0.74	−1.06	−0.96	−1.03	−0.93
25.0	−0.82[c]	−0.83	−0.79	−0.80	−0.77

[a]Initial temperature at 1.5 m set at −3.0°C.
[b]Initial temperature at 1.5 m set at −2.9°C.
[c]Measurement from cable A2 offset from profile above (see Figure 14.7).

Ramped change over 18 years

In response to these considerations, a third simulation was completed with cooler near-surface permafrost. The initial 1.5 m temperature was set at −3.0°C. The results are summarized in Table 14.4. The simulated temperature profile after 18 years is within about 0.1°C of the profile observed in 1990. The results are consistent with points (1) and (2) of the previous paragraph. A slightly better agreement between model and observation was obtained with initial 1.5 m temperature set at −2.9°C (Table 14.4). These simulations suggest near-surface permafrost temperatures in the late 1960s and early 1970s may have been close to −3.0°C.

Sensitivity analysis

Four simulations were run with different cryostratigraphies (excess ice contents, e_i) to determine the effect of variations in thermal properties on model results. The cryostratigraphies comprised (1) the original units, as indicated in Table 14.2; (2) no excess ice within the profile; (3) the upper 13.5 m of the profile at 60% e_i, none below; and (4) 60% e_i throughout the profile. The initial temperature profile was set as in the first simulations (Table 14.3), and a ramped temperature regime at 1.5 m was employed, increasing by 0.074°C yr^{-1} over 15 years. Final temperature profiles are summarized in Table 14.5.

The results indicate that temperature changes during the simulation increase with excess ice content. As mentioned earlier, calculation of thermal properties assumes that the excess ice thaws at 0°C. As e_i increases the unit volume of pore contents $1-e_i$ decreases, and latent heat effects are reduced. The range of temperatures from different simulations is within ±0.1°C of the

Table 14.5 Sensitivity of Temperature Profiles after 15 Years to Changes in Prescribed Excess Ice Content

Depth (m)	Initial Profile	Original Cryo-stratigraphy[a]	No Excess Ice[b]	60% to 15 m 0% to 34 m[c]	60% e_i Throughout[d]
5.0	–2.07	–1.20	–1.31	–1.19	–1.18
7.5	–1.89	–1.28	–1.35	–1.26	–1.25
10.0	–1.72	–1.26	–1.33	–1.26	–1.22
12.5	–1.54	–1.21	–1.27	–1.17	–1.16
15.0	–1.36	–1.14	–1.19	–1.10	–1.08
17.5	–1.18	–1.04	–1.07	–1.02	–0.98
20.0	–1.00	–0.92	–0.94	–0.90	–0.87
22.5	–0.83	–0.77	–0.79	–0.77	–0.73
25.0	–0.65	–0.62	–0.62	–0.61	–0.59

[a]Original cryostratigraphy as described in Table 14.2.
[b]Thermal properties calculated with e_i = 0.0 throughout permafrost.
[c]Thermal properties calculated with e_i = 0.6 for 1.5 to 15 m and e_i = 0.0 for 15 to 34 m.
[d]Thermal properties calculated with e_i = 0.6 throughout permafrost.

profile with original cryostratigraphy, corresponding to a range in response over the simulation period of ±10%.

These results bear out the conclusion of Riseborough (1990) that at sites with warm permafrost, where latent heat effects are significant, climate warming will provide a weak ground temperature signal. However, at sites with large amounts of bulk ice in massive bodies or lenses, the ground temperatures may respond more readily to climate warming.

Finally, the model neglected the change of grain size distribution in the lower portion of the profile. In general, the freezing characteristic of freshwater in sands is a steep curve, with most pore water frozen by −0.3°C (Williams, 1982). The thermal diffusivity of frozen sand may change little at lower temperatures. For frozen, saturated quartz sand, with a porosity of 33%, $\varkappa$ is 2.64×10^{-6} m^2 s^{-1}. (For quartz, K is 8 W m^{-1} K^{-1}, C is 2.1×10^6 J m^{-3}, Williams, 1982.) With 20% e_i, $\varkappa$ is 2.27×10^{-6} m^2 s^{-1}. These values for $\varkappa$ are an order of magnitude higher than those used in the simulations (see Figure 14.10). The implication is that the response of permafrost below 20 m to warming may be underestimated by the simulations. Indeed, the 1990 field observations in this portion of the profile are warmer than the profiles from the 18-year simulations (Table 14.4).

Conclusion

Geothermal modeling and ground temperature profiles from undisturbed sites indicate that the upper 15 m of permafrost in the Stewart River Valley near Mayo, Yukon Territory, has warmed during the past two decades. Near-surface permafrost is about 1.5°C warmer than modeling suggests it may have been in the early 1970s. The ground warming has occurred during a period of rising air temperatures and increasing snowfall.

Acknowledgments

The work has been supported by the Geological Survey of Canada, the Natural Sciences and Engineering Research Council of Canada, the Atmospheric Environment Service, Environment Canada, and the Department of Indian Affairs and Northern Development. Support from A.S. Judge, GSC, and S.R. Morison, DIAND, is acknowledged in particular. Eric Gibson kindly provided drill logs from the Mayo area. Fieldwork at Mayo would not be possible without the continuing generous hospitality of Jim and Shann Carmichael. Field assistance during installation of temperature cables through permafrost was provided by Nick Burn, John Eldon, Joan Ramsay Burn, and Catherine Souch. D.W. Riseborough determined the freezing characteristic of the Mayo soil and has provided many stimulating discussions. Investigations of the near-surface ground thermal regime are in collaboration with M.W. Smith, Carleton University. Helpful comments on the manuscript were received from O.L. Hughes, J.R. Mackay, S.I. Outcalt, H.O. Slaymaker, M.W. Smith, and an anonymous referee.

References

Bostock, H.S., Physiography of the Canadian Cordillera, with special reference to the area north of the fifty-fifth parallel, *Geological Survey of Canada Memoir 247*, 106 pp., 1948.

Bostock, H.S., Notes on glaciation in central Yukon Territory, *Geological Survey of Canada Paper 65–36*, 18 pp., 1966.

Boyle, R.W., Geology, geochemistry, and origin of the lead-zinc-silver deposits of the Keno Hill-Galena Hill area, Yukon Territory, *Geological Survey of Canada Bulletin 111*, 302 pp., 1965.

Brandon, L.V., Groundwater hydrology and water supply in the District of Mackenzie, Yukon Territory and adjoining parts of British Columbia, *Geological Survey of Canada Paper 64–39*, 102 pp., 1965.

Brown, R.J.E., Permafrost investigations in British Columbia and Yukon Territory, Division of Building Research, *National Research Council of Canada Technical Paper 253*, 55 pp., 1967.

Brown, R.J.E., Permafrost, in *Hydrological Atlas of Canada*, Plate 32, Fisheries and Environment Canada, Ottawa, 1978.

Burn, C.R., The development of near-surface ground ice during the Holocene at sites near Mayo, Yukon Territory, Canada, *Journal of Quaternary Science*, **3**, 31–38, 1988.

Burn, C.R., Implications for paleoenvironmental reconstruction of active ice wedges near Mayo, Yukon Territory, *Permafrost and Periglacial Processes*, **1**, 3–14, 1990.

Burn, C.R., and C.A.S. Smith, Observations of the "thermal offset" in near-surface mean annual ground temperatures at several sites near Mayo, Yukon Territory, Canada, *Arctic*, **41**, 99–104, 1988.
Burn, C.R., and M.W. Smith, Development of thermokarst lakes during the Holocene at sites near Mayo, Yukon Territory, *Permafrost and Periglacial Processes*, **1**, 161–175, 1990.
Burn, C.R., F.A. Michel, and M.W. Smith, Stratigraphic, isotopic and mineralogical evidence for an early Holocene thaw unconformity at Mayo, Yukon Territory, *Canadian Journal of Earth Sciences*, **23**, 794–803, 1986.
Cwynar, L.C., and R.W. Spear, Reversion of forest to tundra in the central Yukon, *Ecology*, **72**, 202–212, 1991.
Environment Canada, *Canadian climate normals 1951–1980. Temperature and precipitation. The North-Y.T. and N.W.T.*, Canadian Climate Program, Atmospheric Environment Service, Downsview, Ontario, 55 pp. 1982.
Goodrich, L.E., An introductory review of numerical methods for ground thermal regime calculations, *DBR Paper 1061*, 32 pp., Division of Building Research, National Research Council of Canada, Ottawa, 1982.
Harlan, R.L., Hydrological considerations in northern pipeline development, *Report 74–26*, 35 pp., Environmental-Social Committee, Northern Pipelines: Task Force on Northern Oil Development, Government of Canada, 1974.
Hughes, O.L., Surficial geology and geomorphology, Janet Lake, Yukon Territory, *Map 4-1982*, Geological Survey of Canada, Ottawa, 1983.
Hughes, O.L., Quaternary geology, in *Guidebook to Quaternary Research in Yukon*, edited by S.R. Morison and C.A.S. Smith, pp. 12–16, XII INQUA Congress, National Research Council of Canada, Ottawa, 1987.
Hughes, O.L., and R.O. van Everdingen, Field trip No. 1, Central Yukon—Alaska, *Guidebook, Third International Conference on Permafrost*, 32 pp., National Research Council of Canada, Ottawa, 1978.
Hughes, O.L., J.J. Veillette, J. Pilon, P.T. Hanley, and R.O. van Everdingen, Terrain evaluation with respect to pipeline construction, Mackenzie Transportation Corridor, central part, Lat. 64°—68°N, *Report 73–37*, 74 pp., Environmental-Social Committee, Northern Pipelines: Task Force on Northern Oil Development, Government of Canada, 1973.
Johansen, Ø., *Thermal conductivity of soils*, 231 pp., Institute for Kjöleteknikk, Trondheim, Norway, 1975.
Judge, A.S., The prediction of permafrost thickness, *Canadian Geotechnical Journal*, **10**, 1–11, 1973.
Judge, A.S., A.E. Taylor, M. Burgess, and V.S. Allen, Canadian geothermal data collection—Northern Wells 1978–1980, *Earth Physics Branch Geothermal Series 12*, 190 pp., Energy, Mines and Resources, Canada, Ottawa, 1981.
Lachenbruch, A.H., and B.V. Marshall, Changing climate: Geothermal evidence from permafrost in the Alaskan Arctic, *Science*, **234**, 689–696, 1986.
Lachenbruch, A.H., T.T. Cladouhous, and R.W. Saltus, Permafrost temperature and the changing climate, *Proceedings of the Fifth International Conference on Permafrost*, **3**, 9–17, 1988.
Lachenbruch, A.H., J.H. Sass, B.V. Marshall, and T.H. Moses, Jr., Permafrost, heatflow and the geothermal regime at Prudhoe Bay, Alaska, *Journal of Geophysical Research*, **87(B)**, 9301–9316, 1982.
Osterkamp, T.E., Response of Alaskan permafrost to climate, *Proceedings of the Fourth International Conference on Permafrost*, **2**, 145–152, 1983.
Osterkamp, T.E., Permafrost temperatures in the Arctic National Wildlife Refuge, *Cold Regions Science and Technology*, **15**, 191–193, 1988.

Osterkamp, T.E., and A.H. Lachenbruch, Thermal regime of permafrost in Alaska and predicted global warming, *Journal of Cold Regions Engineering*, **4**, 38–42, 1990.

Outcalt, S.I., Massive near-surface ground ice in arctic Alaska: Description and modelling analysis, *Physical Geography*, **3**, 123–147, 1982.

Outcalt, S.I., F.E. Nelson, and K.M. Hinkel, The zero-curtain effect: Heat and mass transfer across an isothermal region in freezing soil, *Water Resources Research*, **26**, 1509–1516, 1990.

Paine, J.R., Geotechnical investigation (1987), kilometre 52-111, Highway #11, Silver Trail, Yukon Territory, *Report to Highways Branch, Community and Transportation Services, Yukon Territorial Government*, 20 pp., J.R. Paine and Associates, Whitehorse, Yukon, 1988.

Patterson, D.E., and M.W. Smith, The measurement of unfrozen water content by Time Domain Reflectometry: results from laboratory tests, *Canadian Geotechnical Journal*, **18**, 131–144, 1981.

Pike, A.E., Mining in permafrost, *Proceedings of the First International Conference on Permafrost*, 512–515, 1966.

Riseborough, D.W., Soil latent heat as a filter of the climate signal in permafrost, *Proceedings of the Fifth Canadian Permafrost Conference, Collection Nordicana*, **54**, 199–205, 1990.

Smith, M.W., The significance of climate change for the permafrost environment, *Proceedings of the Fifth International Conference on Permafrost*, **3**, 18–26, 1988.

Smith, M.W., Potential responses of permafrost to climatic change, *Journal of Cold Regions Engineering*, **4**, 29–37, 1990.

Smith, M.W., and C.R. Burn, Instrumentation of long-term monitoring of permafrost temperatures in relation to climatic change near Mayo, Yukon, *Report 5114-4320-00001-113-1160(J0055)*, 36 pp., Atmospheric Environment Service, Environment Canada, Downsview, Ontario, 1986.

Smith, M.W., and D.W. Riseborough, Permafrost sensitivity to climatic change, *Proceedings of the Fourth International Conference on Permafrost*, **1**, 1178–1183, 1983.

Smith, M.W., and D.W. Riseborough, The sensitivity of thermal predictions to assumptions in soil properties, *Proceedings of the Fourth International Symposium on Ground Freezing*, 17–23, 1985.

Tukey, J.W., *Exploratory Data Analysis*. 688 pp., Addison-Wesley, Reading, Ma., 1977.

Weast, R.C. (ed.), *CRC Handbook of Chemistry and Physics*, 63rd edition, CRC Press, Boca Raton, Fl., 1982.

Wernecke, L., Glaciation, depth of frost and ice veins of Keno Hill and vicinity, Yukon Territory, *Engineering and Mining Journal*, **133**, 38–43, 1932.

Williams, P.J., *The Surface of the Earth: An Introduction to Geotechnical Science*, 212 pp., Longman, London, 1982.

Williams, P.J., and M.W. Smith, *The Frozen Earth: Fundamentals of Geocryology*, 306 pp., Cambridge University Press, Cambridge, England, 1989.

Index